THE GEYSERS
YELLOWSTONE
of

THE **GEYSERS** of **YELLOWSTONE**
FOURTH EDITION

T. SCOTT BRYAN

UNIVERSITY PRESS OF COLORADO

Published by the University Press of Colorado
5589 Arapahoe Avenue, Suite 206C
Boulder, Colorado 80303

 The University Press of Colorado is a proud member of
the Association of American University Presses.

The University Press of Colorado is a cooperative publishing enterprise supported, in part, by Adams State College, Colorado State University, Fort Lewis College, Mesa State College, Metropolitan State College of Denver, University of Colorado, University of Northern Colorado, and Western State College of Colorado.

∞ The paper used in this publication meets the minimum requirements of the American National Standard for Information Sciences—Permanence of Paper for Printed Library Materials. ANSI Z39.48-1992

Library of Congress Cataloging-in-Publication Data

Bryan, T. Scott.
  The geysers of Yellowstone / T. Scott Bryan. — 4th ed.
    p. cm.
  Includes bibliographical references and index.
  ISBN 978-0-87081-924-7 (pbk. : alk. paper) 1. Geysers—Yellowstone National Park. 2. Yellowstone National Park. I. Title.
  GB1198.7.Y44B79 2008
  551.2'30978752—dc22
                                                              2008020371

Design by Daniel Pratt

17 16 15 14 13 12 11 10 09 08   10 9 8 7 6 5 4 3 2 1

CONTENTS

ILLUSTRATIONS

Maps

It is a pleasure to write this Foreword to the latest edition of what I consider a very important book, *The Geysers of Yellowstone*. I have known author Scott Bryan for thirty-three years. That is longer than the U.S. Army spent on its entire detail to Yellowstone, 1886–1918! During that time, I have become more and more convinced that Scott is "the man" where geysers are concerned and that Yellowstone does not want to return to the "bad old days" of *not* having a book called *The Geysers of Yellowstone*. Since the first edition was published in 1979, it has proven to be not just the only reliable guidebook to the geysers but also a valuable archive of the history of many individual geysers.

I recall clearly my first years of working in Yellowstone, 1969–1972. As a National Park Service maintenance worker I had an immediate interest in the geysers, but later, as a bus tour guide for the park concessionaire, I desperately needed geyser information. Unfortunately, except for some brief pamphlets written by park geologist George Marler, there was little good information available about geyser activity and geyser history at that time. I spent a good deal of time in libraries and archives trying to find Yellowstone's old geyser records. In taking on a large research project about Yellowstone's place names and geyser history, I felt privileged to become a sort of partner to Scott Bryan and Rick Hutchinson, and it seemed I was constantly talking to the two of them. Hutchinson was Yellowstone's thermal geologist then, and he and Scott were the people I always looked up to as my most important geyser authorities. Rick seemed to me a person of deep, if quiet, research, while Scott seemed always on the cutting edge of communicating about geysers with park employees and visitors. Rick's untimely death in 1997 shorted us one seasoned expert, but fortunately we still have the other—Scott Bryan, who I hope will be here for years to come.

Today we also have The Geyser Observation and Study Association (GOSA), founded in 1988. Members of that organization help us monitor Yellowstone's geysers, and that is no small feat. I have learned over my years in the Park that it takes all of us to do a decent job of keeping up with each geyser—the new ones that break out, the old ones that become dormant, and the changing and evolving heights, durations, and intervals of all the rest. With something like 800 to 1,000 actual geysers in Yellowstone, it is nearly impossible to keep up with them all, and it takes time for a geyser enthusiast to become well enough versed about geysers to help us monitor them. There is a lot for all of us to learn, and Scott's book helps us learn it.

Some geysers exhibit remarkable stability and consistency over time, while others change their habits rather quickly. But change in nature is a constant, and keeping up with the changes is the task of all who are interested. In line with this, Scott's book has gotten bigger over time, but that is to be expected. As more persons have been recruited into the "geyser club," more geysers have received attention. We have more eyes and ears on the geysers now. That is good for the Park, good for an ongoing increase in human knowledge, and good for those who want the information preserved for future generations. Information preservation is a big part of what Yellowstone is all about, and we need everyone to help us preserve it. Scott's book serves as a training manual for all who are interested in geysers—from the passing novice to the dedicated expert. In the midst of constant change, this book helps us keep up.

So take this book and walk through the geyser basins, reading and learning as you go. You too may soon be signing up to become a card-carrying "geyser gazer." We'll see you on the boardwalks!

LEE WHITTLESEY
PARK HISTORIAN, NATIONAL PARK SERVICE
YELLOWSTONE NATIONAL PARK, WYOMING-MONTANA-IDAHO

This is the fourth edition of *The Geysers of Yellowstone*. Readers of the earlier editions will find many changes. The entire text has been revisited and most descriptions revised on the basis of activity observed through April 2008. Information about a number of significant new geyser developments has been added, as has recent knowledge about some of the world's geyser fields outside Yellowstone. Every effort has been made to make this the most up-to-date and comprehensive reference to the geysers of Yellowstone National Park ever produced. It is intended to serve those who have been geyser gazers for years as well as those who have never seen a geyser.

The maps here are based on accurate "Thermal Maps" produced by the U.S. Geological Survey (USGS) during the late 1960s and early 1970s at a scale of 1:2,400 (1 inch equal to 200 feet). The locations of roads, trails, streams, and thermal features are precisely shown. Users will have to remember that the scale is small, however, so real distances may be greater than they seem on the map. Exceptions to this are the maps of the Gibbon, Third, and Heart Lake Geyser basins plus portions of the Lower Geyser Basin, for which detailed USGS Thermal Maps do not exist, and the index maps to the Upper and Lower Geyser basins. These are all based on enlargements of standard 7½-minute topographic maps at a scale of 1:24,000 (1 inch equal to 2,000 feet) or on other maps produced by independent researchers. Note also that the maps as they appear in this book have been resized by variable amounts to best fit the published page.

Whether formally named or not, every geyser within each geyser basin has been given a serial number. These numbers relate the text descriptions to the locations on the maps and data in the tables. Most of these numbers are different from those in earlier editions of this book because they are intended only to relate the text to the maps and tables of this edition. Also, they sometimes skip, such as from 56 to 60, reserving numbers 57, 58, and 59 for potential future use.

Special words about the names of the geysers, as they appear in this book, are needed. Geyser gazers (myself included) have always tended to invent names for unnamed features in an offhand way. There is good reason for discouraging this practice. Often, it turns out that the feature does have a name after all, one that has legal priority even if it had been "forgotten" a century ago. Furthermore, a number of people might end up using different names for one spring. In fact, the only persons who technically have the right to apply names to thermal features are Yellowstone's superintendent, chief of interpretation, and research geologist, all of whom may rely on the advice of the Park historian. Even with all this, the ultimate formal approval of any geographical name (within Yellowstone or anywhere else in the United States) rests with the U.S. Board on Geographic Names, which operates under the umbrella of the U.S. Geological Survey.

Nevertheless, numerous informal names have been applied and are now largely accepted because of extensive popular use. Many of these names are included in this book. For example, if a geyser has received a name too recently to be widely recognized, then it is first designated as an unnamed geyser (UNNG) followed by an abbreviation for the hot spring group it is a member of and then a number. The informal name (if any) is given parenthetically and in quotes. An example from the Upper Geyser Basin is:

271. UNNG-PMG-4 ("SECLUDED GEYSER")

271—In the text this is serial number 271 of the Upper Geyser Basin, and it is shown as number 271 on the map that shows this group of springs;

UNNG—The geyser is officially unnamed;

PMG—The geyser is a member of the Pipeline Meadows Group of hot springs;

4—This is the fourth unnamed geyser to have been described in this group;

("Secluded Geyser")—This name is unofficial; although it has received use in a few written reports, it is not considered sufficiently entrenched to have become "established."

When a new name has received only limited use but violates no naming convention, it may be considered "acceptable." In such cases, the "UNNG" designation can be dropped but the name remains in quotation marks. An example is "Aftershock Geyser." Later, if an acceptable name is used repeatedly for a number of years in books or official reports or appears on approved maps, any future attempt to change it would be pointless and confusing. Names of this sort are considered to have become "established" or "entrenched." Established names, of which there are many in this edition, are considered official "unless proved otherwise." Both the "UNNG" part of the designation and the quotation marks then can be eliminated. Uncertain Geyser is a good example of this.

If all this sounds confusing, well, it often is. The numerous name changes that appear in this edition of *The Geysers of Yellowstone* have been made with the approval of the Park historian. We both hope they are all historically correct.

Lee H. Whittlesey, the Yellowstone Park historian, has done an incredible amount of work on Yellowstone's place names. He conducted extensive research in places like the National Archives in Washington, D.C., the U.S. Board on Geographic Names in Reston, Virginia, and numerous regional libraries. His work was significantly added to by independent and cooperative research projects conducted by Rocco Paperiello and Marie Wolf, and Mike Keller. The total result is Whittlesey's compendium titled *Wonderland Nomenclature: A History of the Place Names of Yellowstone National Park*, a typescript volume of more than 2,200 pages published by the

Montana Historical Society; a reformatted version is available from The Geyser Observation and Study Association. Many of the historical asides in this book were extracted from this work.

This edition of *The Geysers of Yellowstone* is also filled with extensive numerical revisions. Geysers are dynamic. Their eruptive activity is seldom stable for very long, and keeping track of all the changes can be difficult. When I wrote the first edition of this book during the 1970s, there were very few geyser gazers. Now more than 400 have banded together as the Geyser Observation and Study Association (GOSA), a fully nonprofit corporation dedicated to studying, understanding, preserving, and simply enjoying geysers. My greatest thanks go to all of them for their valuable information, stimulating discussions, thoughtful ideas, and fun.

The interpretive ranger staffs at the Old Faithful and Grant Village Visitor Centers and Norris Museum keep logbooks of geyser activity. These are gradually being translated into electronic files. Modern electronic temperature monitors are maintained, too, but while they provide continuous, long-term data, they operate on only a few features, and retrieval of their records is occasional. The logbooks, on the other hand, provide daily information that is always available.

Rick Hutchinson kept close tabs on geyser activity in his capacity as Yellowstone's research geologist from 1972 until his untimely death by avalanche in 1997. Although others have followed in his position and there is also the Yellowstone Volcano Observatory staffed by the U.S. Geological Survey, Hutchinson's unique depth of knowledge has been lost forever.

Dr. Donald E. White of the U.S. Geological Survey was responsible for developing many of the modern theories about how geysers and geothermal systems work. He provided ideas vital to this book, and his favorable review of the first edition's manuscript probably assured its publication. Many regret his death in 2002.

Dr. George D. Marler was *the* ranger-naturalist geyser gazer of Yellowstone for several decades before his death in 1978. Had he not kept extensive and detailed notes about the geysers and their dynamic changes throughout the years, relatively little would be known about long-term change in geyser action and its meaning for us today. His numerous publications and reports, although sometimes contradictory, are indispensable references now held in the archives of Yellowstone National Park and the special collections library at Brigham Young University.

We must remember, too, those who were here during the early surveys of Yellowstone. Geologists such as Hayden, Hague, Peale, Weed, Allen, Day, Fenner, and many others set the stage for us by helping to create, maintain, and understand what Yellowstone is really all about. Saxo Grammaticus, a Danish historian and priest, apparently described Iceland's Geysir as early as the thirteenth century and thereby became the earliest "geyser gazer" we know of—although Homer, in *The Iliad*, seemed to describe the small geyser at Ayvacik, Turkey, near the site of ancient Troy, as far back as 700 B.C.

Above all, tremendous appreciation is owed to my wife, Betty. The fact that I wrote *The Geysers of Yellowstone* was her idea in the first place, and this book is a result of her years of patience and encouragement.

Thank you, one and all!

T. Scott Bryan
April 2008

THE GEYSERS *of* YELLOWSTONE

CHAPTER ONE

About Geysers

WHAT IS A GEYSER?

The standard definition of "geyser" in general worldwide use reads like this:

A geyser is a hot spring characterized by intermittent discharge of water ejected turbulently and accomplished by a vapor phase.

It sounds simple enough, but it really is not. The definition includes several "gray areas" that can be interpreted in different ways—how hot is hot; how high must the turbulence be; are there limits as to how long or short the intermittency needs to be? Such questions will probably never be answered to everyone's satisfaction, but there are two similar varieties of hot spring that definitely do not qualify as geyser. *Intermittent springs* undergo periodic overflows but never actually erupt. *Perpetual spouters* (called *pulsating springs* in some parts of the world) may have spectacular eruptions, but their action never stops. In all three of these cases, however, the cause of the eruption is the same—namely, water boiling into steam at some depth below the ground.

WHAT MAKES A GEYSER WORK?

Three things are necessary for a geyser to exist: an abundant supply of water, a potent heat source, and a special underground plumbing system.

The water and heat factors are fairly common. Hot springs are found in virtually all of the world's volcanic regions. The plumbing system is the critical aspect. Its shape determines whether a spring will be quiet or will erupt. It must be constructed of minerals strong enough to withstand tremendous pressure, and it must include a permeable volume so as to hold the huge amounts of water ejected during an eruption.

Nobody really knows what a plumbing system looks like—it is, after all, belowground and filled with hot water and steam. Considerable research drilling has been done in some of the world's geyser areas, and none has yet found any large, open water storage caverns. This fact led to the conclusion that a geyser's main water reservoir is a complex system of small spaces, cracks, and channels in the porous rocks that surround the plumbing system.

In 1992 and 1993 an experimental probe used at Old Faithful Geyser expanded on this idea and resulted in further changes in our understanding of what a plumbing system looks like. Equipped with pressure and temperature sensors and a miniature video camera, the probe was lowered into Old Faithful's vent shortly after an eruption had ended. The result is shown on the left side of Figure 1.1 (p. 3). At a depth of only 22 feet, there was a narrow slot barely 4 inches wide. Just below that was a wider area that had a waterfall of relatively cool, 176°F (80°C), water pouring into it. Then, at about 35 feet, the probe entered a chamber "the size of a large automobile." As Old Faithful refilled, the temperature of the rising water was 244°F (117°C), fully 45°F (25°C) hotter than the normal surface boiling point at that altitude. The researchers reported that the action resembled a seething "liquid tornado" of unbelievable violence. Years ago, in New Zealand, people were able to scramble down into an inactive hot spring. There they found a chamber with smooth walls punctured by numerous small openings. Once upon a time it must have been the scene of wild boiling like that wit-

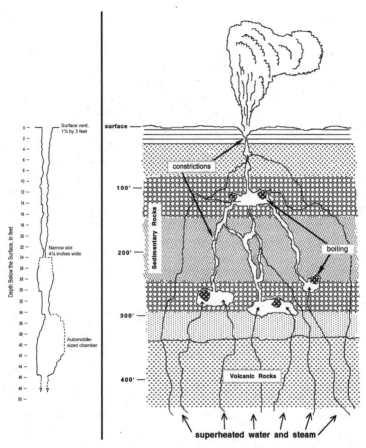

(left) *A video probe lowered into Old Faithful Geyser showed its uppermost 50 feet to be little more than an irregular tube partially filled with violently boiling water. All geysers probably have similar plumbing systems.* (right) *Although nobody really knows what a geyser's deeper plumbing system looks like, research indicates that it is probably similar to this illustration.*

nessed inside Old Faithful, and its existence supports the notion that Old Faithful's inferno is normal rather than unique.

This is quite unlike the standard model of a plumbing system, which has the water rather quietly flowing upward into a single main tube. Still, this revised model probably applies to all geysers. It makes the water supply network more complex but changes nothing about how geysers actually operate. The character and eruptive performance of every geyser are determined by the geyser's plumbing system, and, as in all of nature, no two are alike.

The probe into Old Faithful was never lowered deeper than 46 feet, therefore only into the very top of the plumbing system, but when combined with other data, enough information is available for us to reconstruct the plumbing system of a geyser. An example is shown in Figure 1.1 (right). It consists mainly of tubes extending into the ground, containing many sharp bends and constrictions along their lengths. Connected to the tubes are small open spaces and, especially, layers of water-storing sand and gravel of high porosity. Most of this plumbing is fairly close to the surface, and even the largest geysers extend to a depth of only a few hundred feet. Finally, much of the plumbing system is coated with a water- and pressure-tight lining of *siliceous sinter,* or *geyserite.* This mineral is also deposited outside the geyser and in and about the quiet hot springs. Of course, the geyserite is not magically deposited by the water. Its source is quartz (or silica) in the volcanic rocks underlying the geyser basin.

The water that erupts from a geyser arrives there only after a long, arduous journey. Water first falls in Yellowstone as rain and snow, then percolates through the ground to as much as 8,000 feet below the surface and back again. The round-trip takes at least several hundred years. This is something that can be determined with reasonable accuracy by studying the *tritium* (sometimes called "heavy-heavy hydrogen") content of the geyser water. Tritium is radioactively unstable and decays with age. Young water contains considerable amounts of tritium, while old water contains little or none. It is nearly absent in most Yellowstone waters; in fact, it is believed that the water erupting from Old Faithful today fell as precipitation at least 500 years ago—around the time Columbus was exploring the West Indies—and some geochemical evidence indicates that 1,100 years is more likely.

At depth, the percolating surface water is heated where it contacts a high-temperature brine, which in turn circulates as deep as 15,000 feet where it is heated by the enclosing volcanic rocks. Once heated, it dissolves some of the quartz from the rocks. All this takes place at very high temperatures—over 500°F (200°C) in many cases, and 459°F (237°C) was reached in a research drill hole only 1,087 feet deep. This silica will not be deposited by the water until it has approached the surface and cooled to a considerable extent.

Now an interesting and important phenomenon occurs. Although it was the mineral quartz that was dissolved out of the rocks, the deposit of geyserite is a form of opal (never of gem quality). The mechanisms involved in this process are complex, involving temperature, pressure, acidity or alkalinity of the water, and *time.*

HOW A GEYSER ERUPTS

The hot water, circulating up from great depth, flows into the geyser's plumbing system. Because this water is many degrees above the boiling point, some of it turns to steam instead of forming liquid pools. Meanwhile, additional cooler water is flowing into the geyser from the porous rocks nearer the surface. The two waters mix as the plumbing system fills.

The steam bubbles formed at depth rise and meet the cooler water. At first, they condense there, but as they do they gradually heat the water. Eventually, these steam bubbles rising from deep within the plumbing system manage to heat the water nearer the surface until it also reaches the boiling point. Now the geyser begins to work like a pressure cooker. The water within the plumbing system is hotter than boiling but is "stable" because of the pressure exerted by the water lying above it. (Remember that the boiling point of a liquid is dependent on the pressure. The boiling point of pure water is 212°F [100°C] at sea level. In Yellowstone, the elevation is about 7,500 feet [2,250 meters], the pressure is lower, and the boiling point of water at the surface is only around 198°F [93°C].)

The filling and heating process continues until the geyser is full or nearly full of water. A very small geyser may take but a few seconds to fill, whereas some larger geysers take several days. Once the plumbing system is full, the geyser is about ready for an eruption. Often forgotten but of extreme importance is the heating that must occur along with the filling. Only if an adequate store of heat exists within the rocks lining the plumbing system can an eruption last more than a few seconds. (If you want to keep a pot of water boiling on the stove, you have to keep the fire turned on. The hot rocks of the plumbing system serve the same purpose.) Again, each geyser is different from every other. Some get hot enough to erupt before they are full and do so without any preliminary indications of an eruption. Others may be completely full long before they are hot enough and so may overflow quietly for hours or even days before an eruption occurs. But eventually, an eruption will take place.

Because the water of the entire plumbing system has been heated to boiling, the rising steam bubbles no longer collapse near the surface. Instead, as more very hot water enters the geyser at depth, even more and larger steam bubbles form and rise toward the surface. At first, they are able to make it all the way to the top of the plumbing with no problem. But a time will come when there are so many bubbles that they can no longer freely float upward. Somewhere they encounter some sort of constriction in the plumbing. To get by, they must squirt through the narrow spot. This forces some water ahead of them and up and out of the geyser. This initial loss of water reduces the pressure at depth, lowering the boiling temperature of water already hot enough to boil. More water boils, forming more steam. Soon there is a virtual explosion as the steam expands to over 1,500 times its original, liquid volume. The boiling becomes violent, and water is ejected so rapidly that it is thrown into the air. In fact, people standing near very large geysers sometimes hear and feel a thudding, popping sound. Research indicates that this happens because the superheated water is ejected so quickly and then explodes into steam so violently within the water column that the total speed exceeds the sound barrier—the thuds are caused by small sonic booms within the expanding column of steam and water!

The eruption will continue until either the water is used up or the temperature drops below boiling. Once an eruption has ended, the entire process of filling, heating, and boiling will be repeated, leading to another eruption.

THE DIFFERENT KINDS OF GEYSER

All geysers operate in the same fashion, but they come in three varieties. The differences depend on the size and shape of the plumbing system and its constriction, the depth of a pool, the volume of available water, and so on.

Cone-type geysers erupt steady streams of water that jet from small surface openings. The vent is often, but not always, surrounded by a built-up cone of geyserite. Cone-type geysers are rather uncommon, but because the water is squirted under considerable pressure, they tend to have tall, spectacular eruptions such as those of Old Faithful, Daisy, and Riverside geysers.

Fountain-type geysers look a lot different from the cone type because their eruptions rise out of open pools. Steam bubbles rising through the pools cause a series of individual bursts of water, so the action is more a spraying or splashing than a jetting. The fountain-type makes up the vast majority of the world's geysers. Most are small in size, but there are also large examples such as Grand, Great Fountain, and Echinus geysers.

Some observers do not accept *bubble-shower springs* as true geysers but instead list them as a special case of intermittent spring because no vapor phase can be seen rising from deep within the spring's plumbing system. The eruption consists entirely of violent boiling near the surface of an open pool. Some bubble-shower springs, such as Crested Pool, have eruptions as high as several feet, but in most cases the boiling turbulence reaches up only a few inches. This book considers bubble-shower springs to be geysers.

WHY ARE SOME GEYSERS REGULAR, OTHERS IRREGULAR?

The inflow of water into a geyser system is constant, so it would seem that the activity of any geyser should show little variation from one eruption to the next. However, only a few are classed as regular geysers. To be regular, a geyser must either be isolated from other springs or connected only with springs whose overall activity is so constant that they do not affect the geyser. Old Faithful is the most famous example of regularity. Its eruptions can be predicted with nearly 90 percent accuracy. Some other geysers are even more regular, occasionally operating with almost stopwatch-like precision. (Statistically, there are geysers that yield "coefficients of variability" of less than 3 percent.)

But most geysers are irregular. The time interval between successive eruptions is erratic. One time it may be just minutes between the plays, the next several hours. In no way can these geysers be predicted.

The mechanism behind this has been termed *exchange of function,* as first described by G. D. Marler in 1951. What this basically means is that water and energy can be diverted from a geyser to some other hot spring or hot spring group (another geyser is not necessarily involved). This happens because the plumbing systems of most springs and geysers are intertwined with those of others. The activity of any one of those springs must affect all others in the group.

Just what makes exchange of function take place is unknown. Something must act as a valve so as to shunt the water and heat from one direction to another, but whether this is a vapor lock as a result of a buildup of steam bubbles, a fluidic switch operation, a Venturi effect, or something entirely different has never been determined. Most exchanges are small scale in both extent and time and require a knowing eye to detect. Others can be of great significance. Indeed, one group of active springs may suddenly stop functioning altogether while a nearby group of previously insignificant springs becomes animated with an exchange that might last for years. A good example involved Daisy Geyser and nearby Bonita Pool in the Upper Geyser Basin. For years, Daisy was one of the largest and most regular geysers in Yellowstone, while Bonita overflowed only slightly. Suddenly the energy shifted toward Bonita. Daisy was drained of the energy necessary for eruptions as Bonita overflowed heavily and underwent small eruptions. As a result, Daisy erupted just three times in over thirteen years. Then the energy flow shifted back toward Daisy. Now it is predictable again, while Bonita lies quietly below overflow.

WHY ARE GEYSERS SO RARE?

There are few places on earth where the three requirements for the existence of geysers are met. The requisite water supply poses no great problem; in fact, a few geysers are able to exist in desert areas, places normally thought of as dry. But the heat source and plumbing systems are tied to one another and are much more restrictive.

Nature's heat source is volcanic activity. The heat is supplied by large bodies of molten or freshly cooling rock at great depth. Given a proper water supply, hot springs are possible in any area of geologically recent volcanism.

Most hot spring areas do not contain geysers, however, because there is a catch. Not just any volcanism will do. The plumbing systems must be pressure-tight, and that requires silica-rich rocks to provide the source of the geyserite that lines the plumbing systems.

The answer is *rhyolite*. Rhyolite is a volcanic rock very rich in silica; it is the chemical equivalent of granite. Rhyolite is rather uncommon, though, and large recent fields of it are found in few places. Yet virtually all geysers are found in such areas. Most of the exceptions to this rule are still associated with recent volcanic activity, although their rocks—dacite, andesite, and basalt—are somewhat less rich in silica. And as seems to be usual in science, there are a few anomalous geyser localities, such as Beowawe, Nevada, where the activity is not associated with recent volcanic action or even directly with volcanic rocks of any type or age.

Not only is Yellowstone a major rhyolite field, but it is of very recent origin. Although the last major volcanic eruption was 600,000 years ago, minor activity continued as recently as 70,000 years ago. Yellowstone could well be the site of further volcanic eruptions. That is another story entirely, but for now the Park is incomparably the largest geyser field in the world.

HOW MANY GEYSERS ARE THERE IN YELLOWSTONE?

This chapter opens with a basic definition of "What is a geyser?" and notes that there are a number of gray areas and transitional kinds of springs. When such features as bubbling intermittent springs, variable perpetual spouters, and random splashers are eliminated from consideration, the number of geysers actually observed to be active in Yellowstone in any given year might approach 500, and research indicates that an absolute minimum of 700 springs have erupted since the national park was established in 1872.

The count of 500 active geysers is an amazing number. The second largest of the world's geyser fields, Dolina Geizerov on Russia's Kamchatka Peninsula, contains no more than 200 geysers. El Tatio, remote in the high Andes Mountains of Chile, ranks third, with 85 active geysers. New Zealand's North Island has 70 geysers, and Iceland has about 30. All other areas contain fewer; taken together, the entire world outside of Yellowstone might total fewer than 500 geysers. (See the Appendix for more about the rest of the world's geyser fields.)

Yellowstone, in other words, contains well over half of all the geysers on Earth. The number of geysers is not stable, however.

They are very dynamic features, affected by a wide range of physical factors and processes. The slightest change in the geological environment may radically alter, improve, or destroy the geysers.

Recent studies of historical literature have shown that much about Yellowstone had been forgotten. Many springs taken to be "new" geysers in recent years are now known to have been active during the 1800s. Nevertheless, the number of active geysers appears to be increasing. In 1992, a geochemist with the U.S. Geological Survey, who had conducted studies in Yellowstone for more than fifty years, stated that in 1955 one could easily count the number of geysers on Geyser Hill on two hands. This book enumerates 51 individual or clustered geysers on Geyser Hill, and at least 45 of them were active during 2007. Authors Allen and Day of the Carnegie Institute of Washington counted 33 geysers at the Norris Geyser Basin in 1926; 83 are enumerated here. Similar situations exist throughout Yellowstone.

All this fits with recent studies that suggest that Yellowstone's geyser basins—all of them and all of the geysers and hot springs within them—operate on a "pulse and pause" basis. If this is correct, then decades-long episodes of vigorous geyser activity are separated by longer time spans when few or even no eruptions occur. These pauses are demonstrated by the growth of trees, such as the silicified remains near Old Faithful Geyser, on the rim of Castle Geyser, within the formation of Grotto Geyser, and even the stand of dead trees behind Grand Geyser. The apparent fact is that there really are more active geysers today than perhaps ever before in recorded history. We do not know why this is so, but clearly right now Yellowstone is within a "pulse," and right now is the "good ol' days" of geyser gazing.

CHAPTER TWO

Some Background on the Yellowstone Geysers

Geysers are beautiful and rare. Wherever they are found, they have attracted attention for the duration of their known history. Outside of Yellowstone some geysers have been watched for hundreds, even thousands of years. Within the Park they have been observed for far longer than recorded history. The Indians certainly saw them and wondered about them.

But just *what* the Indians thought about the geysers is uncertain. Few tales have come down to us, and most were probably embellished by the trappers who passed them on to us. We do know that some Shoshoni Indians called the geyser basins "Water-That-Keeps-On-Coming-Out." It was "Many Smokes" to the Blackfeet and "Burning Mountain" to others. But despite popular notions about Native religions, the Indians were obviously not afraid of the geysers in any way. Remains of their campsites have been found in every hot spring area of the Park. In some of these places obsidian chips left over from tool making litter the ground, suggesting that the areas were virtual factories for the production of arrow and spear points.

In at least one place in the Park, an unknown Native piled logs about the crater of a geyser. Why this was done will always remain a mystery, but it further suggests that Indians did not fear geysers. Perhaps they did hold a special reverence for them, though, and the Indians may even have considered the geyser basins to be sacred lands. One story claimed that geyser eruptions were the result of underground battles between spirits but that those wars did not affect mortals. Whatever the Indians thought, they harbored no great myths about either Yellowstone or its geysers.

Their stories about the geothermal wonders, however, attracted many early explorers to the Yellowstone Plateau. Probably the first was John Colter. As a member of the Lewis and Clark Expedition, he was so intrigued with the Montana-Wyoming area that he left the main party in 1806 and moved south along the Yellowstone River. He may have seen the huge terraces of Mammoth Hot Springs and the hot springs near Tower Fall, but he did not encounter any of the geyser basins. In fact, the original "Colter's Hell" referred to hot springs near today's Cody, Wyoming, outside of Yellowstone, but the name was quickly transferred to the park area.

Colter's reports, although factual, were regarded as fancy and largely ignored. But other mountain men soon moved into the region surrounding "Colter's Hell" and emerged with similar stories, always too fanciful and fantastic to be true.

Twenty-eight years later, in 1834, Warren Ferris gained credit as Yellowstone's first tourist, since his visit took place specifically because of the stories he had heard. His journal contains an exquisite description of the Upper Geyser Basin and details about several large geysers—interestingly, *not* including Old Faithful. In the later 1830s, Osborne Russell traveled throughout the Park area. His written descriptions of the geysers and hot springs were so accurate that some individual features can be identified from them today. Yet even these learned observations seemed unbelievable to "civilized" people back East. When mountain man Jim Bridger, notorious for his tall tales, drifted into the Yellowstone country and confirmed the findings of the others, most people listened only for a laugh.

The stories didn't stop, though. In time, the versions of just what Yellowstone really was had become tremendously varied. Clearly, there had to be some honest answers.

Several attempts to lead organized expeditions into the Yellowstone country were made during the 1860s. Because of frequent Indian scares throughout the region or the lack of financing, planned expeditions continually fell through until 1869, when Charles Cook, David Folsom, and William Peterson, all from the mining town of Helena, Montana, decided to go it alone.

They rode down the Yellowstone River, past the Mammoth Hot Springs, and over the Mirror Plateau to the Mud Volcano area. From there they traveled up the Yellowstone River and along the west side of Yellowstone Lake. After spending two days in the West Thumb Geyser Basin, they went on to Shoshone Lake (but not to that geyser basin), then up DeLacy Creek and over the Continental Divide to White Creek, and finally down that drainage straight into the Lower Geyser Basin. They were thrilled by Great Fountain Geyser and amazed at the Midway Geyser Basin. Here, figuring they had seen enough, they moved down the Firehole River and missed seeing the Upper Geyser Basin by only four miles.

Cook, Folsom, and Peterson kept excellent records of what they saw, and their report renewed the enthusiasm of three other men who had thought of making the trip earlier. Bigger names were involved this time: Henry Washburn, the surveyor general of Montana Territory; Cornelius Hedges, a territorial judge; and Nathaniel Langford, later the first superintendent of Yellowstone National Park. Along with a U.S. Army contingent led by Lt. Gustavus C. Doane for protection, these men and others set out in the late summer of 1870 on what has become popularly but erroneously known as the "Discovery Expedition."

It was a long journey that saw hardship and tragedy, but eventually the explorers reached the Upper Geyser Basin. One of their first sights is said to have been Old Faithful in full eruption. Over the next day and a half they saw eruptions of numerous other geysers, large and small. They named several, including Beehive, Giantess, Castle, Grotto, and Old Faithful itself. After recording their findings, they left the basin.

The Washburn Party spent their last night in what is now Yellowstone National Park at Madison Junction. Around a campfire on the evening of September 19, 1870, they discussed the marvels they had seen, and almost to a man they wanted others to be able to see the wonders, too. But how to do it? Popular legend has it that the national park idea was born at that campfire when

The Hayden Survey of 1871 produced the first comprehensive reconnaissance of Yellowstone. Credit, NPS photo by W. H. Jackson.

Judge Hedges supposedly suggested that the Yellowstone country be excluded from settlement and protected for all time by the government.

In fact, that story is a romance; the campfire discussion undoubtedly did not take place. Following their trip, most of the explorers discussed and promoted what they had found. Several wrote articles. One by Langford was returned as unacceptable "fiction," but others were published in various magazines and newspapers, and wonder about the area continued to grow. However, none of those articles except one by Judge Hedges, printed in the Helena, Montana, newspaper, suggested the creation of a park.

One man who was enthralled by the reports was Dr. Ferdinand V. Hayden, the leader of the Geological and Geographical Survey of the Territories. He organized and led a complete survey of the proposed Park area during 1871 and 1872. The Hayden Survey's report, combined with the excellent photographs of William H. Jackson and paintings by Thomas Moran, convinced enough members of Congress to act in favor. On March 1, 1872, quite soon, really, after all those trappers' "tall tales," President Ulysses S. Grant signed into law the bill establishing "The Yellowstone Park."

Many early illustrations of Yellowstone's geysers greatly exaggerated their size. This 1872 woodcut by Harry Fenn (engraved by W. J. Linton) shows a tiny man running away from Giant Geyser, whose cone is really about 10 feet tall.

The entire history of Seismic Geyser—its birth, growth into a major geyser as much as 75 feet high, and decline back into quiescence—was the result of the 1959 earthquake. Credit, NPS photo.

nearly 200 miles from Yellowstone, its magnitude of 7.9 had impressive effects. Exchange of function dramatically changed the performance of many springs on Geyser Hill, and related to that was a temporary and slight slowdown by Old Faithful. Here and there throughout the Park were notable increases in eruptive activity, and many of these have persisted.

Every year in the Yellowstone Plateau, up to 2,000 tremors are recorded by seismometers. Mostly far too small to be felt by people, these quakes are normal events. There is little doubt that Yellowstone has been shaken by a great many major shocks over the ages. The resulting thermal changes must be nearly infinite in number. One day in the future, even Old Faithful will meet its demise, but something new might replace it at the same time.

Of course, not all change and variation are the result of major earthquakes. Yellowstone can be subtle, too, and perhaps nothing shows this as well as the spectrum of colors seen in the hot springs and their runoff channels. Just as each geyser or pool is different, each geyser basin has its own personality. None of the coloration is ever exactly the same, neither from place to place nor from time to time.

The broad flats and cones of the geyser basins are accented by tones of white and gray. These stark colors are caused by the geyserite that, when underground, is so important to the existence of geysers (see Chapter 1). It forms very slowly, sometimes at the rate of only $^1/_{100}$ of an inch per year. Most deposits vary from a few inches to a few feet thick. Obviously, such specimens took many, many years to form. Yet they are delicate and beautiful, too. Pause in the basins to take a close look at the lustrous, pearly beads and compare them with the wafer-thin, artistic laminations of other spots. As you do, remember the tremendous age and rarity of this geyserite. Do not try to take a piece with you. In time it will dry out and crumble into dusty gravel. Leave the formations untouched for others to see.

Minerals other than siliceous sinter are also present, although rarer. Most common are bright yellow sulfur and red-brown iron oxide. These deposits are especially prominent at the Norris Geyser Basin. Here and there throughout Yellowstone the sinter is sometimes stained dark gray or black because of impurities of manganese oxides; smaller amounts of the same mineral cause a fine pink color. Brilliant red and orange-yellow arsenic compounds appear at Norris and Shoshone. Small popcorn-like aggregates of complex sulfate minerals show up as white and light yellow puffs on gravel in barren, acid flats. The minerals are everywhere, always with different combinations and compositions.

The hot pools are often deep blue or green, a coloration that led some early explorers of Yellowstone to believe they had found a bonanza. Copper in solution will turn water blue or blue-green, and that is what they thought they had found—a fabulously rich copper mine without having to sink a shovel into the ground. But any deep body of water will look blue. Water absorbs most of the colors of the rainbow, but not the blues and greens. Those colors are reflected back, giving the water its color. If any other material is present in the water or lining the crater walls, that tint will be added to the blue. Most common is the yellow of either the mineral sulfur or hot water bacteria. This blending of yellow and blue can produce a green of incredible intensity.

An amazing variety of *thermophilic* ("temperature-loving") life thrives in the hot water. The brilliant yellows, oranges, browns, and greens in the wet runoff channels are the result of algae and cyanobacteria (previously known as blue-green algae). These microscopic plants have been studied extensively. You can tell the approximate temperature of a stream by the color of its algae. No cyanobacteria at all and the temperature must be greater than 167°F (74°C). If the color is bright yellow, the temperature is around 160°F (71°C);

brilliant orange, about 145°F (63°C); the dark browns come in at about 130°F (57°C); and pure green shows up at 120°F (50°C) and below. In relatively narrow, fast-flowing channels, one often sees a V-shaped pattern to the colors. The stream cools more slowly toward the center. There the cyanobacteria that thrive in higher temperatures are able to survive at a greater distance from the source of the stream; less tolerant plants hug the edges of the channel where the water cools more quickly. Some runoff channels support a mixture of algae types, which produce different color schemes altogether.

Stringy pale-yellow or pink strands of true bacteria are sometimes visible in very hot runoff channels where the temperature is over 170°F (76°C). One of these is *Thermus aquaticus*, the source of the "Taq polymerase" enzyme necessary for the DNA replication of genetic testing.

Large or small, obvious or subtle, Yellowstone and its thermal features are constantly changing. And now there is another cause of great change on the scene, one able to cause change more extensive and destructive than that of Mother Nature's greatest earthquake. It is called people.

A century ago, some men with vision fought to protect and preserve Yellowstone's environment. A fact often forgotten is that the national park exists not because of its forests and wildlife but rather because of its geysers and hot springs. They are what make it so unique as to have attracted the early explorers and now nearly 3 million visitors each year. It is well that they did so, for we now know that Yellowstone contains well over half of all the geysers in the world and is one of only two geyser fields with any real measure of protection. Yet the remains of an old slat-back chair and other debris were removed from Old Faithful during a 1980s clean-up project. Please, let's not let the Park's founders down.

To leave the boardwalks and trails within the geyser basins is highly destructive of the formations. The slightest bit of trash thrown into a pool can clog it forever. It is also dangerous. Thin crust can look completely solid, but in many places it is only inches thick, brittle, and underlain by a boiling pool. More people have been killed and injured by thermal burns than by all grizzly bear and other wildlife incidents combined. And so, leaving the trails is also illegal. Obey the signs in the geyser basins. Help yourself and others to have safe, enjoyable visits to the greatest geyser field in the world . . . now and in the future.

CHAPTER THREE

Geyser Basins of Yellowstone National Park

Yellowstone National Park, Wyoming, is the home of Steamboat Geyser, the largest active geyser in the world. It is the site of Giant Geyser, which has the greatest total water discharge when it has one of its rare eruptions. Smaller is Bead Geyser, which operates with amazing precision. And so on. Pick nearly any category you wish and you will probably find the "best" example among the geysers of Yellowstone. This should not be a surprise. Yellowstone is by far the largest geyser field on earth. Its more than 700 geysers, most of which are described in the following pages and of which 300 are frequently active, add up to more than 50 percent of all the geysers on our planet.

Geysers have long fascinated humankind. Whenever and wherever they have been encountered, they have received notice—the single geyser (Gayzer Suyu) near the site of ancient Troy in Turkey seems to be mentioned in Homer's *Iliad*, written around 700 B.C., and the first accurate description of a geyser eruption came from a

Danish monk in Iceland in A.D. 1294. Therefore, for all their rarity, geysers are among the most familiar features of our natural world. However, aside from Yellowstone, one must do a lot of traveling to see other good examples. The four largest geyser fields outside of Yellowstone combined—in order, Kamchatka, Chile, New Zealand, and Iceland—contain fewer geysers than does Yellowstone. (For more about these and other world localities, see the Appendix.)

Within Yellowstone are more than 10,000 hot springs of various types. Overall, fumaroles (steam vents), mud pots, and quietly flowing springs are much more common than geysers. The geysers are found only in a few relatively small areas. These are the "geyser basins"—specifically, Upper, Midway, Lower, Norris, West Thumb, Gibbon, Lone Star, Shoshone, and Heart Lake (see Map 3.1).

These geyser basins are not necessarily true basins but instead are simply geographic areas in which hot springs, including geysers, are found. The basins are not large. For example, all of the more than 200-plus geysers of the Upper Basin are located within an area just 2 miles long and ½ mile wide.

The Upper Geyser Basin is the most important. There are more geysers here than in any other area of both Yellowstone and the world. This is the home of Old Faithful, Beehive, Grand, Riverside, and most of the other famous geysers. Interspersed among them are hundreds of other, beautiful hot springs.

The Lower Geyser Basin covers a much larger area—about 5 square miles—but the geyser groups are more widely scattered. Some of the better-known individual geysers here are Great Fountain, Clepsydra, White Dome, and Fountain. It is also the site of the Fountain Paint Pots, among Yellowstone's largest assortment of mud pots.

The Norris Geyser Basin is the third largest, both in area and number of geysers. To many it is the most important basin because it is different from any other. The water at Norris is acidic, and some of it is more caustic than battery acid. All other geyser basins discharge alkaline water. Norris is the site of Echinus Geyser, one of the most beautiful and regular geysers in the Park, and Steamboat, the largest in the world when active.

The Midway, West Thumb, and Gibbon geyser basins contain fewer geysers than the rest. They are, however, accessible areas, with relatively little walking necessary to see the geysers. Definitely, each has its own attractions and is well worth the time it takes to see it.

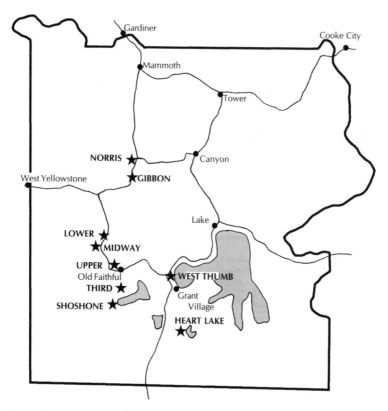

Map 3.1. *Geyser Basins of Yellowstone National Park*

Three other geyser basins lie in the backcountry. The Third Geyser Basin is about 2½ miles by trail and includes two large and several small geysers. The Shoshone and Heart Lake Geyser basins lie at greater distances. Few people visit them, so they remain largely untouched. Marked trails lead to these basins, but none of their features is identified by signs. Although one is free to wander among these backcountry hot springs, remember that all thermal areas are dangerous and extreme caution is necessary.

In the following descriptive chapters, the geysers are detailed according to geyser basin. Each area is further subdivided into groups, and there the geysers are described according to the order in which they lie along the trail. Every set of descriptions is accompanied by a table summarizing the activity of that group.

OLD FAITHFUL GROUP

Only six springs belong to the Old Faithful Group (Map 4.2. Table 4.1, numbers 1 through 5). All are geysers except East Chinaman Spring, just a few feet from Chinese Spring, which boils, overflows, and generally has a geyser-like appearance but has never been known to erupt. In reality, the group may be an isolated portion of the Geyser Hill Group, but because of its location on the opposite side of Firehole River, it seems separate.

1. OLD FAITHFUL GEYSER is the most famous geyser in America and probably in the world; the only likely exception is Geysir in Iceland, the namesake of all geysers. There are geysers that are higher, more frequent, more voluminous, and so on, but no geyser anywhere can match Old Faithful for its combination of size, frequency, and regularity of eruptions. It can be viewed from the many benches near the Visitor Center, Inn, and Lodge, from the mezzanine balcony at the Inn, or from the lobby of the Lodge. However, a much more scenic view is from Geyser Hill, across Firehole River. There, with the greater distance allowing it to be seen to scale and with the hills and sky as a background, Old Faithful provides one of the finest sights in Yellowstone.

Old Faithful was "discovered" by the Washburn Expedition in 1870 (it actually had been seen in eruption at least as early as 1864). The popular story has it that as they entered the Upper Geyser Basin on September 18, the first thing they saw was Old Faithful in full eruption. Unbelievably, here was one of the towering columns of water the mountain men had talked about for so long. Although they spent just 1½ days in the Upper Basin, the expedition members were so impressed by this geyser's frequent activity that they called it Old Faithful. On the basis of their observations, several other geysers could have received the name, but Old Faithful is the only one that has remained true to the name for well over a century.

Numerous rumors about Old Faithful make the rounds. Many claim it once erupted "every hour, on the hour." Others swear that it is smaller than it once was or that it used to play considerably longer than it does now. Following the earthquake at Borah Peak, Idaho, in 1983, dozens of newspaper, radio, and television reports said Old Faithful was dying. That obviously hasn't happened, but Old Faithful is a natural feature and it has undergone some

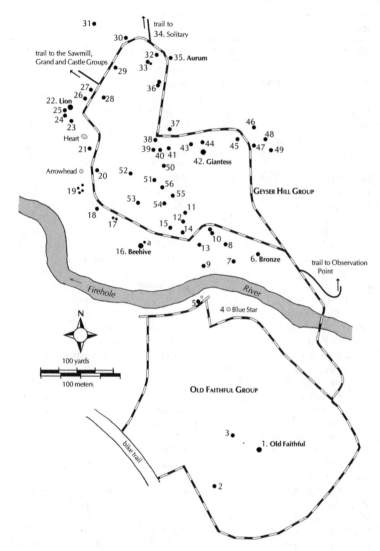

Map 4.2. *Old Faithful and Geyser Hill Groups, Upper Geyser Basin*

changes. One day it will stop playing, but for now it remains much as it has been throughout Park history.

Over the years, more data have been gathered about Old Faithful than any other geyser. The actual average of more than

Table 4.1. Geysers of the Old Faithful and Geyser Hill Groups, Upper Geyser Basin

Name	Map No.	Interval	Duration	Height (ft)
Anemone Geyser, "Big"	10	7–14 min	40–60 sec	6–10
Anemone Geyser, "Little"	10	5–20 min	minutes	2–4
Aurum Geyser	35	2½–14 hrs	1 min	20–25
Beach Geyser	33	frequent	minutes	1–15
Beach Spring	32	minutes	sec–min	boil–10
Beehive Geyser	16	9 hrs–days	5 min	150–200
Beehive's Indicator	16a	see text	0–50 min	6–10
Bench Geyser	55	extinct?	seconds	6
Big Cub Geyser	25	[1998]	4–8 min	40
Blue Star Spring	4	[2002]	minutes	1–2
Boardwalk Geyser	41	hrs–days	8–40 min	20
Borah Peak Geyser	54	[1992]	sec–20 min	3–4
Bronze Spring	6	minutes*	1–3 min	2–15
Butterfly Spring	48	[2003]	2½–3½ min	20–50
Cascade Geyser	9	[1998]	seconds	12–40
Chinese Spring	5	rare	2 min	20
Clastic Geyser	56	infrequent	minutes	2–4
"Coronet Geyser"	41	rare	minutes	5–10
Depression Geyser	18	8–36 hrs	3–10 min	8–10
Dome Geyser	49	min–hrs*	minutes	boil–30
Doublet Pool	36	rare	min	boil–2
Dragon Geyser	51	[1992]	unrecorded	2–5
Ear Spring	29	rare	seconds	2–15
Giantess Geyser	42	0–41/year	1–43 hrs	100–200
Goggles Spring	26	[2002]	2 min	20–30
Infant Geyser	45	see text	–	2
Lion Geyser, between series	22	4–9 hrs	–	–
Lion Geyser, in series	22	1–3 hrs	sec–6 min	30–98
Lioness Geyser	24	[1952]	5–10 min	50
Little Cub Geyser	23	50–90 min	5–10 min	10
Little Squirt Geyser	8	days*	hours	4–8
Model Geyser	50	5–20 min	1–10 min	4
Mottled Pool	46	seconds	seconds	1–10
North Goggles Geyser	27	[2004]	sec–4 min	10–50
Old Faithful Geyser	1	30–127 min	1½–5 min	106–184
"Park Place Geyser"	41	hrs–days	min–hrs	1–2
Peanut Pool	47	[2004]	minutes	1
Pendant Spring	30	[1987]	sec–min	1–6
Plate Geyser	39	1–9 hrs	3 min	15
Plume Geyser	14	27 min–hrs	1–2 min	20–30
Pot O Gold	21	[1988?]	minutes	4

continued on next page

Table 4.1—*continued*

Name	Map No.	Interval	Duration	Height (ft)
Pump Geyser	37	minutes	seconds	2–15
Roof Geyser	52	minutes	seconds	subterra-nean
Scissors Springs	17	irregular	variable	1–5
Silver Spring	7	minutes*	hours	2–15
Slot Geyser	40	min–hrs	seconds	1–4
Solitary Geyser	34	4–8 min	1 min	3–20
Spew Spouter	12	[1959?]	seconds	2–3
Split Cone	2	hrs–days	min–hrs	inches–2
Sponge Geyser	38	1 min	seconds	1–2
Spume Geyser	12	hours*	seconds	5
Surge Geyser	11	[1984]	1 min	8
Teakettle Spring	43	see text	–	–
"Teapot Geyser"	3	unknown	hours	2
UNNG-GHG-2	15	frequent	near steady	1–2
UNNG-GHG-3 ("The Dwarfs")	19	see text	sec–min	1–8
UNNG-GHG-5	31	steady	steady	1–5
UNNG-GHG-7	33	frequent	minutes	bubble–1
UNNG-GHG-8	53	[1986]	1 min	20
UNNG-GHG-10 ("Kitten")	28	7–9 min*	seconds	1–10
UNNG-GHG-11 ("Marmot Cave")	20	infrequent	seconds	4–6
UNNG-GHG-12 ("Improbable")	13	steady	steady	1–3
Vault Spring	44	see text	4–5 min	6–20

* When active.

[ ] Brackets enclose the year of most recent activity for rare or dormant geysers. See text.

46,000 intervals prior to the 1959 earthquake was 64.91 minutes. Since 1959 there has been a tendency toward longer intervals. The 1983 earthquake bumped the average to over 78 minutes. This trend to longer intervals continued into the 2000s. Single intervals longer than 100 minutes are common, while those shorter than 80 minutes now occur only three or four times per week. The average for the 4,189 intervals recorded electronically during January–September 2007 was 90.40 minutes, within a range from 45 to 120 minutes.

In spite of its recent "slowdown," Old Faithful is regular enough to allow the rangers—and you—to predict the time of the

Old Faithful Geyser is one of the natural wonders of the world, reliably reaching over 100 feet high about sixteen times per day.

next eruption. There is a direct relationship between Old Faithful's duration and the following interval. A short eruption means that less water and heat have been discharged, that a shorter time span will be required to regain that water and heat, and therefore that the next interval between eruptions will be shorter than average. For example, if an eruption lasts only 2 minutes, it will be around 65 minutes until the next eruption. A long duration (say, near 4½ minutes) will yield a longer interval; the prediction might be 95 minutes. The historically known range in intervals is from 30 to 127 minutes (plus controversial but probable intervals of only 12, 17, and 18 minutes and another, unlikely interval of 148 minutes). Old Faithful usually has either a long or a short interval. Since the 1990s, the short intervals (defined as 75 minutes or less) have been uncommon, occurring not more than 4 percent of the time. This loss of the shorter play is largely responsible for the increase in average interval.

How tall an eruption might be is subject to variation, too. In June 1985 the average for 211 carefully measured eruptions was 123½ feet, just 3 feet less than a similar calculation done in 1878. Much of the belief that Old Faithful isn't as powerful as it once was may result from different viewing times. Such stories can arise when a person visited the geyser years ago and happened to see an especially fine eruption. The height can exceed 180 feet. Any smaller play on another occasion will make the geyser seem weaker; in fact, on windy days some eruptions barely reach 100 feet. A person's memory can also play tricks; a remembrance of years past is often magnified. All evidence is in contrast. Old Faithful has not changed to any great degree.

Shortly before Old Faithful begins to play, water periodically splashes from the vent. Sometimes this "preplay" will continue for 10 to 20 minutes before the eruption begins, but usually Old Faithful begins to spout after only a few such surges. One splash will be heavier than the others and will be sustained for a few seconds, and the eruption is on. The water column rapidly rockets to its maximum height. After a minute or so, the column begins to shrink and slowly drops. Short-duration eruptions stop abruptly, while those of longer durations end with a minute or two of low jetting and a weak steam phase.

Each of Old Faithful's eruptions is different from every other, and all are magnificent. Old Faithful Geyser is definitely one of the natural wonders of the world.

2. SPLIT CONE is the weathered geyserite cone directly west of Old Faithful Geyser, nearest the Old Faithful Inn and only a few feet from the boardwalk. Small eruptions were given passing notice following the 1959 earthquake but were not recorded again until December 2000. Eruptive episodes have grown more frequent since then. They have occurred at least weekly since 2005, with bubbling and weak splashing up to 2 feet high. Some durations are hours long.

3. "TEAPOT GEYSER" is located on top of a low geyserite mound only about 150 feet northwest of Old Faithful. Although the eruptions may last as long as several hours, they are only 2 feet high. Since the geyser is nearly 400 feet from the nearest trail, they aren't impressive and few people notice "Teapot." The interval has never been accurately determined.

4. BLUE STAR SPRING is the closest blue pool to Old Faithful and is therefore one of the most popular features in the Upper Geyser Basin. The origin of its name is obvious, as is the trash that often litters its crater. Blue Star is a geyser. Eruptive activity was first described in 1925–1926 and then again in 1997 and 2002 when it was occasionally seen having erratic splashes 1 to 2 feet high.

5. CHINESE SPRING. In the early days of Yellowstone Park, there were many more concessions than at present. One of them used Chinese Spring as a laundry. One day the proprietors (who may actually have been Japanese) were doing some wash when the spring they used erupted and scattered clothes across the surrounding landscape. And so the name. Chinese Spring rarely erupts, but when it does the water column surges to 20 feet for about 2 minutes.

The boiling spring a few feet east (right) of Chinese Spring is East Chinaman Spring. It has never been known to erupt. Across the river, the noisy vent at water level is Sputter Spring.

GEYSER HILL GROUP

On Geyser Hill, the white, sinter-clad area directly across Firehole River from the Old Faithful Group, are found over fifty geysers (Map 4.2; Table 4.1, numbers 6 through 56). Some of these are among the largest anywhere, while many others are almost inconspicuously small. Geyser Hill alone, with no other geyser in Yellowstone, would comprise the fourth largest geyser field on Earth.

All of the hot springs on Geyser Hill are connected with all the others. Exchange of function is extremely common. In addition are three mysteries only recently discovered. First is a cycle known as the "Geyser Hill Wave," which causes regular increases and decreases in the intervals of many geysers in the group. Often these variations are subtle, and it takes a lot of observing to become familiar with the changes, but understanding them makes the activity on Geyser Hill anticipatable if not outright predictable. Some geysers additionally but inconsistently exhibit diurnal behavior—that is, short intervals by day and longer intervals at night. Third, first seen during 2002 have been random episodes of strange geyser behavior that apparently have to do with neither the Geyser Hill Wave nor day-night variations. This behavior is marked by odd, often unprecedented eruptions in several geysers at about the same time.

Geyser Hill is traversed by a loop trail. It connects with other trails that lead to Old Faithful, the Sawmill Group, and Solitary Geyser–Observation Point.

6. BRONZE SPRING is the first pool encountered to the left of the boardwalk as one arrives on Geyser Hill from Old Faithful. It is usually a quiet and rather nondescript pool lined with orange-brown cyanobacteria and is often drowned by runoff from springs higher on Geyser Hill. On occasion, though, it undergoes significant eruptions. These are commonly associated with the periodic cycles that take place within Geyser Hill. Most eruptions are only a foot or two high, but bursts higher than 15 feet have been seen.

7. SILVER SPRING lies a few feet west of Bronze Spring (6) and also responds to the cyclic Geyser Hill Wave. Over the course of several days, the water slowly rises in the crater. When the crater is full, intermittent boiling eruptions dome the water as high as 2 or 3 feet. Then, if nearby Little Squirt Geyser (8) erupts, the pool level in Silver will drop rapidly, and eruptions can reach up to 8 feet high (15 feet in October 1988) for several hours. However, if Little Squirt does not erupt, then Silver will remain nearly full with gentle boiling.

8. LITTLE SQUIRT GEYSER was named because of the squirting nature of the small eruptions. Its original name was "Gnome," and later it was called "Spiral." Unless it is playing, the location goes unnoticed by most, since it is only a slight, yellowish depression in the geyserite platform. Before the 1959 earthquake, Little Squirt was irregular and long dormant periods were known. Now it is most likely to erupt at the culmination of the Geyser Hill Wave, but it can skip these opportunities. The intervals can be as short as 2½ days; but at other times, as during 2007, Little Squirt can go as long as 2 weeks without being seen. Most durations are 12 to 18 hours, throughout which the squirting jets reach between 4 and 8 feet high.

9. CASCADE GEYSER was active during the early days of the Park. For part of the 1890s it was a showpiece, with eruptions as high as 40 feet recurring at intervals as short as 10 minutes. Since 1898, however, it has usually been dormant. It was active for brief periods in 1914, 1948, 1950, after both the 1959 and 1983 earthquakes, and again for just three days in January 1988. The most significant activity since the 1890s took place during 1998. Beginning a few hours after a small local earthquake on January 9, Cascade at first had intervals as short as 2½ minutes. The height was as great as 40 feet, but the durations were seldom longer than 10 seconds. The frequency gradually decreased as the year wore on, and the activity ended in mid-October 1998. Cascade has not erupted since then.

10. THE ANEMONE GEYSERS. Anemone is a double spring. Two vents about 10 feet apart are the sites of semi-independent geysers. Most of the time the two act separately, with little apparent relationship between them, but during some periods an eruption of one will invariably follow that of the other, and sometimes they play simultaneously. Both craters are shallow basins lined with pearly beads of sinter.

"Big" Anemone, nearest the boardwalk and the larger of the two, normally shows the greater activity. During most seasons it erupts every 7 to 10 minutes. Some of the spray reaches over 8 feet high, while angled bursts may reach the boardwalk. Most eruptions last about 40 seconds. If "Little" Anemone erupts, then the interval before "Big" does so again may be as long as 15 or 20 minutes.

"Little" Anemone plays from the vent farther from the boardwalk. Its eruptions are smaller than those of its neighbor, but they generally last considerably longer. On rare occasions near-constant activity by "Little" will render "Big" dormant for days at a time.

The first significant variations in Anemone activity began in February 2003 when "Little" fell dormant for its first known time. "Big" simultaneously suffered somewhat longer intervals. Both geysers resumed activity in late 2004.

11. SURGE GEYSER is an old spring or geyser rejuvenated by the 1959 earthquake—the crater was there, but eruptions had never been recorded. Following the earthquake, Surge erupted frequently until 1963. Since then it has sometimes been active in concert with Giantess Geyser (41), proving an underground connection between the two. Such eruptions lasted about 1 minute and reached 8 feet high but have not been seen since 1984.

12. SPUME GEYSER (previously listed as **UNNG-GHG-1**) and Spew Geyser have separate vents within a single crater. Spume may have had occasional eruptions during the early days of Yellowstone, but then it had none until the Hebgen Lake earthquake in 1959. The ragged crater was considerably enlarged by those eruptions, which burst powerfully up to 10 feet high from a vent at the northeast (right-hand) corner of the crater. The action rapidly died down. On a few occasions since, most notably at the time of Giantess Geyser's (41) eruptions, it has played about 5 feet high. Spew apparently never erupted until after the 1959 earthquake. Its vent is in the southwest (left-hand) part of the crater, where eruptions are rare and brief.

A vent informally named "Scuba Geyser" appeared here during the winter of 2004. It and Spume might be the same feature. Active episodes days to weeks apart included intervals as short as 1 hour and durations of about 5 minutes. The lazy splashes were mostly less than 2 feet high, but bursts as high as 6 feet were seen. "Scuba" stopped playing during 2006.

13. UNNG-GHG-12 ("IMPROBABLE GEYSER") was a small, nondescript feature down the slope toward the river from the Anemone Geysers (10). Often appearing to be empty, for years it was known to act as a subterranean spouter that came to be called the "Pathetic Little Hole" (or "PLH"). On October 23, 2005, it erupted with pulsating jets that reached fully 20 feet high. Other major eruptions, as high as 15 feet and with durations longer than 2 minutes followed by a steam phase, were occasionally seen over the next 2 weeks. One interval was shorter than 1½ hours. By early November the small vent had been enlarged into a crater 6 feet long and 4 feet wide, and a new vent made its appearance. With that, the activity declined into perpetual boiling 1 to 3 feet high. Major eruptions have not been observed since early December 2005, and none are expected as long as the constant boiling persists.

14. PLUME GEYSER was created in 1922 when a small steam explosion opened its vent. After about 4 years of irregular activity as "Sinter Geyser," it fell dormant until 1940 and was not frequently active until 1947. Since then it has been active most of the time. For a while, Plume really did seem to erupt "every hour, on the hour." The 1959 earthquake produced a short dormancy, but by 1962 it was again a highly regular geyser. The new average interval was only 27 minutes. Such was the case until 1973.

Plume is a good example of how geysers can change themselves. At the end of December 1972, another steam explosion added an extension to the vent. Much of the eruption now issues through this newer opening. Instead of the slender 40-foot jet of old, the eruptions are massive bursts 20 or 30 feet high.

Plume's modern activity is complex. A model of precision until the late 1980s, its behavior was infrequently modified by outside forces such as eruptions by Giantess Geyser (41). By 1988, observers realized that it was undergoing slight but regular interval variations. During 1992 and 1993, Plume behaved in a strongly diurnal fashion. Daytime intervals could be as short as 25 minutes; at night, it slowed down to intervals as long as 70 minutes and occasionally fell completely "asleep" for several hours. The cause of this is unknown. It clearly is neither a weather nor a tidal effect. Although less extreme, diurnal behavior has also been observed during more recent active episodes, and one overnight interval in midsummer 2007 was of 11½ hours.

After so many years of reliable action, it came as a surprise in late April 2003 when Plume fell into dormancy. For more than a year the only eruptions occurred during the first few days following eruptions of Giantess Geyser and during one case of unexplainably odd events on Geyser Hill. During the dormant periods, the water typically stood several inches below the crater rim and sometimes was cool enough to allow orange cyanobacteria to grow within the westernmost vent. A partial rejuvenation began on July 24, 2004. Active episodes that varied from a few days to several months were separated by equally long dormancies. This behavior continued throughout 2005. Finally, Plume resumed regular action in 2006, and the 2007 average of 6,724 intervals recorded electronically was 55 minutes.

15. UNNG-GHG-2 was a ragged hole a few feet to the left of Plume Geyser (14). During some seasons eruptions were frequent, reaching 2 feet high, but erosion and sedimentation have obliterated the crater. The twin vents of another small geyser lie a few feet beyond GHG-2.

16. BEEHIVE GEYSER is the second or third largest regularly active geyser in Yellowstone. During most of recorded history it was an infrequent performer, with eruptions days to weeks apart. Starting in the early 1970s and continuing into 2005, it was a regular, daily performer. Average intervals usually fell between 16 and 20 hours, but the wide range of 8½ to 25 hours made Beehive technically unpredictable. In 2005 Beehive reverted to erratic activity, with intervals varying between 1 day and a few weeks long, but in 2008 it resumed more frequent activity.

The eruptions issue through a cone 4 feet high and shaped like an old-fashioned straw beehive. The vent within the cone is very narrow and acts like a nozzle so that a slender column of water is shot under great pressure as high as 150 to 200 feet. To observe an eruption of Beehive from the boardwalk near its cone is a unique experience. The awesome display combined with the pounding roar of escaping steam is completely unforgettable. Viewed at a distance the impression is very different. Then the slenderness of the jet becomes apparent, with needle-like rockets of water towering above the surroundings. The entire eruption lasts about 5 minutes, including a short, weak steam phase at the end.

It is difficult to tell when Beehive might erupt. Water spraying out of the cone tells little, although it often attracts a considerable crowd. Such splashing happens throughout much of the quiet interval. Only in the last few minutes before an eruption do the splashes become notably large and frequent. One of these eventually triggers the full display. Knowledge of the Geyser Hill Wave cycle sometimes allows estimates of eruption times to be made, but these guesses will never be more precise than plus or minus a few

Beehive Geyser, on Geyser Hill, is among the largest cone-type geysers in the world. The steady jet reaches between 150 and 200 feet high. Unfortunately, Beehive sometimes goes days or even weeks between eruptions.

hours. Of the greatest use is Beehive's Indicator (16a), whose performances almost always indicate an impending eruption.

16a. BEEHIVE'S INDICATOR plays out of a small vent a few feet to the front-left of Beehive's cone. Historically, when Beehive had days-long intervals, the Indicator was rarely active, but since the early 1970s it has been true to its name. With few exceptions, an eruption of Beehive is preceded by an eruption of Beehive's Indicator. The steady water jet of the Indicator reaches 6 to 15 feet high. It usually starts to play 10 to 25 minutes before Beehive erupts (the known range in lead time is from less than 1 to about 45 minutes). Since it is readily visible from a distance and has such a long duration, the Indicator provides ample opportunity for people to get close to Beehive for the promised show. There are times when Beehive erupts without the Indicator, but they are uncommon.

Unfortunately, the use of Beehive's Indicator is not 100 percent reliable. Sometimes it erupts for long durations (up to an hour) without resulting in play by Beehive. These "false Indicators" are uncommon during most years, but in August 1992 they completely took over the activity from Beehive. Unlike anything seen before, the Indicator erupted every 3 to 5 hours for durations of 50 minutes. Beehive was dormant from August 7 until the Indicator stopped its unusual performances on September 1. A similar episode took place during July 1994, and others have occurred briefly several times since then.

Events called "mid-cycle indicators" are also known. They play meaningless weak and brief jets of water about halfway through Beehive's interval.

There are two other so-called indicator vents for Beehive. One is "Beehive's Second (or Close-to-Cone) Indicator." It lies between Beehive's Indicator's vent and Beehive's cone; most active when Beehive is having days-long intervals, it is too erratic to serve as an indicator. The other is the "West Bubblers," a few feet beyond Beehive's cone; they begin to sputter a few inches high as long as several hours before Beehive.

17. SCISSORS SPRINGS. These two small springs used to flow steadily. The rivulet from each converged a short distance from the pools, then split again so the whole formation resembled a pair of shears. By 1950 one spring had stopped flowing, and the other soon stopped, too. In 1974 they suddenly sprang to life again, not only overflowing but also acting as small geysers a few feet high. The activity has waxed and waned several times since then. Most often only the right-hand spring overflows, while that to the left boils gently.

18. DEPRESSION GEYSER was so inactive before the 1959 earthquake that it hadn't been given a name. The action triggered by those tremors repeated every 3 to 4 hours but lasted only 2 to 3 minutes. Eruptions briefly became even more frequent following the 1983 quake. However, the intervals have gradually been increasing through the years since 1983. Eruptions now range between 8 and 36 hours apart, with an average interval near 15 hours. The bright side is that the durations have also increased, so the 6- to 10-foot splashing sometimes lasts as long as 10 minutes.

19. UNNG-GHG-3 ("THE DWARFS"). In the flat area a short distance to the right (northwest) of Depression Geyser (18) are several spouters and geysers. The number of these springs acting as truly periodic geysers varies, with nine the most recorded at any one time. These geysers were never reported as active before 1947, and then only infrequently until 1972. Now, although most eruptions are less than 2 feet high, they are frequent and vigorous. Beginning in 2005 and continuing through 2006, one of the vents farthest to the right (north), known as "North Dwarf Geyser," had occasional long-duration eruptions that included jetting as high as 10 feet that continued on and off for several hours. In 2007 another, called the "Red Dwarf Geyser," played 6 to 12 feet high with intervals as short as 40 minutes.

Across the boardwalk from the "Dwarfs" are several ragged explosion craters. Although all were formed in prehistoric times, some had eruptions as a result of both the 1959 and 1983 earthquakes. As much as 10 feet high, none performed regularly or frequently, and the only one given a name, Blowout Spring, was active only on the night of the 1959 earthquake.

20. UNNG-GHG-11 ("MARMOT CAVE GEYSER") occupies a jagged crater across the boardwalk from non-eruptive Arrowhead Spring. It was named "Marmot Cave" before any known geyser activity because a yellow-bellied marmot lived in the cavernous opening at the back of the crater. Eruptive episodes are not common but briefly occur during most years. Then the frequent play lasts a few seconds, with splashes 4 to 6 feet high.

In May 2007, a small vent was opened near the boardwalk in front of "Marmot Cave." Eruptions were days apart, but a few reached fine spray onto the walkway.

21. POT O GOLD is, or was, near the boardwalk between Arrowhead Spring and Heart Spring. It was named during the post-1959 earthquake studies when it was a cool, quiet pool lined with orange-yellow cyanobacteria. It first erupted in 1980 and had several eruptions in subsequent years, some of which were simultaneous with eruptions by Giantess Geyser (42). A series of eruptions on July 28, 1988, reached up to 4 feet high. About 12 feet to the right of Pot O Gold was another crater that was only a tiny hole in the geyserite until a series of small but explosive eruptions in 1985 enlarged its crater and littered the boardwalk with chunks of geyserite. Neither of these springs is known to have erupted since the 1980s, and shifting runoff from nearby Heart Spring has inundated both, so they have virtually disappeared.

The Lion Geyser Complex (numbers 22 through 28)

Situated on a high sinter mound rising abruptly above Firehole River at the far northwest end of Geyser Hill are four geyserite cones. Related to them are three other springs, below and northeast of the mound. All seven are geysers, and together they make up the Lion Geyser Complex.

22. LION GEYSER itself issues from the largest cone on the mound, the one farthest to the right (north) and nearest the boardwalk. It is also the largest geyser of the pride. Lion is perhaps the classic example of a cyclic geyser, in which there are distinct series of eruptions. The initial eruption of a series has a duration of around 6 minutes, during which water is jetted to between 50 and 70 feet high; one exceptional eruption was measured at 98 feet in 1988. The subsequent eruptions of a series recur at intervals about 1½ hours (rarely 3 hours) long. With durations of 2 to 4 minutes, they are considerably weaker than the initial eruption and sometimes reach less than 30 feet high. Most series consist of two to four eruptions, although a few have only the initial eruption (a "series" of one) and others see as many as nine eruptions. The longer series often include minor eruptions that look like ordinary activity except for durations of only a few seconds.

The number of eruptions within one series roughly controls how long a "cycle interval" (that is, series start to next series start) will be until the next series begins—the more eruptions within a

Lion Geyser is the largest member of the "Lion Complex," with eruptions ranging between 50 and 100 feet high.

series, the longer the time between series. If there are only one or two eruptions, then the cycle interval can be as short as 4½ hours; longer series produce cycle intervals as great as 19 hours. The average is about 10 hours.

During the quiet between series, Lion occasionally splashes as if about to erupt, and sometimes these splashes seem ready to trigger the geyser several hours before the initial eruption actually takes place. Perhaps it was this action that resulted in Lion's original name, "Niobe." In Greek mythology, Niobe boasted too loudly about her children, only to see them killed by the gods in revenge; bewailing her loss, she turned to stone and remained forever wet with tears. The modern name came about because of the loud roars of steam that gush from the vent just before all of Lion's eruptions, except the initial one.

23. LITTLE CUB GEYSER is the only other member of the Lion Complex to be frequently active. It is the small cone on the far left (south) side of the mound, farthest from Lion. It is larger than it looks from the boardwalk, jetting up to 10 feet high throughout the 10-minute duration. Little Cub historically erupted every 50 to 90 minutes with little variation, but significant changes began to appear in 2004. At times during that summer it underwent diurnal behavior, with mysterious hours-long "sleep" periods at night. Then, in December 2004 it began highly erratic behavior, with intervals that varied from less than 1 to more than 43 hours. It undergoes these changes abruptly and unpredictably and with no known cause. Just off the right-hand side of Little Cub's cone is "The Cubby Hole," which plays a few inches high during Little Cub's eruptions.

24–25. LIONESS GEYSER and **BIG CUB GEYSER** are the two center cones, on the left and right, respectively. These two geysers seldom erupt, but when they do they sometimes play in concert with one another. Lioness can reach 50 feet and Big Cub about 40 feet high.

The Lion Complex clearly shows exchange of function, both between other groups and among its own members. When Lion Geyser is having active cycles, Lioness and Big Cub are dormant. Since Lion has been the dominant geyser during all of recorded history, eruptions by the other two have been rare. Lioness, without Big Cub, might have been active during the 1890s—that is by no means certain—and both geysers exhibited some sort of action during the mid-1920s. There was no activity from 1927 until active phases took place during 1947 and in 1950–1952. The 1947 activity saw at least twelve concerted eruptions, Big Cub starting the action and Lioness following 1 to 2 minutes later. Big Cub erupted at least one time in 1950, and it had a single eruption in 1951. That was followed by at least eight eruptions by Big Cub and six by Lioness during 1952. In this case, most of the eruptions of either geyser were independent of the other.

The Borah Peak earthquake of 1983 produced a dormancy in the entire Lion Complex. When activity resumed in 1985, it seemed as though both Lioness and Big Cub were about to enter a new active period, as they frequently boiled and splashed. The only eventual eruptions, however, were single ones by Big Cub alone on August 6, 1987, and October 20, 1998. Both geysers boil and sputter vigorously at times and seem to be on hair triggers. Someday they are sure to be seen again.

26. GOGGLES SPRING is the irregular crater a few feet north of Lion's mound. It was named during the late 1910s for uncertain reasons but at the time when automobile drivers wearing goggles became common in Yellowstone (the first car entered the Park in 1915). The shallow, round, orange pool next to Goggles' crater used to contain a small vent of its own, but it has been choked with debris for many years. Goggles Spring was probably active in some fashion during the 1920s. It erupted one time in 1952, when Big Cub and Lioness were also active, but then not again until 1985. Following the rejuvenation of the entire Lion Complex from its 1983–1984 dormancy, there was an increase in activity by the entire complex. Goggles shared in this by having several eruptions in concert with those of nearby North Goggles Geyser (27) and Lion. The eruptions were brief and not more than 6 feet high, but the bursts were jetted with enough force and at such a low angle out of an empty crater that they reached the boardwalk more than 20 feet away. The only additional known activity was a single, very different eruption that took place near dark on April 26, 2002, a day when other strange events took place on Geyser Hill. This bursting eruption reached at least 30 feet high, lasted more than 2 minutes, ended with a long and noisy steam phase, and was in concert with a major eruption by North Goggles Geyser.

27. NORTH GOGGLES GEYSER was evidently the site of frequent action before recorded times, as it has large runoff channels leading away from its small cone. Historically, though, it was rarely, if ever, seen prior to the 1959 earthquake. It has been erratically active since then. Whether active or not, North Goggles undergoes quiet overflow every few minutes. It is strong bubbling and boiling during a long overflow period that triggers an eruption. North Goggles has minor, intermediate, and major eruptions. During most years the minors are the only variety seen. These are very brief and seldom more than 10 feet high. Studies indicate that they are most likely to take place during the quiet cycle interval of Lion Geyser. During 2003, these eruptions occurred almost daily. The intermediate type of eruption was seen only during 1994. They resembled major eruptions except for shorter durations and the lack of a concluding steam phase. Major eruptions, the dominant form of activity only during 1985 and rare ever since, are spectacular. Almost always occurring during or shortly after a Lion series, they can reach more than 50 feet high. The steady water jet pulsates throughout the eruption and only begins to decline near the end of the 4-minute duration. These eruptions are followed by short but briefly powerful steam phases. North Goggles has not erupted in any fashion since 2004.

28. UNNG-GHG-10 ("KITTEN GEYSER") is a small pool located about 30 feet across the boardwalk from North Goggles Geyser. Although seldom noticed, "Kitten" is always active as an intermittent spring, repeating quiet overflows every 7 to 9 minutes. Sometimes, usually shortly before Lion's initial eruption, it has true eruptions with the same 7- to 9-minute frequency. Most bursts are a foot or less high, but splashes up to 10 feet high have been seen.

29. EAR SPRING is a prominent small pool right next to the boardwalk. Its water is usually superheated (the water temperature is hotter than the local 198°F [93°C] boiling point), so there is a constant sizzling and bubbling about the rim of the crater. True eruptions are extremely rare. One reported as 15 feet high took place during 1957; the 1959 earthquake produced a few eruptions perhaps 2 feet high, and additional action of that size was observed in 1986, 1992, and 2004. Also, Ear Spring sometimes undergoes heavy surging without actually splashing in association with eruptions by Giantess Geyser (42) and as was seen independently in 2002. In the opposite fashion, Ear is occasionally found cooler than boiling, resting quietly several inches below overflow.

30. PENDANT SPRING is usually a quiet pool. Years ago it was known as "Algous Pool" when it was cool enough to house a thick growth of cyanobacteria, but more often it has been too hot for that. Geyser eruptions are known from just three episodes. During the 1870s and 1960s, the eruptions had durations of hours and heights of 1 to 2 feet. Following the 1983 earthquake, the plays were just seconds to a few minutes long, but some bursts reached up to 6 feet high. Pendant Spring has been dormant since 1987.

31. UNNG-GHG-5 lies among some old craters near the woods beyond Pendant Spring (30). Active as a perpetual spouter of very small size prior to the 1983 Borah Peak earthquake, it had frequent eruptions up to 5 feet high during 1984. GHG-5 then gradually declined, and it is again a weak perpetual spouter, nearly invisible from the boardwalk.

32. BEACH SPRING played as a 10-foot geyser during 1939 and on a smaller scale during 1947. Now it acts as a cyclically boiling intermittent spring with enough vigor to qualify it as a bubble-shower spring. Much of the time the water lies low within the crater. Every few minutes it fills, sometimes high enough to cover or even overflow the beach-like terrace around the pool. As the water rises in the crater, it undergoes vigorous superheated boiling and rarely a few 1-foot surges. The duration is a few seconds.

33. UNNG-GHG-6 and **UNNG-GHG-7.** These two small geysers lie a few feet beyond Beach Spring (32). The larger, right-hand (northern) geyser of the two (UNNG-GHG-6) is probably the Beach Geyser of the early 1900s. These two springs are intimately related to one another, filling, erupting, and draining in concert. Beach Geyser had frequent eruptions as much as 15 feet high during 1939, the same year nearby Beach Spring was also active as a notable geyser. Now eruptions are infrequent and seldom more than 2 feet high. Smaller GHG-7 usually does little more than bubble.

Across the boardwalk from Beach Spring and GHG-6 and 7, a trail leads through the forest and up the hill to Solitary Geyser (34). Along the way is a lush forest meadow with some of the finest wetland wildflowers anywhere in Yellowstone. At their best in midsummer, the dominant flowers are columbine, monkshood, Indian paintbrush, chives, sticky geranium, and a rare veronica. From Solitary Geyser this trail continues back across the mountainside to Observation Point and then down to the starting point of the Geyser Hill loop.

34. SOLITARY GEYSER was, contrary to many accounts, active as a small geyser during the early years of Yellowstone, but it was dramatically altered in 1915. In that year permission was given to use the water of the seldom-visited spring for a swimming pool. Solitary Spring immediately became Solitary Geyser. A deepened runoff channel had lowered the water level enough to allow boiling to take place at depth, and, consequently, eruptions began. The play was frequent and powerful, reaching 25 feet high. The Old Faithful Geyser Plunge was closed in 1948 and the runoff channel was repaired, but eruptions continue despite the renewed high water level. Modern eruptions recur every 4 to 8 minutes and last about 1 minute. Most bursts are less than 3 feet high, but a few that reach 15 to 20 feet can be seen with a little luck or a lot of patience.

35. AURUM GEYSER was named because of the golden color of iron oxide that stains the inside of the little cone. Aurum boiled and splashed within its vent at almost all times, but eruptions were rare prior to the 1983 Borah Peak earthquake. It has been continuously active since 1983. The steady jet is angled toward the boardwalk. Reaching 20 to 30 feet high, almost every eruption inundates the walk at the peak of the 1-minute play. Interestingly, Aurum is one of few Upper Basin geysers—perhaps the only one—known to be directly affected by short-term weather changes. During winter into early summer, when the nearby meadow is wet, the intervals tend to be relatively short and regular, ranging between 2½ and 4 hours. Once the meadow has dried out in the heat of summer, the intervals grow erratic, varying unpredictably from 3 to 14 hours. Heavy rain, as falls during summertime thunderstorms, can cause Aurum to revert to the winter mode, but this is always temporary.

36. DOUBLET POOL is one of the most beautiful blue pools in the Upper Geyser Basin. Its water level and overflow are slightly intermittent, and it bubbles and pulsates at the times of high water. On two occasions associated with eruptions by Giantess Geyser (42) and once following the 1959 earthquake, Doublet Pool had true bursting eruptions. They were only 2 feet high.

Near the runoff from Doublet Pool is a small geyserite cone that is sometimes active as a geyser. It has been given the very informal

Aurum Geyser was a rare sight before the 1983 earthquake. Now it erupts as often as eight times per day.

name of "Singlet Geyser," and the play is less than 1 foot high. Several other bubbling vents lie in the same area.

37. PUMP GEYSER was a perpetual spouter 2 to 3 feet high that never seemed to vary its appearance until about 1990. It has since become a true geyser, albeit with both intervals and durations only a few seconds long. The water is sprayed into the sloping basin, with most draining back into the vent. Every few minutes a more powerful eruption jets water under considerable steam pressure, so it fans outward from the long vent to as high as 15 feet. Pump Geyser was named in allusion to The Pump, a perpetual spouter a few feet to the north that has disappeared.

38. SPONGE GEYSER is the second smallest officially named geyser in Yellowstone (Tiny Geyser, #13 in the Gibbon Geyser Basin, is considerably smaller). It got its name because the geyserite decorations on the outside of its cone resemble a sponge. The cone is remarkably big given the size of the eruptions, and Sponge might once have been a more significant feature. When active, water stands near the top of the cone, and eruptions a few inches high recur every minute or so. Dormant periods are known.

39. PLATE GEYSER was named because of plate-like sheets of geyserite found scattered about its crater following the 1959 Hebgen Lake earthquake, which triggered the first known eruptions. It remained an infrequent performer, however, and most eruptions were in association with Giantess Geyser (42) and lasted only a few seconds. There has been much more action since the 1983 earthquake, and Plate was a regular geyser during many recent seasons, with average intervals of about 3½ hours. However, during 2005, Plate developed a cyclic nature, and now eruptive episodes just a few hours long are separated by quiet periods of up to 2 weeks. With durations of 3 to 10 minutes, the highest bursts can reach 15 feet.

40. SLOT GEYSER occupies a crack-like vent between Plate Geyser (30) and Boardwalk Geyser (41). Historical eruptions were described as 1 or 2 feet high but had not been seen for many years until 2007, when it began to have tiny eruptions on a highly erratic basis. The intervals range from 10 minutes to 20 hours. The durations are rarely as long as 10 seconds.

Almost immediately adjacent to Slot is an unnamed and much larger round crater that appeared when Slot reactivated in 2007. Apparently never previously seen to erupt, it splashed 1 to 4 feet high in unison with Slot but also had independent action. Like Slot, the duration is only a few seconds.

41. BOARDWALK GEYSER (formerly known as **"ABRUPT GEYSER"**) made a dramatic appearance in May 1992. It played from an old crater along a fracture; several other geysers, including "Park Place Geyser" (next paragraph), Slot Geyser (40), and Plate Geyser (39), occupy other craters on this same rift. Boardwalk's play began without warning, with water gushing instantly at a sharp angle to as high as 20 feet. It then slowly declined into a weak steam phase at the end of durations that ranged from 8 to 40 minutes. Seen only during 2000 were a few minor eruptions with durations shorter than 1 minute. The intervals were always highly erratic, varying from a few hours to many days. Boardwalk entered complete dormancy in 2006.

Several other geysers have appeared in the general area of Boardwalk since 1992. One, informally called "Park Place Geyser," reaches only 1 to 2 feet high but pours a flood of water into Boardwalk's vent. It is probably the same as the "Second New Thing" that was active along with Boardwalk in 1992. Another member of this complex is "Coronet Geyser," whose rare eruptions splash outward in all directions to perhaps 3 feet high. Like Boardwalk, these geysers have been dormant since 2006.

42. GIANTESS GEYSER is one of the largest and most powerful geysers anywhere. You must be very lucky to see it erupt, though. During most of recorded history it has averaged only two or three active episodes per year; the all-time record for activity was set in 1983, when it had forty-one eruptive episodes (all before the Borah Peak earthquake on October 28). Given the infrequency, it is amazing that members of the Washburn Expedition saw Giantess during their brief visit in 1870. They described it as "the grandest wonder" of their trip.

During the long quiet phase, the large pool periodically boils around the edge of the crater. An active episode begins during one of these "hot periods." One study indicated that the hot periods might become slightly stronger and more frequent during the last few hours of the quiet phase, but this idea is very tentative. The action starts when sudden surging pours a tremendous flood of water over all parts of the surrounding sinter platform. After several minutes this pauses, and the water level drops within the crater. Then the eruption begins in earnest. Bursting, mostly less than 50 feet high, leads to a second pause in the action. Usually 50 to 70 minutes after the initial surging, Giantess starts playing again. If

Giantess Geyser averages only two or three eruptions per year, but its water jets may approach 200 feet high shortly before it enters a steam phase that is sometimes audible more than a mile away.

there is to be a steam phase—and there usually is—jets of water as high as 200 feet blast from the crater.

The entire active phase may last anywhere from 1 to 43 hours; the 1959 earthquake caused one of more than 100 hours. Each of these phases consists of a number of separate eruptions and can be classified into four types. The water-phase type will jet water for 5 minutes on intervals of 25 to 50 minutes, over a total of about 24 hours. In the steam-phase type, the water gives way to powerfully roaring steam, usually during the second water eruption. This roaring has been heard more than a mile away, but then it steadily declines and ends after about 12 hours. First observed in 1959 and the most common eruption type in recent years is the mixed-phase type. This begins like the steam phase but reverts to water phase after 3 to 6 hours; this action lasts as long as 43 hours. Finally, first observed in 1981 and never since 1983 is the short- or aborted-phase type. This is either a water- or a mixed-phase eruption that ends after a very short and weak duration. Following any of these varieties, the crater will refill in anything from a few hours to 3 days. Nothing more than the occasional boiling occurs until the next active phase begins weeks to months later.

A geyser as great as Giantess would be expected to have an effect on other springs in the area, and indeed it does. Dome Geyser (49), Vault Spring (44), Infant Geyser (45), Teakettle Spring (43), Plate Geyser (39), Doublet Pool (36), Ear Spring (29), Pot O Gold (21), Beehive Geyser (16), Plume Geyser (14), Spume Geyser (12), and Surge Geyser (11) are all directly connected with Giantess. This is known because of these springs' reactions to the eruptions of Giantess. In fact, they are all part of one subsurface system, and no doubt it is these numerous connections that make Giantess an infrequent and irregular performer.

43. TEAKETTLE SPRING is not a geyser and perhaps never was. It is mentioned here because it does have a large geyser-like cone. The crater was full to the rim until 1947. In that year nearby Vault Spring (44) began erupting after a long dormancy. The water immediately began ebbing in Teakettle. It can now be heard sloshing deep within the crater.

44. VAULT SPRING was known as a geyser in the early days of the Park, when it occasionally erupted several hours *before* nearby Giantess Geyser (42) began an active period, but no consistent action was known until 1947. In that single year it had remarkable activity, with intervals of 40 to 50 hours, durations of 12 to 24 hours, and heights of 15 feet. More recently it has exhibited two modes of behavior. Most commonly, it begins to play about 6 hours *after* Giantess starts and is usually done erupting before Giantess has quit. Vault can also have independent active phases, as during 1998 when every 2 to 6 days it underwent short series of eruptions. Vault has never completely filled its crater since that episode. Whether with or independent of Giantess, the intervals are around 1 hour. The play reaches 6 to 20 feet high over durations of 4 to 10 minutes.

45. INFANT GEYSER is a spring that seems to have somehow gotten partially disconnected from the rest of the Geyser Hill Group. Prior to the 1959 earthquake, the water level stood well below overflow, and small eruptions took place only in concert with Giantess Geyser (42). After the tremors, Infant began having eruptions on its own several times per day. Gradually, the water level rose, and the eruptions ceased when the crater began to overflow in 1964. During all this time the water was clear and alkaline like the other Geyser Hill springs. Infant has since dropped to the water level of old, but now it is murky gray and strongly acid. It still shows a connection with Giantess, however, by varying its water level whenever Giantess plays, as it also did during the 2003 activity by Butterfly Spring (48).

46. MOTTLED POOL lies uphill, across the boardwalk from Infant Geyser (45). Nothing of it can be seen from the boardwalk (and, of course, you must stay on the boardwalk). There isn't much to see, anyway. Deep within the crater, eruptions rise from a small pool. Frequent but brief, most of the splashes are only 1 to 2 feet high. Mottled probably had some substantial eruptions during the 1800s, but in 1927 it was described as "extinct." The modern activity might date to the 1959 earthquake, and it has continued without change since the early 1970s. Despite its proximity to Butterfly Spring (48), those major eruptions in 2003 had no evident effect on Mottled Pool.

47. PEANUT POOL was nearly forgotten until 1991, when it began to have eruptions. The play was never more than 1 foot high, and Peanut had reverted to an empty hole by the end of 1992. In 2003 it was active again, and it showed a relationship to Butterfly Spring (48) by dropping below overflow whenever Butterfly had strong minor eruptions.

48. BUTTERFLY SPRING, named for the original shape of the crater, had episodes of geyser eruptions as much as 6 feet high during the 1880s and in 1936, but otherwise it always acted as a 1-foot perpetual spouter—until May 2, 2003. During that single month, Butterfly had a series of spectacular major eruptions that converted the butterfly-shaped crater into a deep jagged hole. Jetted at an angle toward and onto the mound of Dome Geyser (49), the muddy, chocolate-brown water reached as high as 50 feet. The play recurred as often as every 4½ hours and lasted 3½ minutes. The last major eruption was on May 31, 2003. Minor activity a few feet high continued thereafter, sometimes on a scale that seemed to fall just short of triggering more major eruptions, but even that had ceased by the end of 2005. Butterfly now splashes continuously, but most bursts are too small to be seen from the boardwalk.

49. DOME GEYSER occupies the large geyserite mound at the upper east end of Geyser Hill. It was a rare performer until after the 1959 earthquake. The November 1959 rejuvenation may have been a delayed response to the August tremors. Even with that, Dome was seen only infrequently until 1972, when active phases began to take place several days to a few weeks apart. Curiously, after the major activity of Butterfly Spring (48) ended on May 31, 2003, Dome did not play again until February 4, 2006. Active episodes last as long as several days, during which eruptions recur every 20 to 40 minutes. The first eruption of a series can reach 30 feet high, but the remaining action consists mostly of vigorous boiling with little actual splashing.

Other Geysers of the Geyser Hill Group

Seven other geysers of this group are worthy of mention. Because of their small sizes and locations well away from the boardwalk, they are not easily viewed.

50. MODEL GEYSER is a small spouter of regular and frequent activity. Best viewed from the boardwalk near Sponge Geyser (38), it plays from a fairly obvious yellow-brown crater in the central portion of Geyser Hill. The eruptions reach about 4 feet high. A smaller unnamed geyser lies a few feet beyond Model, and another is located just to Model's front-left.

51. DRAGON GEYSER is a beautiful blue pool, again best seen from the vicinity of Sponge Geyser (38). Dragon is known to have had occasional eruptions throughout Park history, most recently during 1992 when play 2 to 5 feet high was recorded. Washed areas around the crater following the 1959 earthquake implied that some sort of voluminous eruption took place that night.

52. ROOF GEYSER can rarely be seen at all except for its intermittent steam cloud a short distance northwest of Dragon Geyser (51). It lies completely below ground level, in a deep hole over which is a "roof" of geyserite. The small pool erupts every few minutes for a few seconds. Infrequently, Roof is seen to spray fully 4 feet above the ground.

53. UNNG-GHG-8 is actually just one among a collection of craters atop the hill above Scissors Springs (17) that erupted following the 1983 Borah Peak earthquake, but it was the only one of these craters that continued its activity for more than a few days. Eruptions up to 20 feet high recurred on an irregular basis into 1986. No eruptions are known from any of these craters since then.

54. BORAH PEAK GEYSER was in existence prior to the October 1983 earthquake, but it was much more active for several years following those shocks. It had both minor and major eruptions. All intervals were a few minutes long. The minor eruptions had durations of only a few seconds, whereas the occasional majors lasted as long as 20 minutes. With either type, the play was vigorous bursting 3 to 4 feet high. Borah Peak has been nearly dormant since the early 1990s. (Between 2003 and 2006, an unnamed geyser that plays only inches high from a fracture at the crest of the sinter hill above Beehive Geyser [16] was incorrectly recorded as Borah Peak Geyser, which lies atop the sinter platform farther from the boardwalk.)

55. BENCH GEYSER was once very active, but the runoff from the frequent eruptions by Giantess Geyser (42) during the early 1980s completely filled its crater with gravel. The site is now invisible except as a spot of steam in cold weather.

56. CLASTIC GEYSER is little-known. The name appears on a map drawn shortly after the 1959 Hebgen Lake earthquake. Starting in about 2000, a geyser has occasionally been seen playing 2 to 4 feet high about 75 feet southeast of Dragon Geyser. This is probably Clastic.

CASTLE, SAWMILL, AND GRAND GROUPS

The Castle, Sawmill, and Grand groups of geysers (Map 4.3, Table 4.2, numbers 60 through 103) are often considered to comprise a single "Castle-Grand Group," which is how they were handled in the previous editions of this book. However, historically they were separated into the three separate groups given here. This is the most extensive area of contiguous geysers in the Upper Basin. Most

Table 4.2. Geysers of the Castle, Sawmill, Grand, and Orange Spring Groups, Upper Geyser Basin

Name	Map No.	Interval	Duration	Height (ft)
Belgian Pool	83	[1930s?]	unrecorded	unre-corded
Bulger Geyser	84	frequent	1–12 min	6–12
"Bush Geyser"	99	see text	–	–
Castle Geyser	63	12½–13½ hrs	1 hr	40–100
"Chimney Cone Spouter"	70	steady	steady	inches
Churn Geyser	74	5–20 min*	1½–2 min	10–15
Crack Geyser	100	[1962]	minutes	6
Crested Pool	65	minutes	seconds	boil–10
Crystal Spring	83	[1987]	seconds	2–20
Deleted Teakettle Geyser	73	2–10 min	1 min	2–15
Dishpan Spring (Terra Cotta "B")	66	infrequent	seconds	12
East Economic Geyser	97	[1959]	2–3 min	35–40
East Triplet Geyser	86	[1993]	minutes	5
Economic Geyser	96	[1999]	seconds	10–25
Four Bubblers	108a	rare	unrecorded	1
Gizmo Geyser	63	minutes*	minutes	15
Grand Geyser	93	6–20 hrs	9–13 min	150–200
Liberty Pool	69	[1887]	seconds	75
Limekiln Springs	103	steady	steady	1–2
North Triplet Geyser	85	extinct	–	–
Old Tardy Geyser	82	min–hrs	sec–min	10–15

continued on next page

Table 4.2—*continued*

Name	Map No.	Interval	Duration	Height (ft)
Orange Spring	106	[1986]	seconds	3–15
Orange Spring Geyser	107	[2002]	minutes	4
Oval Spring	81	see text	sec–min	1–20
Penta Geyser	79	2½ hrs–days*	35 min–2 hrs	25
Percolator Geyser	88	hours	min–hrs	1–2
Rift Geyser	90	2–20 hrs	5 min–4 hrs	4
Sawmill Geyser	75	1–3 hrs	9 min–hrs	3–35
Scalloped Spring	72	see text	–	–
Slurp Geyser	83a	see text	min–hrs	1
Spanker Geyser	67	steady	steady	1–6
Spasmodic Geyser	80	1–3 hrs	min–hrs	inches–10
Spatter Geyser	68	[1962]	minutes	6–10
Sprinkler Geyser	60	20–30 min	40 min	10
"Sputniks, The"	89	erratic	minutes	inches–4
Tardy Geyser	77	min–hrs	sec–hrs	10–20
Terra Cotta "A" Geyser	66	2 hrs	minutes	10
Terra Cotta "D" Geyser	66	20 min–days	5 min	1–3
Terra Cotta "E" Geyser	66	frequent	minutes	1
Tilt Geyser	64	45–100 min*	1½–3 min	6
"Topsoil Spring"	94	rare	hours	inches
Tortoise Shell Spring	62	steady	steady	1–6
Turban Geyser	91	17–22 min	5 min	5–20
Twilight Spring	78	see text	minutes	2–5
Uncertain Geyser	76	hrs–days	2–5 min	10–15
UNNG-CGG-6	69	irregular	sec–min	1–2
UNNG-CGG-8 ("Snake Eyes")	71	[1993]	seconds	4
UNNG-CGG-9 ("Key Spring")	95	[1988]	seconds	10
UNNG-CGG-10	66a	hours	hours	inches–2
UNNG-OSG-1 ("Pulsar Spouter")	105	steady	steady	1
UNNG-OSG-2 ("UNNGOSG")	108	[1988]	minutes	15
UNNG-OSG-3 ("South Orange")	109	[1986]	1½ min	2–6
Vent Geyser	92	with Grand	30 min–2 hrs	35–70
Washtub Spring (Terra Cotta "C")	66	20 min*	5 min	20
Wave Spring	98	[1990]	unknown	few feet
West Triplet Geyser	87	hours	20–40 min	6–10
Witches Caldron	102	steady*	steady	2

* When active.

[ ] Brackets enclose the year of most recent activity for rare or dormant geysers. See text.

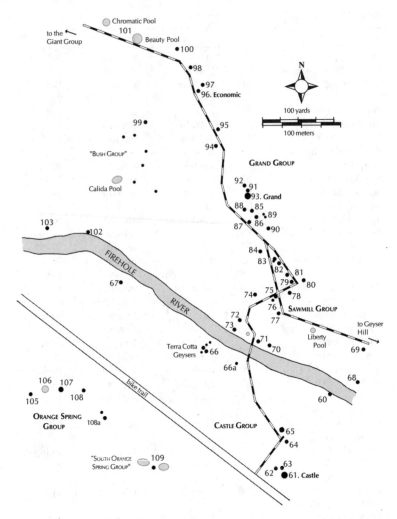

Map 4.3. *Castle, Sawmill, Grand, and Orange Spring Groups, Upper Geyser Basin*

of the forty geysers are active to some degree all of the time, and two—Castle and Grand—are among Yellowstone's largest as well as predictably regular. Besides the numerous geysers, these groups also contain several beautiful pools.

These groups can be approached by two routes. A paved trail leads directly from the Visitor Center, the Inn, and the lower store to Castle Geyser. The other route leaves the Geyser Hill loop near Lion Geyser and leads through the Sawmill Complex directly to Grand Geyser.

THE CASTLE GROUP

The Castle Group (numbers 60 through 67) consists of those springs that lie between the paved trail and Firehole River, and for convenience it also includes the Terra Cotta Geysers (#66) and Spanker Geyser (#67).

60. SPRINKLER GEYSER is difficult to see well; the best viewing spot is from the boardwalk across the river, near Liberty Pool (see #69). Named in the 1870s, it has probably been active at all times since, but no detailed study of its behavior was performed until 1993. Sprinkler turned out to be quite regular in its performances, actively erupting nearly 60 percent of the time. Most intervals were between 20 and 30 minutes long, while some durations exceeded 40 minutes. The height is about 10 feet. A subsidiary vent at river level, known as River Sprinkler, simultaneously erupts a foot or two high when not drowned by the river.

A short distance up the slope above Sprinkler is an unnamed geyser that erupts infrequently as high as 3 feet; it was most recently seen during 2005. About 200 feet upstream from Sprinkler Geyser is Heartbeat Spring, named because of rhythmic pulsations of the pool's water surface. Heartbeat has rarely been known to splash 1 or 2 feet high.

61. CASTLE GEYSER always attracts a lot of interest. It was named in 1870 by Nathaniel P. Langford and Gustavus C. Doane, who felt the cone resembled the ruined tower of an old castle. Over 12 feet high, the huge geyserite structure is one of the landmarks of the Upper Basin, spectacular even when the geyser is not erupting.

Castle's large cone suggests a great age, but that appearance might be deceiving. Age dating of the broad geyserite mound beneath the cone, using a dating process only recently developed in New Zealand, implies that the cone itself formed in not more than 10,400 years and likely in less than 8,800 years. Furthermore, there are indications that Castle has been active only intermittently during that time, often dormant for time spans long enough to allow soil to form and trees to grow—trees that are now silicified and incorporated within the cone's geyserite. Castle's current episode of activity probably started about 1,000 years ago. (There is evidence for this on-again, off-again "pulse and pause" aspect to the geothermal activity throughout the geyser basins, where silicified wood is found in many of the geyserite deposits.)

When people talk about the personalities of geysers, Castle heads the list. It has been known to undergo at least four different types of eruptions. Common through most of the Park's first century was steady steaming punctuated every few minutes by water jets 40 feet high. When that type of action took place, Castle had an extremely powerful eruption once every several weeks. Well over 100 feet high, the 10-minute water eruption was followed by a steam phase that lasted several hours. At other times, Castle did not give off such frequent splashes of water. Instead, on relatively short intervals, it had eruptions lasting only 4 minutes not followed by the steam phase. On still other occasions, this same sort of action resulted in durations of just 2 minutes. This kind of play still takes place. Known as minor eruptions, they are becoming a more frequent part of Castle's modern activity (discussed later).

Typical of Castle since the 1959 earthquake is yet another variety of activity. Steam wells out of the crater quietly until shortly before an eruption. Then occasional splashes rise above the top of the cone until an especially heavy surge initiates the eruption. This preliminary play is quite variable—sometimes the splashing is seen as long as 3 hours before an eruption, yet many eruptions start with virtually no warning. For about 10 minutes, water is jetted into the air as an almost continuous column. Some jets are only 30 feet high, but many reach 70, 80, or even 100 feet. As the eruption progresses, water is slowly lost from the system and the steam phase begins. At first it is very powerful, producing a deep, thunderous roar while darting sharp water jets through the billowing steam cloud. Then, over the next hour or so, Castle slowly calms down. The interval between major eruptions is extremely regular, averaging about 13 hours with little variation—if there is no minor eruption.

Minor eruptions have become increasingly common in recent years. During the first few minutes of its play, Castle normally has occasional brief pauses. Minor eruptions happen when the geyser somehow does not restart from such a pause. The 2- to 4-minute durations of before are no longer the rule, and minors now last

The huge sinter cone of Castle Geyser, probably the most massive free-standing cone of geyserite anywhere, is topped by eruptions that commonly reach around 80 feet high.

from 1 to 14 minutes. Very approximately, the duration of a minor eruption controls the length of the interval until the eventual major: every 1 minute of minor duration typically results in about 1 hour of interval. Castle tends to do a lot of splashing throughout a minor interval, action that erroneously leads many observers to think a full eruption is about to begin. Two consecutive minor eruptions were never seen until 1992, but they now occur frequently, and three consecutive minors were recorded one time each in 2006 and 2007. So, it is possible that Castle is slowly progressing into a mode of behavior in which major eruptions will be the exception rather than the rule.

The major eruption that follows a minor interval is identical to any other major eruption, but it will be followed by an interval not of 13 hours but of 14 to 18½ hours.

Castle's changeable nature makes it one of the most interesting geysers in Yellowstone. At any time it could revert to some form of activity from the past, and it probably has new tricks to spring on us in the future.

62. TORTOISE SHELL SPRING has a remarkably large geyserite cone of its own, but lying as it does immediately at the base of Castle Geyser's (61) cone, it seems hardly noticeable. In spite of this proximity, there is little, if any, connection between the two. Even Castle's most powerful eruptions do not visibly affect Tortoise Shell. This spring is one of the hottest in Yellowstone, with superheated water temperatures as great as 207°F (97°C) having been recorded. The result is a constant, violent boiling strong enough to throw considerable water out of the open crater. Rarely, brief eruptive bursts send spray as high as 6 feet, but because the overall action is constant, Tortoise Shell is not considered a geyser.

63. GIZMO GEYSER was a series of quietly overflowing, slightly bubbling vents until early 1988, when it began having strong eruptions. The play rises from several openings in the geyserite platform at the base of Castle Geyser's (61) cone. Studies indicate that Gizmo's activity might be weakly controlled by Castle, with slightly shorter intervals and longer durations as the time for Castle approaches. The most noticeable part of an eruption comes when one of the vents temporarily produces a loud steam jet that sends spray up to 15 feet high. Unfortunately, since early 2006 Gizmo has failed to produce the loud steam phases, and it splashes almost constantly.

64. TILT GEYSER is often dormant because of runoff from Crested Pool (65) flowing into its crater, and its future has been significantly altered by the resulting changes. The original vent is now invisible, thoroughly buried beneath mats of sinter-laden cyanobacteria and probably permanently clogged. However, the vents of "Tilt's Baby" still exist. To the left of the original crater and immediately next to the corner of the boardwalk, these openings formed as a result of a small steam explosion on July 6, 1976. When active, the vents splash from a few inches to 6 feet high and are sometimes accompanied by angled jets that reach up and out as far as 12 feet for durations of 1½ to 3 minutes—amazingly, "tilted" in almost identical fashion to the original, now-buried Tilt. Intervals are known to vary between 45 and 110 minutes but tend to be fairly regular at any given time. Dormant periods weeks to months long are also common.

65. CRESTED POOL is the classic example of a bubble-shower spring. At least 42 feet deep and intensely blue, the water is superheated.

Boiling around the lip of the crater is constant but variable. On frequent occasions a sudden surge of hot water will rise to the surface, causing violent boiling that may dome the water up as much as 2 to 6 (rarely 10) feet high. This sort of action often triggers true, bursting eruptions in other boiling springs, but none is ever known to have happened at Crested Pool.

Crested Pool is not only beautiful but also very dangerous. The spring is surrounded by a railing because it has taken human life—a young boy who was running through the steam cloud and couldn't see where he was going. Yet you see people climb the railing and try to balance on top while taking "that special picture." It could be special, all right! Don't climb the railings.

66a. UNNG-CGG-10 is a small spring along the runoff channel near the boardwalk. Unlike most new features, its development has been gradual. It first appeared in the late 1990s as a small spot of bubbles within the runoff stream. Now it is a true geyser that erupts from a shallow, gravel-filled crater that is all but invisible during the quiet intervals. Much of the activity is only inches high, but splashes that reach 2 feet are seen on occasion. Both the intervals and durations appear to be hours long.

66. TERRA COTTA GEYSERS is the collective name for a number of separate hot springs, all of which are geysers. In practice, they have come to be called Terra Cotta "A," "B," "C," "D," and "E"—awkward since it is now known that the original Terra Cotta Spring of 1878 is a small brick-red pool about 200 feet farther downriver. In fact, two of these geysers have official names of their own.

Most often, Terra Cotta "A" is the most active. Spouting from several vents, the major part of the play reaches 5 to 10 feet high at a slight angle from the vertical. Intervals are usually around 2 hours and durations a few minutes.

Dishpan Spring (Terra Cotta "B") is much less active than "A," but its strongly angled eruptions can reach 12 feet high. The durations typically are only a few seconds long.

Terra Cotta "C" bears the name Washtub Spring. It and Terra Cotta "D" usually erupt simultaneously. Washtub plays from a small square pool whose sinter has a slight pinkish cast. Uncommon in most years, the eruptions last as long as 5 minutes and can reach 20 feet high at an angle so as to reach into the river. The play by "D" is much smaller.

Lastly, Terra Cotta "E" is a small geyser, hardly visible from the boardwalk but quite frequent in its action.

67. SPANKER GEYSER has had remarkably constant activity, with no known dormant period. Rising from the left bank of Firehole River far downstream from the boardwalk bridge (which provides as good a view as any), it is actually a perpetual spouter. Most bursts are less than 6 feet high, so steam is often all that is visible of Spanker.

THE SAWMILL GROUP

The Sawmill Group (numbers 68 through 83a) consists of those springs and geysers directly across Firehole River from the Castle Group. The boardwalks leading from Geyser Hill, the Castle Group, and the Grand Group pass close to all the members of the Sawmill Group.

68. SPATTER GEYSER is directly across Firehole River from Sprinkler Geyser (60). Spatter's only known activity was a result of the 1959 earthquake, when its behavior made it a near twin of Sprinkler. Spatter returned to dormancy in 1962, and its cavernous crater is now partially filled with debris slumped inward from the slope above.

69. UNNG-CGG-6 is a small pool on the southwest side of the board-walk, directly across from a railing. CGG-6 mostly acts as an inter-mittent spring with short intervals and durations, but when over-flowing it occasionally splashes a foot or two high.

The yellow-green pool next to the boardwalk railing is called "Rubber Pool" because of the way the sides of the crater seem to bend and wiggle during infrequent episodes when the pool's water pulsates slightly. It is not known to have erupted except, possibly, on the night of the 1959 earthquake.

Liberty Pool (see Map 4.3) has an uncertain history. A spring by this name was active only during 1887, when it was first seen on the Fourth of July. The play was as high as 75 feet, but the dura-tions were only 15 to 20 seconds long, and the intervals were always erratic. The question is: Was this geyser the modern Liberty Pool? Some of the descriptions of the activity better fit the location of Dark Algal Pond, one of the numerous cool springs between Rubber

Pool and today's Liberty Pool. Collectively, this entire assortment of small springs is known as the Frog Pools. Several of them—but not Liberty Pool—erupted the night of the 1959 earthquake.

70. CHIMNEY CONE, a prominent sinter cone a few feet east (upstream) of the footbridge, is not a geyser and probably never has been. Such a tall and narrow cone is more typical of springs that flow only a slight volume of water very steadily over many years. A small spring at the base of the cone, "Chimney Cone Spouter," constantly bubbles and splashes a few inches high.

71. UNNG-CGG-8 ("SNAKE EYES GEYSER") has infrequent eruptions out of a pair of small vents near the boardwalk in front of Chimney Cone (70). The first known eruptions took place during 1992. More eruptions, often in series with intervals of only a few minutes and durations of seconds, occurred during 1993 but were not seen again until May 2008. The maximum height of the spray was about 4 feet. Just to the right of "Snake Eyes" is another small spring. It, too, interrupts its normal bubbling with infrequent eruptions less than 1 foot high.

72. SCALLOPED SPRING has probably never been a natural geyser. There are ways of making hot springs erupt unnaturally—all illegal in Yellowstone—and Scalloped was a victim of such an event in 1955. The induced eruption was evidently rather powerful. No details about the play were recorded, for the people who caused it did not report their act. The pool never recovered, and the water level is now several feet below the ground surface.

As a historical aside, it is now known that Scalloped's original and better name was "Gargoyle Spring." The name "Scalloped" was originally applied to today's South Scalloped Spring, where the term is much more fitting.

73. DELETED TEAKETTLE GEYSER got its name when it was decided that too many springs in Yellowstone had been named "Teakettle." Apparently the intention was that the name of this particular geyser would be deleted entirely from the records; instead, a "deleted" notation became part of the now-official name. The geyser erupts from a small cone right on the brink of the steep Firehole River bank. A 15-foot eruption is visible in a photograph taken in 1915, but no additional eruptions were recorded until after the 1959 Hebgen Lake earthquake. For several days thereafter, it frequently played up to 10 feet high. Then the activity declined. By 1964 it only boiled up about 1 foot high, and eruptions of this sort continue. Recurring every few minutes and lasting around 1 minute, the initial surges may reach 2 to 3 feet high. Observers with sharp eyes will notice that the water level in South Scalloped Spring drops slightly when Deleted Teakettle erupts and then quickly recovers to the higher level during the quiet interval.

Sawmill Geyser Complex (numbers 74 to 83a)

The Sawmill Geyser Complex is one of the most active groups of geysers in Yellowstone. Every one of its springs has a history as a geyser. The entire Sawmill Complex is cyclic in its action. It is usually Sawmill, Tardy, and Spasmodic geysers that have significant eruptions. This is known as the "Sawmill mode" activity. Less common in most years is the "Penta-Churn mode," in which Sawmill does not play at all during long durations by Tardy and Spasmodic that can result in otherwise rare play in Penta and/or Churn geysers. Whichever of these modes is in force, when the eruptions end (within a few minutes of one another in most cases) the water of almost every spring in the group drops from a few inches to several feet below its full level. This often leads visitors to comment about the dead springs that no longer erupt, but if they return in a short while they will see a very different set of hot springs. Sometimes people find themselves surrounded by activity, with eruptions rising from five or six geysers at once. All the springs refill simultaneously, and the next eruptive cycle usually begins shortly before Sawmill reaches overflow.

No place in Yellowstone better expresses exchange of function. What is "normal" at one time may be replaced by an entirely new pattern of activity at another time as the energy shifts back and forth between the members of the complex. Also, although the evidence is weak, the entire complex may be connected with the neighboring Grand Group.

74. CHURN GEYSER is one of the less active members of the Sawmill Complex. Although it was known as a 10-foot geyser in 1884, it was named because most of its known activity was little more than a surface commotion of the water. After the 1959 earthquake it would sometimes boil up a foot or two and overflow, but no modern eruptions were recorded until 1971. It is now known that Churn is likely to erupt only when the Sawmill Complex is active on the less common "Penta-Churn mode." The active periods tend to occur shortly after Tardy (77) and Spasmodic (80) geysers quit playing. As the water level drops in other members of the complex, Churn remains nearly full and may then begin a series of one to seven eruptions. Intervals range from 5 to 20 minutes. The bursting play reaches 15 feet high for durations of about 1½ minutes.

75. SAWMILL GEYSER is the namesake and largest member of the Sawmill Complex. During the eruption, water appears to spin about in the crater, resembling a large, circular lumber mill blade. Sawmill is a very interesting fountain-type geyser. The eruptions are a series of separate bursts of water; some are no more than 3 feet high, but others easily exceed 35 feet. Throughout the eruption there is a copious discharge of water.

Studies during the past few years have shown that Sawmill has eruptions of three distinctly different lengths. By far the most common are durations of 30 to 50 minutes; others last between 9 and 20 minutes *or* are longer than 80 minutes (occasionally as long as 4 hours). Since no eruptions have durations of other lengths, it seems clear that these three different types involve progressively greater portions of the plumbing system, almost as though Sawmill is three geysers that share one crater. The longer the eruption, the deeper the drainage of the entire Sawmill Complex when the eruption ends. A "deep drain" following exceptionally long eruptions happens about once a day and may be associated with otherwise unusual eruptions by some of the smaller members of the complex.

During the "Sawmill mode" activity, Sawmill is in eruption about 30 percent of the time, and a typical interval is between 1 and 3 hours long. In contrast, Sawmill does not erupt at all (or only infrequently and briefly) when the group functions on the "Penta-Churn mode."

76. UNCERTAIN GEYSER was named at a time when its activity seemed to bear little relationship to the surrounding geysers. The small, round vent lies nearly hidden within the deep sinter shoulders of the far (southern) side of Sawmill Geyser's (75) crater. Uncertain is most likely to play when the Sawmill Complex undergoes a "deep drain" following an exceptionally long-duration Sawmill eruption, typically 1 to 1½ hours after the start of the drain. Eruptions are preceded by several minutes of small bubbling and splashing within the vent. Activity can occur at other times, and Uncertain has been seen playing in concert with Sawmill. The play is a steady jet of mixed spray and steam reaching 10 to 15 feet high. It lasts from 2 to 5 minutes and is followed by a long, weak steam phase.

77. TARDY GEYSER looks and acts somewhat like the much more impressive Sawmill Geyser (75), and indeed it was once known as "Little Sawmill." The present name may have been given because of Tardy's tendency to have a series of brief, steamy eruptions near the end of Sawmill's exceptionally long plays. When the Sawmill Complex is operating on the "Sawmill mode," Tardy undergoes eruptions of relatively short durations while Sawmill is playing. Tardy's jets reach up to about 10 feet high. If there is a "deep drain" of the system, then Tardy may send explosive bursts of steamy spray over 20 feet high. Under the "Penta-Churn mode," Tardy gets a jump on Sawmill. In company with Spasmodic Geyser (80), these eruptions may have durations as long as a few hours and sometimes result in eruptions by Churn (74) and/or Penta (79) geysers.

78. TWILIGHT SPRING lies to the right of the boardwalk as one walks from Sawmill Geyser (75) toward Spasmodic Geyser (80). Water is constantly rocking about in the twin craters, occasionally splashing over the edge. When the water level drops during a "deep drain" of the Sawmill Complex, Twilight sometimes erupts. The splashes reach up to 2 to 5 feet high.

79. PENTA GEYSER (once called "The Handsaw") is a very enjoyable geyser. Its eruption, jetting up to 25 feet high from the main vent and splashing 1 to 4 feet high from four other openings, rises from a small cone only about 7 feet from the boardwalk. Penta is a complex and irregular geyser. During most years, the Sawmill Complex operates almost exclusively on the "Sawmill mode," and eruptions by Penta are rare at best. Many seasons have passed with only two or three eruptions recorded. During other years, the "Penta-Churn mode" is more common. Then Penta is seen more often, even daily, and intervals as short as 2½ hours are known. As the long eruptions of Tardy (77) and Spasmodic (80) geysers progress, the water level in Penta slowly rises. If it reaches near or to overflow *and* begins to bubble, then an eruption is possible although by no means certain. Many such cycles sometimes pass before Penta finally responds. The play ordinarily lasts 35 to 50 minutes, but longer durations are known.

Two other types of eruptions occasionally take place in Penta, and these might explain the relatively large geyserite cone at a geyser normally very inactive. Both are uncommon. One is best described as a "steam-phase eruption." Usually seen during a "deep drain" by the Sawmill Complex, these eruptions briefly spray mixed steam and water a few feet high and do little more than moisten the cone. Intermediate "mixed-phase eruptions" are similar, except that they occur while a "deep drain" is just starting and jet water 10 to 12 feet high with enough volume to form small pools next to the cone during durations of 1 to 5 minutes.

80. SPASMODIC GEYSER plays from at least twenty separate vents. When erupting, the activity is constant or nearly so from the numerous small openings near the boardwalk. Every few seconds to minutes there will be a momentary increase in intensity, and then all the vents spout strongly. Meanwhile, there is occasional bursting and boiling in the two deep, blue pools. A last vent is located on the far side of the crater, somewhat between the pools. When Spasmodic as a whole is erupting, this geyser plays every 1 to 2 minutes up to 10 feet high, the action lasting a few seconds. Spasmodic is active during both modes of Sawmill Complex activity. Occasionally during "deep drain" episodes of the entire Sawmill Complex, the south (right-hand) crater will undergo tremendous superheated boiling and the north (left-hand) pool will simultaneously have bursts easily 10 feet high, both of these eruptions rising from water levels several feet below the crater rim.

81. OVAL SPRING had only one recorded episode of geyser eruptions prior to the Hebgen Lake earthquake. That was in 1931. Otherwise it was never more than a quiet, greenish pool. The 1959 tremors induced an eruption, and Oval boiled heavily for the next several weeks. By 1960 it had resumed its pre-quake state, and only a few further eruptions were seen until the 1980s.

The modern activity is of at least four different sorts. All are uncommon. Three kinds take place during "Sawmill mode" activity, when Oval Spring's water level has dropped during a "deep drain" of the Sawmill Complex. These eruptions prove that Oval is really three geysers in one. Rising from in front of the large cave-like opening that is exposed at depth, bursting eruptions sometimes occur in series with intervals of a few minutes, durations of seconds, and heights of 1 to 20 feet. Other eruptions can take place in a second vent to the northeast (right-front) of the cave; these are seldom more than 1 to 2 feet high. The most powerful sort of activity is very rare, observed only a time or two each during 1983, 1985, 1989, 1990, 2001, and 2006. Those eruptions blasted white, muddy water at an angle out of the cave itself, jetting 12 feet high and outward as far as the boardwalk.

The fourth type of eruption occurs only during the "Penta-Churn mode." This lazy splashing 1 to 2 feet high occurs when the pool is completely full. It is accompanied by a pulsing motion of the pool's surface that apparently led to an early name, "Jellyfish Spring."

82. OLD TARDY GEYSER is fairly irregular in its performances. A member of the Sawmill Complex, it follows the same cyclic rises and falls in water level and eruptions as the rest of the group. Intervals range from a few minutes to many hours. Corresponding durations are from seconds to hours, but in all cases the height is 10 to 15 feet. Eruptions are most common during "Sawmill mode" activity, and a "deep drain" by the system can result in long-duration eruptions in which Old Tardy is the group's only active geyser for an hour or more. On the other hand, Old Tardy is nearly dormant when the complex is operating on the "Penta-Churn mode."

Between Old Tardy and Oval Spring (81) is a narrow opening in one of Old Tardy's runoff channels. It is a geyser, but its eruptions seldom rise far enough above a subterranean pool to reach ground level. This "UNNG-OTO" (Old Tardy–Oval) developed from a steamy patch of ground during 1987, as did an all-but-invisible fracture on the west side of Old Tardy.

83a. SLURP GEYSER plays from a round hole in the runoff channel at the edge of Crystal Spring (83). It is inserted here, before Crystal Spring, because despite only 1 foot of surface distance between the two, they are decidedly separate from one another at depth. Slurp Geyser is a member of the Sawmill Complex. Its activity is very erratic, occurring infrequently and only during a "deep drain" of the Sawmill Complex. The play is a long-duration chugging within the vent, the small splashes reaching no more than 1 foot high and producing limited runoff.

THE GRAND GROUP

The Grand Group (numbers 83 through 100) is located along the boardwalk that leads north from the Sawmill Group. The majority of these springs—at least fifteen from Crystal Spring at the south to Crack Geyser 800 feet to the north—are known to be interconnected at depth, forming a single extended complex. As would be expected, the relationships between these springs are, well, complex.

83. CRYSTAL SPRING is the shallow, nearly colorless pool at the top of the formation north of Old Tardy Geyser (82). Its rare geyser activity indicates that it is a member of the Grand Geyser Complex. On most occasions, water rocks about within the crater, and every few minutes it rises to cause a brief overflow. Crystal is thus an intermittent spring. A few major eruptions have been recorded. Some took place in 1931–1932, when it was known as "Gusher Geyser." It was active again in 1973, 1977, and 1983. Some of these eruptions reached 20 feet high. A series of smaller, 2-foot eruptions occurred in late 1987.

Just north of Crystal Spring is Belgian Pool. Only appearing to be rather cool (it is actually about 180°F [81°C] in the hottest part), this spring received its name because of the death of a Belgian tourist who fell into the water in 1929. There are records from the early 1930s of a Belgian Geyser. It is likely that this and Belgian Pool are the same, but it is also possible that the geyser name refers to Crystal Spring, which is known to have been active at the same time. The water in Belgian Pool drops slightly during eruptions by Rift Geyser (90).

84. BULGER SPRING may be a remote member of the Sawmill Group. The term *bulger* was commonly applied to small geysers and spouters by Yellowstone's early geological explorers, and this fact alone implies that Bulger Spring was a performer during the 1870s. It apparently remained so through the 1890s but then was nearly dormant until the 1959 earthquake. Now it is again a frequent geyser, exhibiting both minor and major eruptions. As with most geysers, the minor activity is more common. These eruptions have durations of only a few seconds and produce little or no runoff away from the cone. Major eruptions have durations as long as 12 minutes and produce substantial runoff. Each kind of eruption sends bulging bursts of water about 6 feet high, with occasional jets double that.

The Grand Geyser Complex (numbers 85 through 93)

The Grand Geyser Complex includes at least ten springs, all geysers, as members of one close-knit complex. In addition, the complex is known to be connected with other hot spring clusters. Grand is clearly related to the Economic Geysers more than 400 feet to the north, and there might be weak connections to the Sawmill Group to the south and to the "Bush Group" of springs in the meadow to the northwest. Grand Geyser itself was unpredictable through much of its history, and the fact that it can now be predicted with considerable accuracy is a wonder.

85. NORTH TRIPLET GEYSER was the closest of the Triplets to Grand Geyser (93). At some point the Park Service removed numerous rocks from within and about Grand's crater, and this apparently altered the drainage in the area. The water from Grand's eruptions was partially diverted into North Triplet. Gravel completely filled its crater, and now nothing is visible at the site of North Triplet.

86. EAST TRIPLET GEYSER is, along with North Triplet, the least important existing member of the Grand Complex. Prior to 1947, East Triplet erupted fairly often. Most seasons since then have seen no activity, and the open crater that still existed in the early 1990s has disappeared. In the general area of East Triplet are the miscellaneous vents known collectively as "The Sputniks" (89). These springs and spouters show a close relationship to West Triplet (87) and Rift (90) geysers and may represent East Triplet's energy.

87. WEST TRIPLET GEYSER lies in a symmetrical funnel-shaped vent near the boardwalk. In the years prior to 1947, it erupted regularly about every 3 hours, nearly always in concert with the East and North Triplets. Since then it has been more irregular and changeable. Now, one or two eruptions are commonly seen during Grand

Geyser's quiet interval. The bursting play reaches up to 10 feet high.

The nature of West Triplet's action relates directly to that of both Rift Geyser (90) and Grand Geyser (93). It used to be said that whenever West Triplet had an eruption, that of Grand would be delayed by at least two interval cycles of Turban Geyser (91), that is, by about 40 minutes and sometimes much longer. (See the description of Grand Geyser for an explanation of these cycles.) More recently, such a relationship has not been clear. Indeed, rising water in West Triplet's crater has been taken as a sign that Grand's time is near, and it often begins to play while Grand is in eruption. However, if it starts before Grand, then a delay is likely. Furthermore, West Triplet invariably precedes eruptions by Rift Geyser, and activity by Rift is definitely not beneficial to Grand. That relationship has also changed in the past few years, but it is clear that an average eruption by Rift can delay Grand by as long as 2 hours. The possible influence of West Triplet therefore cannot be ignored.

88. PERCOLATOR GEYSER was felt by some to be an expression of old North Triplet's energy, but a photograph shows the three Triplets as well as Percolator in eruption during the 1890s. Evidently, then, it was simply too small to be considered important, because the name was not applied to this geyser until about 1970. Percolator is most active when West Triplet (87) is playing and for a short period of time before Grand Geyser (93) erupts. The 1- to 2-foot eruptions usually stop shortly after Grand begins its play.

89. "THE SPUTNIKS" are two clusters of small geysers in slightly separated locations that behave as a single unit. These geysers first appeared during the early 1980s. In some seasons they have been virtually invisible, while in other years they have been vigorous enough to play several feet high and develop prominent craters along the edge of the grass beyond the various Triplets. For part of 2004, one of the "Sputniks" was regularly active, with intervals of 11 minutes and durations of 80 seconds.

90. RIFT GEYSER is a very important geyser, yet one would little suspect that a geyser of any sort lies at its site. The crater is a slightly depressed, sandy area at the base of some rhyolite boulders. Because the eroded remnant of a large geyserite rim is visible between Rift and the boardwalk, it is possible that Rift used to be an open pool that was buried by a landslide. The vents, which may total more than a dozen, are actually nothing more than cracks in the rocks beneath the sand. When described in 1924, it was called "Six Fissures Geyser." Most of the water jets are only a few inches high, but two of the central vents play to about 4 feet. Intervals can range from as little as 2 hours to several days, but they usually take place about twice per day. Rift's eruptions almost invariably begin 30 to 40 minutes after the start of West Triplet's eruption and while that geyser is still playing. Throughout the duration, which lasts from 5 minutes to 4 hours, there is a copious discharge of water.

That is the "problem" with Rift Geyser. Because it is a member of the Grand Complex, this large volume of discharge must affect the rest of the complex, and it can have a substantial delaying effect on the start of the next eruption of Grand Geyser (93). An old rule of thumb said that one could expect a 2-hour delay in Grand's eruption for every 1 hour of action in Rift. Like the similar rule at West Triplet Geyser (87), this no longer seems to rigidly apply. Now we know that Grand can erupt with Rift but that if it doesn't do so within about half an hour of Rift's start (little or no delay), then it probably won't begin playing until after Rift has stopped.

91. TURBAN GEYSER, lying as it does within a prominent sinter bowl, is often mistaken for Grand Geyser (93). Because of this, many visitors think Grand is starting when actually Turban is undergoing its ordinary small eruption. Grand, however, is the less obvious large pool with no rim just to the right-front of Turban.

Turban is a very important part of the Grand system, however, because Grand will begin its eruption *only* about the time of the start of Turban's action. There is no exception. Turban normally erupts on intervals of 17 to 22 minutes, although "delay intervals" as long as 37 minutes are known. The duration is about 5 minutes, throughout which the water bursts 5 to 10 feet above the rim. During Turban's quiet intervals, the water level in Grand rises to overflow, then drops as Turban plays. Eventually, the rising water level in Turban acts as a trigger, sending Grand into play. Turban then erupts in concert with Grand and intermittently for 1 to 2 hours after Grand, with powerful bursts 10 to 20 feet high.

92. VENT GEYSER is an unexpected bonus to an eruption of Grand (93). It issues from a small crack-like vent on the left side of Turban's massive geyserite shoulder. Vent normally erupts in concert with Grand, starting 2 to 3 minutes after Grand begins. At the beginning of the eruption, the slightly angled water column slowly builds in force until it is fully 70 feet high—a major geyser in its own right. Thereafter, it dies down to about 35 feet. Vent continues to erupt in company with Turban for an hour or more after Grand has stopped.

On a few known occasions during the 1960s, once in 1978, and once in 1982, Vent began erupting by itself 1½ to 2 hours *before* Grand. These eruptions were weak but steady. As the durations progressed, the water levels in both Grand and Turban slowly dropped about 1 foot. Then Grand and Turban went into normal eruptions. Even more rarely and not at all since 1993, Vent had eruptions completely independent of Grand. Occurring about 4 hours after the previous Grand and thus about the time when Grand reached its first overflow, Vent simply joined Turban during its normal 5-minute duration. The height of these plays was around 20 feet.

93. GRAND GEYSER. If any geyser anywhere is worth seeing, it is Grand. Countless people have waited for hours, commenting later that it was well worth the time. With its massive water column sparkling in the sun, a rainbow captured in its steam, and accompanied by the slender arching jet of Vent Geyser (92) and massive bursting of Turban Geyser (91), Grand is unlike any other sight in the world.

During the long quiet period after an eruption, Grand slowly fills with water. The first overflow is about 4 hours following the previous eruption. There is little variation to this figure. From then until the time of the next activity, the water slowly rises and falls in sympathy with the eruptions of Turban Geyser. Each cycle, from high water through low water as Turban plays and back to high water, takes exactly the same amount of time as that particular cycle of Turban—usually between 17 and 22 minutes. This water level variation allows the geyser gazer to judge how near Grand is to an eruption.

Stand where you can see the outermost part of the basin where small ridges of sinter project through the water. Use them to gauge the level of the pool. If the water level drops enough to clearly expose the ridges and stop almost all overflow when the time for

Grand Geyser is the tallest predictable fountain-type geyser in the world, reaching 150 to 200 feet high about three times each day.

Turban to play is approaching, then you know there will be at least one more cycle of Turban before Grand erupts. Finally will come a cycle when the water level does not appear to drop at all. Now is the time to watch closely. Very small at first, waves begin to wash across the surface of the pool because of pulsations over Grand's vent. Soon the waves become obvious as the water level continues to rise and discharge becomes heavy. Only rarely will the wave action start without a consequent eruption; usually this is accompanied by slight, bubbling overflow by Vent Geyser (92). If that does happen there will probably be a substantial "Vent delay"—at least two and perhaps several more Turban cycles—before there is another chance for an eruption. Remember, though, that any prediction is only educated guesswork; anything might happen, and appearances can be deceiving. Grand's pool *can* drop but then recover its water level in seconds, and some eruptions start explosively with little wave action.

The eruption begins when the water of the pool suddenly domes over the vent. The bubbling, frothing, and surging will continue for several seconds. It may even stop an agonizing time or two. But Grand soon rockets forth, sending massive columns of water to tremendous heights. Some of the early bursts can reach 150 feet, but the best is yet to come.

Grand's eruption consists of a series of "bursts." The first burst normally lasts 7½ to 9 minutes. During this time Grand averages "only" about 100 feet high, but the activity is continuous and a great amount of water is discharged. Suddenly, this burst ends and the geyser is quiet. After a few moments of rapid refilling, Grand jets forth again. This and any succeeding bursts usually top out at around 170 feet, and sometimes they approach 200 feet high. If the second burst is short, there may be another pause leading to a third burst, and so on. An exceptional eruption in 1983 had eleven bursts. Unfortunately, sometimes there is only the one initial burst, as was the case the majority of the time in 2007.

Since 1950, Grand's intervals have been very regular considering its many connections with other geysers. During most seasons the average is near 8 hours, and the eruptions can be predicted with considerable accuracy. However, other years find Grand erratic and much less frequent, with average intervals longer than 12 hours, and Grand is then nearly unpredictable. Changes in the overall pattern can literally happen overnight, as in early 2003, when for a few weeks Grand averaged less than 7 hours between eruptions before one day reverting with no transition whatsoever to intervals near 9½ hours.

Whatever the nature of its current activity, Grand is the largest frequently active geyser in the world. No geyser anywhere has ever approached it for a combination of size, frequency, and predictability. And the consensus never changes: no geyser can match Grand in beauty either.

94. "TOPSOIL SPRING" is a small, round pool immediately next to the west side of the boardwalk. It is often cool enough to support a growth of orange cyanobacteria, at least around its edges, but on infrequent occasions it heats to near boiling. Then eruptions a few inches to 1 foot high may take place.

95. UNNG-CGG-9 ("KEY SPRING") is normally a rather ugly feature right next to the boardwalk, shaped something like an old skeleton key. On August 21, 1988, however, "Key Spring" underwent a series of eruptions. Although all durations were less than 1 minute long, the spring sprayed muddy water across the boardwalk and up to 10 feet high. The moral to this story and that of Wave Spring (98) is that virtually any Yellowstone hot spring has eruptive potential, requiring only a slight change in the status quo to trigger activity.

96. ECONOMIC GEYSER was once a geyser of considerable interest and importance. It was named because of the economical manner in which the erupted water drained back into the crater. It played as often as every 4 minutes and was one of those spouters that could always be counted on to perform. So it continued until sometime in the 1920s. Economic has been almost totally dormant since then. Five eruptions were witnessed one day in 1957, and it was active for a few weeks following the 1959 earthquake, when rather frequent eruptions reached up to 25 feet high. Economic is at least indirectly connected with the Grand Geyser Complex, more than 450 feet away. When Economic was active during 1959, Grand was dormant for the first time in decades. Grand reactivated within a few days of Economic's renewed dormancy. Since 1959, the only known action by Economic took place in August 1997 and July 1999, when lucky geyser gazers waiting for Grand witnessed solo 10-foot eruptions. Now, Economic's water temperature is usually cool enough for the crater to be lined with orange-brown cyanobacteria.

97. EAST ECONOMIC GEYSER, located just 20 feet to the back-left of Economic (96), was a rather frequent geyser from 1888 through 1901, when it was sometimes referred to as "The Wave." However, although nearby Economic was famous, this pool did not receive its modern name until after the 1959 earthquake. East Economic's post-quake activity was closely tied to that of Economic—an eruption of one would result in a lowering water level in the other, and both returned to dormancy on the same day. The eruptions of East Economic were considerably more powerful than those of its neighbor, sometimes reaching 35 or 40 feet high. It has not played since 1959.

98. WAVE SPRING is normally a quiet pool. Lined with a thick coating of orange-brown cyanobacteria, its surface pulsates slightly so as to produce a constant train of tiny waves outward from the center of the spring. In late summer 1990 it erupted. The eruption was not seen by reporting witnesses, but pieces of bacterial mat blown from the crater landed several feet away. The height may have been near 10 feet. In early 2004, a small spring a few feet east (to the back-right) of Wave underwent eruptions a few inches high. As small as that was, Wave's water level responded by dropping more than 6 inches, then recovered whenever the small geyser stopped playing.

99. UNNG-CGG-4 ("BUSH GEYSER"). In the open flat across the boardwalk west of the Economics and Wave Spring is an assortment of small features, informally called the "Bush Group." Several are rarely active as geysers. "Bush Geyser" (the shrubbery that marked its location is gone) is among the larger of these, infrequently playing up to 4 feet high. Another, seen only during the afternoon of July 19, 1986, exceeded 10 feet in height, and a small pool was caught in the act of splashing 3 feet high on the single day of May 9, 2004.

100. CRACK GEYSER became active on September 21, 1959, the same day the Economic geysers (96 and 97) went dormant, inferring that subsurface connections extend north as well as south from Economic. The eruptions rose from an earthquake-caused crack in the sinter platform. Through 1961, there were frequent eruptions from Crack. The 6-foot-high spout would last for several minutes. There has been no activity except bubbling since 1962.

101. BEAUTY POOL and **CHROMATIC POOL** are not geysers, but no springs provide a better example of exchange of function than these. A periodic energy shift from one pool to the other causes one to overflow while the other declines, then the reverse. The time interval between the shifts ranged from a few weeks to a year or more before the 1959 earthquake. Since then the flow has almost always been from Chromatic Pool, with brief shifts to Beauty usually months apart. When high and hot, Beauty is truly among the most beautiful pools in the Upper Geyser Basin. Unfortunately, most of the time since 2002, both pools have been at intermediate water levels, and both have cooled and lost much of their previous beauty.

102. WITCHES CAULDRON lies right beside Firehole River at such a level that high water will cover and drown it. At low water it roils and boils, seemingly toiling for the proverbial trouble, but the highest surges are only 2 feet tall. Witches Cauldron is all but invisible from any trail, but it can be seen through the trees from the paved trail on the west side of the river.

103. LIMEKILN SPRINGS is a set of perpetual spouters (one is separately named The Seashell) that play from small cones perched atop a large sinter mound. The eruptions are 1 to 2 feet high. Like Witches Cauldron, Limekiln is best seen from the paved trail across the river.

ORANGE SPRING GROUP AND "SOUTH ORANGE SPRING GROUP"

These two groups of related springs (Map 4.3, Table 4.2, numbers 105 through 109) consist mostly of quiet pools, and no geyser activity has ever been of long life. The Orange Spring Group is the larger of the two, lying farther along the trail from Castle Geyser. Its largest spring, Orange Spring itself near the highest elevation in the group, has played as a significant geyser. The "South Orange Spring Group" consists of just two small pools with one geyser vent between them, all of which are difficult to see from the trail.

105. UNNG-OSG-1 ("PULSAR SPOUTER") is barely visible from the paved trail, erupting from a vent on the far side of the highest geyserite mound in the group. The constant activity is only about 2 feet high.

106. ORANGE SPRING, at the summit of the broad geyserite mound, has had a few short episodes of eruptive activity, most recently during 1986. The bursting eruptions were brief but were up to 15 feet high. Usually, however, it is cool enough to support thick mats of orange cyanobacteria.

107. ORANGE SPRING GEYSER is a small pool on the slope on the near side of Orange Spring. It was active during the 1960s and on the single day of August 30, 2002, when bursts 4 feet high were seen. Throughout 2007 it underwent intermittent overflow, but no actual eruptions were seen.

108. UNNG-OSG-2 lies near the front (east) base of the geyserite mound topped by Orange Spring, where a cluster of springs includes a single pool of size. Referred to as the "UNNGOSG" (pronounced "un-GOZ-gee"), the pool frequently had eruptions up to 15 feet high in the early 1970s and then a few of smaller size in 1988.

To the left (south) of "UNNGOSG" are two springs (108a) that act together. One of these is a small pool where a long fracture extends through a crater; the other is normally only a damp depression colored by orange cyanobacteria. Probably the springs called the Four Bubblers during the 1880s, they are rarely seen to play less than 1 foot high.

109. UNNG-OSG-3 ("SOUTH ORANGE GEYSER") is the only geyser ever recorded as part of the South Orange Spring Group, and it was active only during the summer of 1986. It was remarkably regular, playing every 20 minutes with durations of 1½ minutes. Early in the season some of the play reached 4 to 6 feet high, but the force gradually declined, and the last observed eruptions reached no higher than 2 feet.

GIANT GROUP

The Giant Group (Map 4.4. Table 4.3, numbers 110 through 123) includes everything near Firehole River between the Grand Group and the Grotto Group. Although rather widespread, these springs are known to comprise a single interfunctional group that includes two of the most important geysers in the Upper Geyser Basin. Oblong, although not very high, discharges a tremendous

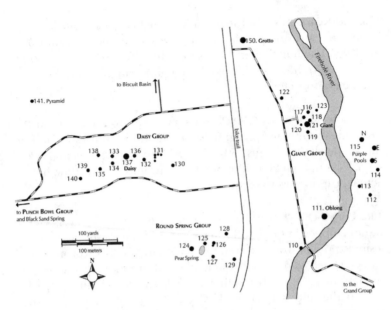

Map 4.4. *Giant, Round Spring, and Daisy Groups, Upper Geyser Basin*

amount of water; and Giant, with the sole exception of Steamboat Geyser (Norris Geyser Basin, Chapter 7), is the tallest geyser in the world.

The Giant Group lies in the area where the two Upper Basin trails begin to merge into one. The best views are from the boardwalk, but all of the springs can also be seen from the paved trail that runs directly between Castle Geyser (63) and Grotto Geyser (150).

110. INKWELL SPRING sits on a geyserite mound immediately next to Firehole River, where several small craters constantly discharge water. Although nearly a perpetual spouter, there is some variation to the play and infrequent brief pauses. Some of these vents used to jet as high as 2 feet, but the activity has weakened in recent years, and the largest pulses now reach only an inch or two high. This spring is called Inkwell because of the black coloration in and about the vents, which apparently is caused by non-crystalline iron sulfide minerals.

Table 4.3. Geysers of the Giant, Round Spring, Daisy, and Punch Bowl Groups, Upper Geyser Basin

Name	Map No.	Interval	Duration	Height (ft)
Bank Geyser	130	1–3 min	seconds	4–10
Bijou Geyser	116	minutes	min–hrs	5–30
Black Sand Pool	146	[1959]	minutes	boil–20
Bonita Pool	136	[1978]	days–years	2
Brilliant Pool	134	with Splendid	10 sec	20
Catfish Geyser	117	see text	minutes	15–100
Comet Geyser	133	steady	steady	1–10
Daisy Geyser	137	85 min–5 hrs	3½–4 min	75–150
Daisy's Thief Geyser	135	[1972]	min–hrs	8–15
Demon's Cave	147	infrequent	minutes	3–6
East Round Spring	129	[1961]	3 min	20
Giant Geyser	121	days–months	85–105 min	150–275
Inkwell Spring	110	steady	steady	inches
Mastiff Geyser	118	see text	minutes	30–125
Mud Pool	140	steady	steady	1–2
Murky Spring	139	[1960s?]	10–15 min	1
New Geyser	112	[2004]	minutes	2–10
Oblong Geyser	111	2½–11 hrs	5–7 min	10–25
Pear Geyser	125	[2004]	2–5 min	12–20
"Platform Vents, The"	120	see text	minutes	1–8
Purple Pool, East	115	[1986]	unrecorded	2–4
Purple Pool, North	115	[1986]	unrecorded	2–4
Purple Pool, South	115	[2000]	minutes	40
Pyramid Geyser	141	3–8 min*	1 min	8
Radiator Geyser	132	infrequent	sec–min	sub–10
Round Spring	127	[1990]	10 sec–3 min	5–60
Round Spring Geyser	126	seconds	seconds	8–20
Splendid Geyser	138	[1997]	1–10 min	50–218
Trefoil Spring	128	[1990]	seconds	1
Turtle Geyser	119	see text	sec–min	inches–20
UNNG-GNT-1	114	[2000]	minutes	15
UNNG-GNT-2 ("The GIP")	122	[2006?]	unknown	unknown
UNNG-GNT-3	123	frequent	sec–min	4
UNNG-GNT-4 ("Solstice")	113	[2006]	near steady	2–15
UNNG-PBG-1	145	infrequent	1 min	1–4
UNNG-RSG-2	126	minutes	seconds	2–6
West Round Spring	124	[1990?]	seconds	6–15
Zig Zag Spring	131	hours	min–hrs	2

* When active.

[ ] Brackets enclose the year of most recent activity for rare or dormant geysers. See text.

111. OBLONG GEYSER is not a major geyser in terms of the sheer size of its eruptions, but because of its tremendous water discharge it has always been high on the list of important geysers. The actual amount discharged has never been accurately determined because it spreads out in all directions across a broad sinter platform and almost immediately cascades into Firehole River, but it probably amounts to at least 10,000 gallons.

During the quiet intervals the water periodically rises and falls in the crater, resulting in light overflow. Each such cycle lasts about 20 minutes. It is during a period of overflow that the eruption begins. The water level rises a bit more so that a sudden flood of water leaves the crater, and soon the ground begins to pound as steam bubbles deep within the plumbing system form and collapse. Slowly, the boiling within the crater increases until the entire pool is involved. The water is never really jetted into the air but instead is domed upward by a massive boiling that can reach 25 feet high. The entire play lasts between 5 and 7 minutes. About 2 hours are required for the crater to refill and resume the intermittent overflow.

At times during the 1980s and early 1990s, Oblong sometimes had series of eruptions on short intervals. Often, the first of these intervals would be around 2½ hours; the second and third intervals (if any) could be as short as 30 minutes. These secondary eruptions rose from a partly empty crater and had bursts as much as 50 feet high. Such activity has not been witnessed since 1992, but at least one case was electronically recorded in 2005.

For many years Oblong was regular enough to be predicted, with intervals near 8 hours, but that ended in the early 1990s. Oblong was dormant most of the time from July 1993 until June 1996. On the infrequent dates when it did erupt during those years, there were often other unusual events in this part of the Upper Geyser Basin, an indication that the deep plumbing system roots of the Giant, Daisy, Grotto, and Chain Lakes groups are intertwined at depth. From 1996 into 2006, Oblong's action was unpredictably erratic, with intervals ranging from less than 3 hours to almost 2 days. However, when Giant began frequent action in April 2006, Oblong's intervals abruptly decreased to an average near 4 hours. Exceptionally short intervals often occur shortly before an eruption by Giant and aberrantly long intervals within a few hours after Giant.

112. NEW GEYSER was initially active for only a few weeks in 1970 and apparently not again until 2004. Since nothing is "new" for very long and the term has been overused, some people object to this name. Nevertheless, it appeared in U.S. Geological Survey publications and is therefore "accepted." The first known eruption was on September 1, 1970, and lasted several minutes. Pieces of jagged sinter strewn about the crater indicated that a small steam explosion had enlarged an old vent. In the succeeding two months, several other eruptions were seen. The intervals could not be determined, but they might have been several days long. The maximum height of the eruptions was about 15 feet. In December 2000 a few spraying eruptions up to 10 feet high were seen and recorded as New Geyser. These most likely were by "Solstice Geyser" (113). However, New definitely did have some small, splashing eruptions during the summer of 2004, activity no doubt related to the intense action of "Solstice."

113. UNNG-GNT-4 ("SOLSTICE GEYSER") was seen for the first known time on the day of the winter solstice, December 21, 2000, when a 10-foot geyser "across the river from Oblong" was reported. Initially described as being New Geyser (112), which may have had some activity at about the same time, "Solstice" became active again during the winter of early 2004. Eruptions continued into the following summer. In fact, "Solstice" is a reactivation of an old spring never previously seen to erupt, located downslope and much closer to the river than New. Intervals and durations were highly variable. Most eruptions were just 2 or 3 feet high and lasted only a few minutes, but in early August 2004 "Solstice's" action became more vigorous and nearly perpetual. Some bursts sent spike-like jets of water 12 to 15 feet high. That activity persisted for four months. In December 2004 "Solstice" regressed to weak, intermittent eruptions, and it was soon dormant. Activity resumed during October 2005 and continued into March 2006. "Solstice" has been dormant since then, and not a trace of steam from it is visible even on frigid days.

114. UNNG-GNT-1 plays from a small pool about 30 feet from South Purple Pool (115). It was possibly active during the 1950s, too, but the records kept about that activity are insufficient to determine its exact site. It broke out on May 1, 1970, as a perpetual spouter. Then, after an early-summer dormancy, the play resumed on a periodic basis. The eruptions were about 15 feet high. It is believed that the 1970 activity ended on September 1, the same day New Geyser began to play. Since then, during the episodes of major eruptions by South Purple Pool in 1987 and 2000, GNT-1 joined in with a few eruptions of small size.

115. THE PURPLE POOLS—SOUTH, EAST, and **NORTH**—are all directly connected with Giant Geyser (121). It has long been suspected that variations in the water levels and/or boiling action in the Purple Pools might serve as an indicator for Giant Geyser, but nò such relationship has been proven. However, any activity in Giant can affect the Pools—as an eruption of Giant progresses, the water levels in the Pools drop until they are about 3 feet below overflow.

All three of the Purple Pools are geysers in their own right. The first known activity was in 1971, when the East and North Purple Pools played frequently to 2 to 4 feet high. At the same time, South Purple boiled vigorously. Some people thought this activity was a precursor to rejuvenation by Giant, but the eruptions soon stopped without anything out of the ordinary happening. On the other hand, for a short time before Giant returned to limited activity in 1986, the Pools underwent action similar to that of 1971. At times since 2004, both the North and East Purple Pools have been seen to boil up a foot or two high.

Following the Giant eruption of September 12, 1987, South Purple Pool had a series of powerful eruptions. Voluminous bursts reached over 30 feet high and killed trees on the nearby hillside. A solo eruption by South Purple was seen in 1998. Most recently, a brief series of eruptions similar to those of 1987 took place in August 2000. Some bursts reached over 40 feet high, with both intervals and durations only a few minutes long.

Giant Geyser Complex

The Giant Geyser Complex (numbers 116 through 122) comprises the geysers on the "Giant Platform," a flat, raised geyserite structure capped by the cones and vents of several geysers—Bijou, Catfish, Mastiff, Turtle, the "Platform Vents," and Giant. The complex also includes "The GIP" a short distance away. These geysers are intimately connected belowground, and they all participate in events called "Giant hot periods."

Triggered by interruptions known as "pauses" in Bijou Geyser's jetting play, a hot period starts when the water level rises in all other members of the complex. The "Platform Vents" usually begin to erupt first, but the key is Mastiff, which will overflow and may boil inches to several feet high. During a long-duration hot period, there may also be a steam phase in Bijou and minor eruptions by Catfish and Turtle. Eruptions by Giant itself can begin only during a hot period. However, the great majority of hot periods, which can take place as often as every hour, do not produce eruptions.

The Giant Complex is a pivotal and highly important group of hot springs. It is positively known to be connected with the Grotto Group, a few hundred feet to the northwest. Much of what happens within the Giant Complex depends on the activity of the Grotto Group. Giant is also related to Purple Pools across Firehole River, to Oblong Geyser to the south, and possibly to the Daisy Group to the west.

116. BIJOU GEYSER (pronounced "BEE-zhoo") used to be called "Young Faithful Geyser," apparently because of its frequent action. Bijou is the most active member of the Giant Complex. It is the highest cone on the left (north) side of the Giant Platform. For much of its history, it has played almost constantly 5 to 15 feet high, wetting all sides of its cone. During the 1980s it was discovered that the details of Bijou's action may serve as an indicator of "Giant hot periods," the only times in which Giant Geyser (121) can start an eruption. Although Bijou's spouting is nearly constant, it has occasional quiet intervals. These usually last only a minute or two, but when they continue for several minutes, the rest of the group responds with higher water levels. Then and only then is it possible for there to be a hot period; therefore, Bijou's action is closely watched. Note, however, that long pauses in Bijou's action also take place during and for a few hours after the long "marathon" eruptions by Grotto Geyser (#150, which see). Exceptionally strong hot periods often take place about the time these "marathon pauses" finally end.

117. CATFISH GEYSER erupts from a ragged cone near the front base of Bijou Geyser's (116) cone. At the end of those "Giant hot periods" that fail to produce an eruption of Giant (the usual case), Catfish erupts 6 to 15 feet high for a few seconds. It frequently sends up thin jets of water at other times, too. During 1951, the thermal energy within the Giant Complex shifted to the north, away from Giant and to Catfish, Mastiff (118), and Bijou (116) geysers. During this activity, known as the "Mastiff function," Catfish became a major geyser, erupting to heights ranging between 75 and 100 feet. Such activity ended in January 1952. Some of Giant's recent eruptions have been similar to the Mastiff function, but Catfish has not matched the power of 1951.

118. MASTIFF GEYSER seems to be the key to eruptions by Giant Geyser (121), with its action during hot periods serving as the trigger for Giant. During hot periods, Mastiff fills and overflows, boils vigorously, sometimes domes its water a few feet high, and, ideally, erupts as high as 30 feet or more. The stronger Mastiff's hot period action, the more likely is an eruption of Giant.

Mastiff was reported to erupt 20 to 30 feet high just before an eruption of Giant Geyser in 1873. Smaller play was seen during the 1880s and 1890s and was described in passing in 1919. These reports probably all refer to Mastiff's normal hot period action (although hot periods were not described as such until 1951). After 1919, however, Mastiff was not again reported in action until 1951, when an exchange of function shifted the energy within the Giant Complex to the north side of the platform. Catfish (117) erupted to 75 feet. And, unexpectedly, Mastiff joined in, one massive column of water reaching well over 100 feet high and a smaller secondary column jetting at an angle to perhaps 75 feet. Such action was totally without precedent, for just a few minutes later Giant itself began playing. Catfish quit, but Mastiff continued on equal terms with Giant for 5 minutes, and only when Mastiff stopped did Giant assume its usual stupendous eruption. When the Giant Complex behaves in this fashion, it is called the "Mastiff function." Some of the eruptions since the 1980s have been referred to as Mastiff function plays, but perhaps they should not be. In these cases Mastiff usually has reached only 30 to 50 feet high with durations as short as a minute or two, and it has seldom been joined by powerful eruptions of Catfish. For these reasons, the modern action is sometimes called "North function" rather than "Mastiff function."

In 2006, Mastiff underwent a few solo eruptions of a sort never seen before. They jetted water 20 to 35 feet high for durations of a few seconds to several minutes. Although they took place during otherwise normal hot periods, they were not accompanied by significant action in Catfish or Bijou and did not result in eruptions by

Giant. Evidently, Giant was just not quite "ready" to erupt, because on each of these occasions it did so about a day later.

119. TURTLE GEYSER is little known. It possesses a highly eroded and somewhat detached cone on the far right (south) side of the Giant Platform. Apparently, Turtle is active only at the time of a hot period or a Giant eruption, and not always even then. It might be the 20-foot geyser referred to in 1925 as having regular eruptions at intervals of 40 minutes, and it might also have been active in 1931 and 1933. However, the first eruptions positively known took place in 1951. Restricted to that one year, they reached up to 20 feet high for durations of just a few seconds. An eruption reported in 1956 is considered questionable, since the Giant Complex was inactive at that time. Turtle bubbles and overflows during most modern hot periods, and splashing bursts 1 to 3 feet high are occasionally seen during eruptions by Giant.

120. "THE PLATFORM VENTS" are two clusters of geysers that are active only at the time of Giant Geyser's (121) hot periods or eruptions. A hot period takes place when, during an exceptionally long quiet pause by Bijou Geyser (116), the water level rises within the entire Giant Platform. As the water level visibly rises in Mastiff Geyser (118), small holes in the platform directly in front of Giant's cone begin to overflow and then spout. The tallest is "Feather Vent." It can reach as high as 15 feet and used to be known as the "Christmas Tree." A second cluster of vents toward the far right-hand side of the platform, called the "Southwest Vents," erupts at the same time. In total, as many as nine individual vents can be simultaneously active—the more, the better, and a strong hot period is a very impressive sight. It is known that Giant can begin an eruption *only* during a hot period. There are no exceptions. Unfortunately, the huge majority of hot periods, which can recur on intervals as short as one hour, do not result in eruptions. Nevertheless, all are eagerly awaited because of their potential.

121. GIANT GEYSER was appropriately named. When in eruption there is nothing like it. A huge tower of water is thrown far into the air, and because it is a cone-type geyser, this is a steady column rather than the intermittent bursts seen in most other large geysers. Giant erupts with two quite different modes of play, but it is a tremendous geyser even at its weakest. With the single exception of Steamboat Geyser (see Norris Geyser Basin, Chapter 7), Giant is the tallest and most voluminous geyser in the world.

Most of Giant's eruptions are of the type called "Normal (or Giant, or South) function." In these, Giant is not accompanied by significant play by Mastiff Geyser (118), and the water column typically reaches "only" 150 to 220 feet high. In the early references to Yellowstone, Giant was commonly listed as erupting to 250 to 300 feet. Probably in those days, certainly during 1951, and occasionally in recent years, it has played on what is called the "Mastiff function." In this function, Giant's eruptions are triggered by powerful activity in Mastiff Geyser; Catfish Geyser (117) may join in with eruptions stronger than usual, and Bijou Geyser (116) often goes into steam phase. Giant does not join the action until several minutes later. Some of these eruptions have been measured by triangulation to be at least 275 feet tall. (Since the modern action by Mastiff is significantly weaker than that described in the 1950s, the preferred term for current activity of this sort is "North function" rather than "Mastiff function.")

When active, the Giant Complex is characterized by well-defined hot periods. Most hot periods do not trigger major eruptions of Giant. On the other hand, eruptions begin *only* during hot periods. Marked initially by rising water within Mastiff and then by jetting from the "Platform Vents" (120), the water in Giant's cone may surge and lift a few inches. Most often this is all that happens, and there is no eruption. But if conditions are right, the surging becomes jetting that fills the cone and pours water onto the platform. This may happen repeatedly before the eruption is triggered. Only seconds are required for Giant to reach its maximum height. A single eruption will last as long as 105 minutes, and even when it is more than half finished, the eruption can still be over 100 feet tall. The total volume of water expelled has been estimated at 1 million gallons (4 million liters).

When Giant was active between 1950 and 1955, the intervals ranged from 2 to 14 days; during much of 1952 to 1954, the average was only 54 hours and Giant was nearly predictable. However, any geyser the size of Giant may be adversely affected by exchange of function. Given that the Giant Complex is connected with the Purple Pools, Oblong Geyser, the Grotto Geyser Group, and possibly the Daisy Geyser Group, it is no surprise that Giant has not been frequently active. When it went dormant in 1955, many presumed it would reactivate within a few years. Indeed, during early

Although the intervals between its eruptions range unpredictably from a few days to many months, Giant Geyser can reach over 250 feet high.

1959, Giant did show signs of impending activity. And then the 1959 earthquake struck.

Some references state that Giant has been dormant since 1955 and that the earthquake somehow altered or destroyed the deep plumbing system connections with the surrounding hot spring groups. This is not so. Although some eruption intervals have been years long, there has always been some level of activity, and Giant did erupt six times between 1963 and 1987. Finally, in 1988, Giant began a slow progression into its most recent phases of vigorous activity.

As of April 1, 2008, Giant had had 269 eruptions since 1955. It was at its best from late 1996 into 1998, when the intervals were nearly predictable with an average of about 4 days. Then, from April 1998 until late 2005, the intervals varied erratically from just 7 to over 450 days. Another episode of consistent action began in October 2005 and continued into 2008, with intervals of 4 to 16 days; a short series of intervals only 3 to 6 days long took place during April–May 2006. But such is Giant's known history going back to the establishment of Yellowstone: for every one or two years of frequent eruptions, there have been several years with minimal or no activity.

Giant Geyser bears additional mysteries. During those long years of little activity from 1955 into 1996, when Giant did erupt it almost invariably did so during the late-summer–early-fall seasons, specifically between late August and late October. Was this chance? Probably not; only 5 of 25 eruptions in those years occurred during the other nine months. In fact, seasonal changes are seen in other geysers, and they almost always happen during late summer or early fall. The best-known case of this is in the Norris Geyser Basin, where it is commonly called the "disturbance" (see Chapter 7). The causes of these changes are not clearly understood, but they probably involve variations in the level and/or volume of non-thermal groundwater.

The second mystery is just why Giant Geyser should undergo such frequent periods of years-long dormancies. Giant is always superheated and there is always some degree of activity in its complex. The key probably involves the volume of water lost from the system via Grotto and other related hot springs. But while those springs certainly show variations of their own, nobody has yet been able to use such changes in any predictive way. So it is that well-named Giant is, and no doubt will remain, one of the rarest geyser gazing treats.

122. UNNG-GNT-2 ("THE GIP") lies within a weathered hole near the boardwalk between the Giant and Grotto groups. "GIP" stands for "Giant's Indicator Pool." In spite of its distance from the Giant

Platform, the water level within The GIP rises and falls along with that in the rest of the Giant Complex; indeed, the rising water that portends a hot period is seen here as easily as it is at the Giant Platform. As "Giant #1," The GIP was reported as a small geyser in 1878, and a photograph taken in 1921 shows it either erupting or undergoing a powerful steam phase during an eruption of Giant Geyser. Washed areas indicated that a few small eruptions (or possibly just exceptionally high water levels) took place between 1988 and 1991 and once in 2006.

123. UNNG-GNT-3. In addition to all the miscellaneous geysers of the Giant Complex, a number of other geysers and perpetual spouters exist on the platform that, in spite of their very close proximity to the rest, seem to be a separate group of springs. These are hidden low on the east side of the geyserite mound and are not visible from the boardwalk. In total, there are at least a dozen springs here. One is a geyser that historically has played up to 6 feet high.

For people following the Upper Geyser Basin boardwalk beyond the Giant Group, the next group of geysers is the Grotto Group (beginning with Grotto Geyser, #150).

ROUND SPRING GROUP

The Round Spring Group (Map 4.4, Table 4.3, numbers 124 through 129) is a small cluster of springs of little overall significance. However, most have records of geyser activity, and Round Spring itself proved to be a major geyser during May 1990. The group is not threaded by any trail, but it is close enough to the paved Upper Basin trail that most of its features can be easily seen from there. There has been some confusion over the years about the names in this group; the versions given here should set things straight.

124. WEST ROUND SPRING, along with Trefoil Spring (128) and several small, unnamed features within the group, erupted the night of the 1959 earthquake. Little is known about that activity except that it continued into 1960. West Round did not erupt again until August 1971, when a brief active period saw frequent eruptions as high as 15 feet. Again it apparently was quiet until May 1990, when West Round was seen to have at least two eruptions perhaps 6 feet high at the same time Round Spring (127) was active. West Round Spring is difficult to see from any trail, so it is possible that small eruptions are more common than is recognized.

125. PEAR GEYSER (not to be confused with Pear Spring, the large pool visible in front of the trees and just to the left of the geyser) had its first known activity in 1958–1959, *before* the Hebgen Lake earthquake. Then in 1961 it was highly regular, erupting every 5 minutes to a height of about 12 feet. Additional brief episodes of activity have been seen since then, the most recent being several series of eruptions in May and June 2001, when some play reached over 20 feet high, and two cases of small eruptions witnessed in July 2004.

126. ROUND SPRING GEYSER and **UNNG-RSG-2** are located between Pear Geyser (125) and Trefoil Spring (128). They have a confused history but are probably identical to the "small geysers" reported as early as the 1870s but then not again until 1937. Definitely active during 1956, they reactivated in 1982 and now play frequently. The northernmost (right-hand) of the two, Round Spring Geyser (previously RSG-1), is the larger and more active, reaching up to 8 feet high on both intervals and durations only a few seconds long. The smaller geyser plays only after an exceptionally long interval by the other and, in fact, may be the trigger for a new episode of short-term cyclic action. While Round Spring was active in May 1990, both of these geysers were more vigorous than usual, and one jetting eruption by Round Spring Geyser was easily 20 feet high.

127. ROUND SPRING lies near Pear Spring, to the left of UNNG-RSG-2 (126). The fact that it played as a small geyser in 1895 had been forgotten until it had a series of powerful eruptions in late May 1990. The geyser was active for only about 30 hours, and most intervals were just a few minutes long. The eruptions had durations of only seconds, but they reached between 5 and 20 feet high. A few so-called major eruptions exceeded 30 feet, and one, which had an extraordinarily long duration of 3 minutes, reached an estimated 50 to 60 feet high. After this episode, Round Spring returned to dormancy, and it is now a cool pool lined with dark brown cyanobacteria.

128. TREFOIL SPRING (also known as **"NORTH ROUND SPRING"**) was named for its threefold shape. It apparently was always a quiet pool until May 1990, when nearby Round Spring (127) was active. Trefoil was then seen to bubble vigorously, and on several occasions the action broke the surface with splashes 1 foot high. Since 2000, it has had a water level several inches below overflow.

129. EAST ROUND SPRING (incorrectly identified in previous editions as "Round Spring") is the member of the Round Spring Group closest to the trail. It was active as a geyser in 1940–1942 and in 1951, when eruptions 20 feet high had durations of 3 minutes. East Round might be the "Round Spring Geyser" that reportedly had weak eruptions from the time of the 1959 earthquake into 1961. It has not erupted since then, although it does act as a long-period intermittent spring.

DAISY GROUP

The terms "Daisy Group" (Map 4.4, Table 4.3, numbers 130 through 141) and "Daisy Geyser Complex" are nearly synonymous. Of all the geysers here, only two—Bank and Pyramid—are not known to be connected with Daisy. Aside from the geysers and numerous small spouters, the group includes only three quiet pools. The group is thoroughly separated from other hot spring clusters at the surface, but there is evidence of a subsurface connection with the Giant Group.

The Daisy Group lies west of Firehole River and the Giant Group, up on a low hill. There are four approaches to the group. From the paved Upper Basin trail, two smaller paved trails lead up the hill; they pass on opposite sides of the springs and then merge to continue westward toward the Punch Bowl Group. From there the trail continues on to Black Sand Basin and so, reversed, provides a third access to the Daisy Group. Fourth is a trail that leads through the forest from near Biscuit Basin.

Daisy and Splendid geysers are very large and among most gazers' favorites. Splendid is seldom active, but when it is it may rival Grand Geyser in size and beauty. When Splendid is not active, Daisy usually is. During some years Daisy has been the most regular major geyser in Yellowstone. That unfortunately began to change in the 2000s, and Daisy is now only marginally predictable.

130. BANK GEYSER might have been formed, or at least was substantially enlarged, by a steam explosion in 1929, and for a time during 1933 it had bursting eruptions up to 20 feet high. Even though it lies within a few feet of springs that definitely are members of the Daisy Group, Bank is believed to be isolated and unrelated to any other spring. It lies in a small alcove down the slope toward the Giant Group. It presently erupts every 1 to 3 minutes. The play is a series of splashes over the course of a few seconds. Most eruptions are minor and only 4 to 6 feet high. Major eruptions with wide bursts up to 15 feet high have infrequently been seen since 2004. During some years the action degenerates into periods of mild boiling. The small erupting vents adjacent to Bank's overflow area are collectively known as "The Bubblers."

The Daisy Complex (numbers 131 through 140)

The Daisy Complex is a very interesting assortment of springs. It includes at least ten geysers, two of which are among the largest in the Park, and almost no other springs except for a miscellany of small "sputs." All of the geysers are directly connected to one another at depth, and frequent exchange of function leads to irregularity among them all. Any description of one geyser demands mention of the others.

Daisy is, in spite of its many relatives, a predictable geyser most of the time. Signs showing current predictions are maintained both at the Visitor Center and at the start of the southernmost of the two paved trails that lead to the group.

131. ZIG ZAG SPRING (previously **UNNG-DSG-1**) consists of two sets of several vents apiece that lie along jagged fractures in the geyserite crust between Bank (130) and Radiator (132) geysers; where the cracks cross one another is another spring crater. Although the most distant of these vents are separated by 20 feet, they all erupt simultaneously. Details about the activity are uncertain. Durations can be as long as several hours. Zig Zag is definitely more frequent during active episodes by Splendid Geyser (138) than it is when Daisy Geyser (137) is active, as is the case with nearby Radiator Geyser (132). Most of the vents sputter only a few inches high, but the central vent can splash up to 2 feet.

132. RADIATOR GEYSER was mapped in 1878, but years later it was covered by road work and forgotten. When it sprang to life in the old Daisy-area parking lot, people thought their cars' radiators were boiling, hence the name. Eruptions were seldom more than boiling until the time of the 1959 earthquake. Radiator has been active in company with Daisy (137) and Splendid (138) geysers ever since, on most occasions playing a few minutes after the end of the larger nearby eruption. When Daisy is the active major geyser, Radiator's eruptions are mostly subterranean and seldom noticed, but when Splendid is active, Radiator can easily reach 10 feet high. It has also had a few active episodes in which the action seemed to be independent of the other geysers. At these times, the play is not more than 2 feet high.

133. COMET GEYSER is located directly between Daisy (137) and Splendid (138) geysers. Its cone is the largest in the complex, yet the geyser is among the smallest—usually. The cone has been built to its size by nearly constant splashing over a long period of time. At best, this play reaches about 6 feet high. Every few minutes the surging becomes heavier and large amounts of water are ejected, but even this discharge is never sufficient to form a surface runoff. The only time Comet seems to stop erupting is following full major eruptions of Splendid and, rarely, of Daisy. Even then, though, the play has not actually quit; instead, the water level has simply been drawn down so far that the eruption is entirely confined to the subsurface. Occasional reports of large eruptions by Comet are controversial. Almost always they are later proven to have really been play by Daisy, but there are exceptions. One day in 1992, for example, reports of two different 30-foot eruptions were turned in separately by people familiar with the Daisy Group, and several times during June 2004 Comet was seen jetting well over 10 feet high. Thus, one day it might prove to be a third major member of the Daisy Group.

134. BRILLIANT POOL was evidently never more than a calm spring prior to 1950. Before then it always ebbed a few inches at the time of Daisy's eruptions and then refilled simultaneously with Daisy. In 1950 Brilliant occasionally began to overflow heavily. Whenever this happened, eruptive action in Daisy would stop. This interplay was observed many times during 1950 and might have been a prelude to renewed activity by Splendid Geyser (138) in 1951.

The response of Brilliant Pool to eruptions by Splendid is quite different. Following any such eruption, the water level can drop as much as 4 feet. After a series of several eruptions by Splendid, Brilliant's crater may be completely empty. Only then can Brilliant commence its own activity, with sharply angled jets that reach up and outward as far as 20 feet. This play lasted as long as 2 minutes in 1951, but more recently the eruptions came in series with short intervals and durations usually shorter than 10 seconds.

135. DAISY'S THIEF GEYSER was known as "Dewey Geyser" (named for Admiral George Dewey, famed for defeating the Spanish fleet in the Battle of Manila Bay during the Spanish-American War) during 1898–1899. A reference to a "Daisy's Indicator" in 1932 might also be this geyser, and it definitely was active during 1936. Starting in 1942, Daisy's Thief was seen a few times in most years through the 1940s. Moments before Daisy Geyser (137) was ready to begin spouting, the Thief started jetting a steady column of water to 15 feet. The eruption lasted about 25 minutes without pause. Throughout, the water level in Daisy dropped slowly; at the end of the Thief's activity, it had been lowered to the same level that followed a normal eruption of Daisy. And because Daisy finally did erupt after a normal interval following the eruption of Daisy's Thief, it appeared that the latter had discharged an amount of water and energy exactly equivalent to one eruption by Daisy. The origin of the name is therefore clear.

Following a few years of quiet, Daisy's Thief reactivated in 1953, but the nature of the action was very different. Instead of a 25-minute eruption, the spouting lasted several hours. It was less forceful than before, and when it ended Daisy erupted immediately. This type of function continued until the time of the 1959 earthquake. Since then there have been few eruptions. In July 1968 one occurred that lasted more than 7 hours, during which Daisy had three brief, weak eruptions. Later in 1968, in the days following a swarm of small earthquakes, Daisy had a series of nine eruptions, seven of which were preceded by Daisy's Thief acting as an indicator rather than a robber. Those were the only eruptions by Daisy during an eleven-year dormancy. As Daisy rejuvenated from dormancy in the early 1970s, the Thief had a few additional very small eruptions. Because Daisy was then having long and erratic intervals, no clear relationship between the two geysers could be discerned. Daisy's Thief is not known to have erupted since 1972, and the small round crater on a geyserite mound is difficult to see.

136. BONITA POOL was used as an indicator of Daisy Geyser's (137) eruptions—the time of first overflow was related to the priming of Daisy—until 1937. In that year the water level rose higher than before, and whenever overflow was achieved, all eruptions in Daisy were stopped. Note the similarity between this and the occasional overflow by Brilliant Pool (134) during 1950. On one occasion the high water persisted for several months, and Daisy was dormant that entire time. This happened several additional times prior to the 1959 earthquake.

At first, following the big tremors, Daisy continued with normal activity. Then, in early 1960, Bonita Pool became very active, overflowing steadily and experiencing frequent small eruptions. Daisy went dormant, and Bonita did not stop overflowing until 1967. During that summer it would sometimes stop. Immediately, Daisy began filling and heating, but just when an eruption seemed imminent, Bonita would refill. Except for two occasions in 1968 when activity was initiated by Daisy's Thief Geyser (135), Bonita's action was enough to bring about a complete dormancy in Daisy from 1960 into 1971. Even then Daisy often tried to play without quite succeeding, but a few eruptions did occur. The time required for a complete exchange of function back to Daisy was long, but by 1978 Daisy could finally be predicted with confidence. Bonita has remained quiet and low, even during the early 2000s when Daisy began having longer intervals. A little spray plays from "Bonita's Sputs" on the edge of the crater near the time of Daisy's eruptions, but Bonita seems finished for now, not remotely meriting its name (Spanish for "pretty").

137. DAISY GEYSER is the most important member of the Daisy Complex. Although Splendid Geyser (138) is larger, Daisy is far more active. This has apparently been the usual case, for Daisy's runoff channels are wider, deeper, and more extensive than Splendid's.

Daisy erupts from a crater partially surrounded by a heavy sinter rim. At two points this margin is accented by small cones that begin to spout shortly before an eruption. When Daisy is predictably regular, the activity in these cones can be used to give an accurate time of eruption. Usually, the larger of the two begins to splash

Daisy Geyser can erupt as often as, and more regularly than, any other major geyser in Yellowstone, including Old Faithful.

about 20 minutes before the eruption, while the smaller begins a few minutes later. Meanwhile, the water in the crater boils and surges constantly over the main vent. The eruption begins when this splashing suddenly grows higher and stronger. Daisy rockets forth within a few moments. The maximum height of 75 feet is reached within the first half minute of play, with the water column sharply angled toward the northwest. This height is maintained for most of the duration of 3½ to 4 minutes.

When the activity of Bonita Pool (136) is such that it does not overflow and erupt, Daisy can be extremely regular. During some seasons it has been the most predictable of Yellowstone's major geysers. Until 2001, the average interval ranged between 85 and 110 minutes. Although there was some variation from year to year, during any one season the actual range between the longest and shortest intervals tended to be small. For reasons that are unclear, starting in 2001 Daisy became less frequent and more erratic. Since 2006 the intervals have averaged near 3 hours; several have been as long as 5 hours, and at least one interval reached 2½ days. No other member of the Daisy Complex has responded with increased action, so this change may represent an exchange of function away from the group as a whole—perhaps to the more active Giant Complex. As a result, Daisy is now only marginally predictable.

Whatever the nature of Daisy's performances, three things can further affect it. One is strong wind, especially out of the south, which may delay Daisy's eruptions by as much as half an hour. Another is earthquake tremors, even from shocks hundreds of miles away, which are known to temporarily decrease Daisy's intervals. The third is Splendid Geyser (138).

During recorded history Daisy has been much more active than Splendid. But Splendid has had occasional active episodes, and these times have always been periods of great irregularity in Daisy. The longest such period was during the years before 1900, and others occurred discontinuously during the 1950s, 1970s, 1980s, and 1990s. One of Splendid's best years ever was 1985, which was one of Daisy's worst. Yet Splendid plays a beneficial role in Daisy's activity, too. All this can best be shown with the story of events during the 1970s.

Following the 1959 earthquake, Daisy was active as before. Indeed, as time passed it became more vigorous until the average interval had shortened to only 48 minutes. Then, in February 1960, Bonita Pool began to overflow and erupt. Daisy rapidly declined and was soon dormant. Except for July 31 and September 18, 1968, when eruptions were triggered by Daisy's Thief Geyser (135), Daisy remained dormant until July 22, 1971. A couple of eruptions on that day were the first signs of a shift back toward Daisy, away from Bonita. By 1972, Daisy would have two or three eruptions during the course of a few hours, then lapse back into dormancy for several days. These brief active phases were often initiated by eruptions of Splendid Geyser, and by 1973 it became clear that not only did active phases begin with Splendid but that continuing action by Splendid was required to keep Daisy going. So it was until 1978. By then, there had been such a complete shift of energy to Daisy that not only was Splendid no longer needed, but it could not erupt at all.

From this it seems that maybe eruptions by Splendid do not occur so much because of energy shifts to it but rather because of energy shifts away from Daisy. The two are not the same thing. In any case, it is only at such times that Daisy and Splendid can erupt simultaneously. Known as concerted eruptions, such plays were first seen during the 1880s but then not again until 1972. Never were concerted eruptions seen with such frequency as during 1985 and 1986, when single active phases of Splendid included as many as four concerts. Similar activity was seen in 1996–1997. When in concert, Daisy is usually much stronger than normal. Most such eruptions exceed 100 feet, many surpass 120 feet, and the measured record is 152 feet. Daisy also has increased durations, playing for as long as 6 minutes. (It should be noted, though, that some in-concert Daisy eruptions are very weak, reaching less than 30 feet high for durations of less than 2 minutes.) Clearly, Daisy owes a lot to Splendid's existence.

Situated on a hill, Daisy can be seen from much of the Upper Geyser Basin. It is always spectacular. It is best to see it erupt close at hand from a position where the sharp angle of the column is visible. Other geysers might be higher, more sharply angled, longer lasting, of greater volume, and so on, but somehow Daisy is a particularly special sight.

138. SPLENDID GEYSER has seldom been active during recorded times. It is possible that Warren Ferris saw and gave an accurate description of Splendid in 1834, and it was certainly active through most of the early years of the Park. It went into dormancy in 1898. During 1931 it had a few eruptions, but no vigorous active episodes took place until 1951. Splendid then had a number of eruptions, often in series, during each year until the 1959 earthquake. Splendid rejuvenated along with Daisy Geyser (137) beginning in 1971, was dormant from 1978 until 1983, underwent many intense series of eruptions during 1985 and 1986, and was once again a major performer during the mid-1990s. There have been no eruptions since 1997.

Splendid is always on a hair trigger, requiring little to set off a series of eruptions. For years it has been known that Splendid is most likely to begin to erupt during a storm, when the barometric (atmospheric) pressure drops quickly as a weather front moves across the Park. This has the effect of slightly reducing the boiling temperature in Splendid's plumbing system. It appears that such falling pressure is required for Splendid to be active, but surely other factors are involved as well. Once the initial eruption of an active episode has taken place, Splendid can continue to play regardless of further changes in the pressure.

The water in Splendid's crater is always agitated, frequently boiling up several feet. The strongest action is usually about the time of, or within a few minutes after, Daisy's eruption. If an eruption by Splendid is to occur, the surging will abruptly build to as much as 15 to 30 feet high. It would seem that such a tremendous output of water would have to trigger an eruption, but this is not so. "False starts" are common. When it is to be an eruption, the surging will hold its height for several seconds, then explosively jet to the maximum. Few eruptions are less than 150 feet high, and nearly all those that small are in concert with Daisy. Many eruptions approach 200 feet high, and one was measured at 218 feet, making Splendid perhaps the third tallest geyser in Yellowstone. Once started, an eruption may last from 2 to 10 minutes and, if it is a series of eruptions, the intervals range between 1 and 12 hours.

At times the activity of the 1970s and 1980s was without precedent. Most of the time, if Daisy or Splendid has been active, the other has been dormant. Starting in 1972, there were probably more eruptions by Splendid than in all its previous active phases combined. During 1985 it played 44 times; there were another 99 eruptions in 1986 and 26 in 1987. Roughly 30 percent of those eruptions were in concert with Daisy. With only two known exceptions, every concert began with Splendid. Within the first 2 minutes of its start, Daisy joined in. The two geysers would nearly match one another for size, each reaching between 100 and 150 feet high. The spectacle cannot be adequately described. The refilling of the craters was rapid, and Daisy often erupted alone scarcely an hour after the concerted play. After 1987, Splendid was nearly dormant for 9 years. It played three times in 1988, not at all in 1989, and twice in 1990.

Concerted eruptions by Daisy Geyser (left) and Splendid Geyser are always rare and have not been seen since 1997.

A different mode of activity first appeared in September 1992. Taking place a few minutes after an eruption by Daisy, this play reached only 50 to 70 feet high for durations of just 1 to 1½ minutes. Definitely not full eruptions but certainly more than false starts, these were described as minor eruptions. During 1993 they seemed to occur about once every 2 to 4 weeks, but they have not been seen since that year.

Splendid's next active episode began in 1996 and lasted into 1997. Most of the play was comparatively weak, only 70 to 80 feet high, and there were few concerts with Daisy. Splendid has not erupted in any fashion since 1997.

The fact that Splendid has not been frequently active is shown by the surrounding sinter platform. The recent cycles of activity caused considerable erosion in the immediate area, yet the run-off channels are smaller and less extensive than Daisy's. Very many eruptions would cause a considerable change in the crater and its surroundings. It may be that Splendid is of relatively recent origin and that much more can be expected of it in the future. For now, it is a rare treat.

139. MURKY SPRING, somewhere west of Daisy's Thief Geyser (135), was active during the 1950s and again during the 1960s. Fairly regular, the eruptions recurred every 6 to 8 hours, lasted 10 to 15 minutes, and reached 15 feet high. It is amazing, then, that the descriptions do not allow a positive identification as to which spring is Murky. Near Splendid Geyser (138) and northwest of Daisy's Thief Geyser are two sinter-lined craters that show signs of having erupted in the past. They are definitely part of the Daisy Complex, as shown by their dropping water levels whenever Splendid erupts. One of them is probably Murky Spring.

140. MUD POOL is located within a stand of trees on the far southwest side of the Daisy Complex. Since 1982 it has been active as a perpetual spouter, the steady play only 1 to 2 feet high. It is slightly possible that Mud Pool is identical to Murky Spring (139).

141. PYRAMID GEYSER is located at the base of White Pyramid Geyser Cone (better known simply as the White Pyramid or White Throne), across the meadow northwest of the Daisy Complex. Little was known about this spring prior to the 1959 earthquake. Some eruptive episodes with intervals of 3 to 4 hours were recorded, but the evidence is that Pyramid was minimally active. The quake caused a dormancy that lasted until 1971. Since then it has been continuously active, but on a cyclic pattern. During most years, a quiet period about 3 hours long is followed by an eruptive episode consisting of several closely spaced eruptions. With intervals of 3 to 8 minutes and durations of around 1 minute, the height of the steady jet is 8 feet.

PUNCH BOWL SPRING GROUP AND BEYOND

The Punch Bowl Spring Group (see Map 4.1, Table 4.3, numbers 145 through 147) is a small collection of springs in the vicinity of Punch Bowl Spring, along the trail between the Daisy Group and Black Sand Basin. Punch Bowl itself is a boiling, intermittent spring. Vague reports say it was active as a geyser during the 1870s and 1880s. The spring immediately to the right of the cone is also an intermittent spring, with overflow periods that show no relationship to the action by Punch Bowl.

145. UNNG-PBG-1 is the nearest of the pools west of Punch Bowl Spring. Active episodes are infrequent during most years, but then eruptions may recur every few minutes. The biggest bursts are just 1 foot high. More impressive were eruptions occasionally seen during the 1980s, when superheated boiling was punctuated by surging up to 4 feet high.

The two small pools beyond PBG-1 also have histories of infrequent geyser activity, and they probably play only when PBG-1 is also active. Both erupt 1 to 2 feet high.

146. BLACK SAND POOL. Down the trail that leads from Punch Bowl Spring toward Black Sand Basin are two springs, which properly comprise their own group. Black Sand Pool alone was the original "Black Sand Basin" of the 1880s. (At that time, the present Black Sand Basin was called "Sunlight Basin.") Black Sand Pool has a history of infrequent but rather powerful eruptions. It may have been active when the Park was established, as its original name read "Black Sand Geyser," but the first eruptions specifically recorded were during 1895, when it played up to 20 feet high. It was not active again until 1950, when there were several eruptions about 12 feet high. The only other known activity, more of a vigorous intermittent boiling, took place during the first few weeks after the 1959 earthquake.

147. DEMON'S CAVE is located across the trail a short distance southeast of Black Sand Pool. It was originally named "Cave Geyser" in 1872 and was referred to as a "boiling cauldron" in 1881, facts that imply eruptive activity. The record is then blank, however, until the 1980s. Based mostly on signs of splashing and washing within the deep crater, occasional eruptions a few feet high still take place. Demon's Cave is a dangerous feature; the approach directly from the trail leads to a wide geyserite ledge that overhangs the pool. To view it, swing wide around the crater to its back side.

GROTTO GROUP, RIVERSIDE GEYSER, AND THE CHAIN LAKES GROUP

The Grotto Group and the area a short distance to its north consist of three functional hot spring clusters—the Grotto Group, Riverside Geyser, and the Chain Lakes Group (Map 4.5). Several of the geysers are large. Riverside is the only one that receives much publicity, but the overall activity is vigorous and, at least in part, related to that of the Giant Group.

The two main trails through the Upper Geyser Basin merge at Grotto Geyser. This point is about 0.9 mile from the Old Faithful Visitor Center and 0.2 mile from the end of the improved trail at Morning Glory Pool.

The Grotto Group (Map 4.5. Table 4.4, numbers 150 through 159) includes ten geysers, all intimately related to one another. Grotto Geyser is a never-ending source of pleasure for visitors because of its frequent activity and strange cone. It is also closely monitored because of its relationship to the Giant Complex— Giant hot periods often occur about the time Grotto starts an eruption, and a substantial proportion of Giant's eruptions take place as the entire system recovers from the long "marathon" eruptions of Grotto.

Table 4.4. Geysers of the Grotto Group, Riverside Geyser, and Chain Lakes Group, Upper Geyser Basin

Name	Map No.	Interval	Duration	Height (ft)
"Central Vents, The"	152	irregular	minutes	1–3
Culvert Geyser	162	[1988]	minutes	2
Grotto Fountain Geyser	155	with Grotto	7–17 min	15–83
Grotto Geyser	150	2–49 hrs	6 min–40 hrs	15–40
Indicator Spring	154	[1988]	seconds	6
Link Geyser, major	165	days–years	1 min	20–75
Link Geyser, minor	165	1–4 hrs	15–30 min	4
Marathon Pool	159	rare	minutes	6
North Chain Lake Geyser	166	[1984]	seconds	35
"Persistent Spring"	163	steady	steady	inches
Riverside Geyser	160	5½–8½ hrs	21 min	75
Rocket Geyser	151	with Grotto	2–12 min	10–50
South Grotto Fountain Geyser	156	irregular	5–30 min	10
Spa Geyser	158	2–5 min*	seconds	6–72
Square Spring	161	5–10 min*	seconds	1–10
"Startling Geyser"	157	rare	2–4 min	10–20
UNNG-CLC-1 ("Clasp")	167	[1974]	seconds	20
UNNG-CLC-3 ("Victory")	164	[2000]	minutes	7
Variable Spring	153	see text	hours	boil–2

* When active.
[ ] Brackets enclose the year of most recent activity for rare or dormant geysers. See text.

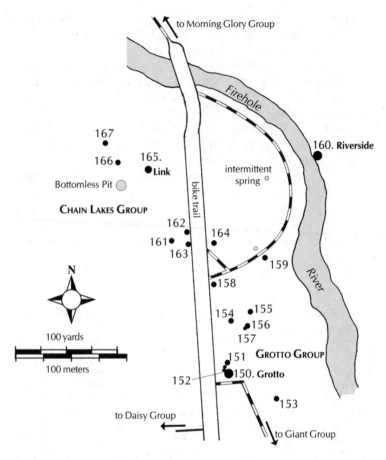

Map 4.5. *Grotto Group, Riverside Geyser, and Chain Lakes Group, Upper Geyser Basin*

150. GROTTO GEYSER was given its name by the Washburn Expedition in 1870. The cone's interesting projections and caverns apparently formed as siliceous sinter was deposited about a stand of dead trees. The remaining stumps are now thoroughly coated and petrified by the geyserite.

Grotto's eruptions almost invariably begin a few minutes after Grotto Fountain Geyser (155) has begun to play. The eruption of Grotto resembles a series of large splashes. At the very beginning of the activity, the initial surges may reach over 40 feet high, but

The odd geyserite cone of Grotto Geyser is believed to have formed as geyserite was deposited on the remains of pine trees.

the play quickly dies down to 15 feet or less, which is held for the remainder of the eruption. The duration of an eruption roughly controls the length of the following interval and is bimodal. Most eruptions, called "short mode" or "normal," last from 50 minutes to 3½ hours and recur on intervals of 5 to 8 hours. Interspersed among the normal eruptions are "long mode" or "marathon" eruptions. Most of these have durations that exceed 12 hours, and the average is near 19 hours. Marathons lasting a full day are known, and a few have exceeded 40 hours in length. These are followed by intervals of complete quiet as long as 49 hours. Several days (and numerous normal eruptions) used to pass between marathons, but they have become more common since 2000. Sometimes only one or two normal eruptions take place between them, and seven back-to-back marathons were confirmed by an electronic recorder during 2005. In addition, and with increasing frequency, the first Grotto eruption following a marathon may have a duration as short as 6 to 15 minutes. These changes might be a function of the recent activity by Giant Geyser, as a significant percentage of its eruptions have been associated with Grotto marathons.

Grotto's discharge of water is very high. A typical marathon eruption will release at least 700,000 gallons of water, about as much water as Old Faithful erupts in three days. Because of this high rate of water loss as well as its relationship to Giant, Grotto has always been numbered among the most important geysers. Grotto has never been known to fall completely dormant.

151. ROCKET GEYSER nearly always erupts in concert with Grotto (150). During Grotto's activity, Rocket steadily churns and splashes about in its crater, and most bursts reach no more than 3 feet above the crater rim. Rocket seems to be nothing more than an additional outlet for Grotto. On occasion, however, Rocket will take over the entire activity with a major eruption. Water jets in a steady stream as high as 50 feet while Grotto falls nearly silent. Major eruptions usually begin about 1½ hours after the start of Grotto. They last from 2 to 12 minutes. At the end of a Rocket major, Grotto resumes its normal splashing action but then often stops within a few minutes. Interestingly, and showing that Grotto's normal and marathon eruptions are quite different phenomena, Rocket majors are fairly common during the short eruptions but less frequent during marathons. In 2007 there were three known cases in which Grotto's activity was initiated by Rocket majors rather than by Grotto itself, one of which also included no action by Grotto Fountain Geyser (155). Extremely rare, observed only a handful of times, are independent eruptions by Rocket that begin around half an hour after Grotto has quit.

152. "THE CENTRAL VENTS." Between the cones of Grotto (150) and Rocket (151) geysers is an assortment of small vents that play about 1 foot high throughout Grotto's activity. On occasion, these "Central Vents" erupt before Grotto starts, at about the time Grotto Fountain Geyser (155) appears ready to start playing. The result is a "Central Vent Delay" in which Grotto's start is postponed for at least 20 minutes. Infrequently, the "Central Vents" take on the role of Grotto Fountain Geyser (155), triggering the start of Grotto, which then progresses without any eruption by Grotto Fountain.

153. VARIABLE SPRING is the first pool east of Grotto, below the boardwalk leading toward the Giant Group. It is possible that this spring is the "Algae Pond" of post-1959 earthquake studies, but if

Major eruptions by Rocket Geyser most often take place near the end of adjacent Grotto Geyser's activity.

so, the spring has changed dramatically since then. Eruptions by Variable were rare prior to 1983. They occur only during and shortly after "marathon" eruptions by Grotto Geyser (150)—the longer the duration of the marathon, the more extreme is Variable's action. At these times the clear water drops as far as a foot within the crater and becomes murky to outright muddy as superheated boiling splashes the remaining pool as high as 2 feet.

154. (GROTTO'S) INDICATOR SPRING is a pool lying between Rocket (151) and Grotto Fountain (155) geysers. During the last couple of hours before Grotto erupts, the spring intermittently fills to near overflow and then drops a few inches. The water level gets a bit higher with each 20-minute cycle and is synchronized with similar changes in nearby Grotto Fountain Geyser (155). Back around 1911, an "Indicator Spring" was said to "spurt up" just before the start of Grotto, but that was probably Grotto Fountain. That being the case, the eruptions observed on July 5, 1988, are the only ones known for Indicator Spring. That activity, which reached as high as 6 feet above an empty crater, was associated with other "unusual" events within the Grotto Complex, including a simultaneous eruption by Grotto Fountain that had an extraordinarily long duration of 53 minutes.

155 and 156. GROTTO FOUNTAIN GEYSER and **SOUTH GROTTO FOUNTAIN GEYSER** are difficult to describe separately. They are closely related to each other as well as to the other members of the Grotto Complex. Their histories are somewhat unclear because of a surfeit of names historically used for Grotto Fountain. It is reasonably certain that Grotto Fountain was mapped as a "spouter" in 1872, and eruptions 30 feet high were described in 1886. It might be the "Indicator Spring" of 1911. Major eruptions, said to have reached over 100 feet high, were infrequently seen between 1922 and 1932. Grotto Fountain was active under the names "Strange Geyser" and "Grotto Drain Geyser" during the 1940s, and in spite of the fact that the present name was officially approved in 1949, it was also called "Surprise Geyser" from the early 1950s into the late 1960s. It is probable that South Grotto Fountain has been active along with Grotto Fountain on all occasions.

The small cone of Grotto Fountain does not look like the source of a major geyser. Only a few inches high, it sits in the center of a broad, depressed runoff channel leading away from Grotto and Rocket. Yet when Grotto Fountain plays, the highest jets often reach between 30 and 50 feet high for most of the 7- to 41-minute duration. Grotto Fountain is capable of extraordinary eruptions, too—one in 1987 was measured as 83 feet high, while another in 1988 had a duration of 53 minutes. The steady cone-type jetting is beautiful, especially when joined by the bursting of South Grotto Fountain.

South Grotto Fountain's crater is about 30 feet to the south (right) of the other. When the two geysers are active, Grotto Fountain usually plays first, closely followed by its southern neighbor. If the reverse is true, then Grotto Fountain may not play at all. South Grotto Fountain's bursting usually does not exceed 10 feet high but may last for more than 30 minutes.

Grotto Fountain serves as an indicator for Grotto in two ways. First, during the last hour or two before Grotto erupts, Grotto Fountain periodically fills, boils, and drops, completing one cycle in roughly 20 minutes in synchrony with the filling in Indicator Spring (154). The boiling becomes stronger with each cycle, and experienced observers usually can accurately judge when Grotto will start on this basis. This is the case because Grotto Fountain can erupt *only* immediately before Grotto starts—thus, by "predicting" the time of Grotto Fountain, one is also predicting Grotto. Its head start ranges from less than 1 to more than 20 minutes; in general, the longer this lead time, the stronger will be both Grotto Fountain and the initial surges by Grotto. As noted, if South Grotto Fountain begins playing before Grotto Fountain, the latter might not play at all.

157. "STARTLING GEYSER," previously referred to by the long-winded name "South South Grotto Fountain," plays from two small holes immediately in front of South Grotto Fountain's (156) crater. It is a rare performer not seen during most seasons. The vents are not visible from the trail, so to see the impressive eruption rising from an apparent nothing is "startling" to say the least. "Startling" is a pretty geyser. One of the vents sends a steady jet of water 15 feet high (the other splashes weakly) throughout durations of 2 to 4 minutes. The eruptions take place about the time when Grotto Fountain (155) would normally be expected to start and thus a few minutes before an anticipated Grotto (150) eruption. Rather than replacing Grotto Fountain, however, "Startling" plays an adverse role, invariably delaying both Grotto Fountain and Grotto. Rare series of eruptions yield one delay for each eruption of the sequence. The intervals are about 20 minutes long, corresponding to the cycles shown by Indicator Spring and Grotto Fountain Geyser.

158. SPA GEYSER looks something like a large, oval bathing pool, but nearly every eruption by Grotto (150) results in overflow at near-boiling temperatures in Spa. Eruptive episodes are relatively infrequent and mostly associated with Grotto's "marathon" eruptions. Evidently, the greater discharge of water by the marathon versus a normal eruption is required to trigger Spa's action. The activity consists of a series of brief eruptions, usually at intervals of 2 to 5 minutes during episodes as long as 3 hours. Very explosive, the play bursts large masses of water between 6 and 72 feet high. Not seen prior to 2004 were occasions where, rather than filling and erupting during a marathon, Spa boiled steadily and vigorously, with a water level more than 2 feet below overflow. During the 1950s Spa was said to have erupted powerfully after every eruption by Giant Geyser (121), but such a relationship has not been seen since then.

159. MARATHON POOL is the circular pool located on the right (east) side of the trail leading between Spa (158) and Riverside (160) geysers. Although it is now known that eruptions were seen during 1941, it was only named in 1988 because of its reaction to long-duration ("marathon") eruptions by Grotto Geyser (150). Marathon's water level drops gradually throughout a marathon. Although extremely rare, Marathon Pool can erupt as high as 6 feet around the time Grotto finally quits.

Across the trail from Marathon Pool is an unnamed spring that varies its water level by several inches. The times of high water are accompanied by vigorous bubbling, but true eruptions are unknown.

160. RIVERSIDE GEYSER is one of the least variable geysers in Yellowstone. On a statistical basis, Riverside is far more regular than Old Faithful and most other major geysers. Its large cone proves a very long history of activity, but it might have been dormant when Yellowstone Park was established and first thoroughly explored—the expeditions of the 1870s describe Grotto at length, but although Riverside was named in 1871, their reports give it little more than passing mention. Even after it became regularly active in 1883, the geyser was seldom seen by Park visitors. Indeed, it was years before maps and guidebooks were in full agreement as to just which geyser really is Riverside; some used the name for what is now called Mortar Geyser (173).

Riverside is an isolated spring, probably not connected with any other. As a result, the flow of water and energy into its plumbing system is constant, resulting in extreme regularity. Only gradual, long-term changes in the flow rate can alter Riverside's performances. Historically, the average intervals have varied between 5½ and 8½ hours, but recently they have consistently held to about 6½ hours.

Riverside's action is bimodal. Unlike the eruptions of most regular geysers, few of Riverside's eruptions occur at the time of the average interval. Instead, most take place either a few minutes before *or* after the average. Posted predictions are based on the simple average, because one never knows whether the next interval will be short or long. Beyond that, though, eruptions seldom occur more than 30 minutes off the average.

The setting of Riverside is superb. The crater rises directly out of the far bank of Firehole River against a background of grassy meadowland and forest. The cone is shaped somewhat like a chair. The main vent is in the shallow basin near the front part of the "seat." On the far side of the seat are two minor vents. The large hole on the chair "back" may once have been the main vent, and some play still issues from it during the early stages of an eruption.

Between 1½ and 2 hours before an eruption, the main vent begins to overflow. The discharge is variable and punctuated by boiling spells. About 1 hour before the play, the minor vents behind the main vent begin to bubble, spouting to a few inches (they can be difficult to see). During this pre-play, the old vent will occasionally splash directly into the river. There can be "false starts" when it appears that the eruption is starting but then fails to continue. When this happens it will often be another 15 to 20 minutes before Riverside makes another attempt to play. A particularly heavy splash along with main vent surging initiates the eruption. Boiling over the main vent becomes heavy, and within a few seconds Riverside is arching out over the river, sometimes nearly spanning it. The maximum height of 75 feet is held for several minutes. Then the geyser slowly dies down. It doesn't stop spouting for 21 minutes, and that is followed by a short, weak steam phase.

Riverside's average interval is practically the same now as it was in the 1880s. Riverside is a very stable geyser, and it will probably continue to play about as it does now for a long time to come. With its column arching over the river and a rainbow in its spray, Riverside Geyser is one of the beauties of Yellowstone.

THE CHAIN LAKES GROUP

The Chain Lakes and surrounding springs (Map 4.5, Table 4.4, numbers 161 through 167) on the west side of the paved trail are all connected belowground. Geyser activity among the Chain Lakes themselves is rare but is major in size. There is some evidence that the activity of the cluster can be altered by that of the Grotto Group a short distance away.

161. SQUARE SPRING and the two smaller springs beyond it had episodes of eruptions during the 1950s, but no details were written about that activity. Square alone had a brief series of eruptions up to 10 feet high following the 1959 earthquake. Reactivated in 1982, it performs mostly as an intermittent spring. The times of high water are accompanied by bubbling and occasional splashes 1 to 4 feet high. The intervals are 5 to 10 minutes long.

162. CULVERT GEYSER probably did not exist as such prior to the early 1950s. When the highway was realigned through the Upper Basin, a number of small springs blocked the engineer's plans. One of these near Spa Geyser (158) was buried, and another at the edge of the new road had a retaining wall built around its crater. Now known as Culvert Geyser, the latter was certainly not the deep pool of today. Apparently, the energy of the buried spring was diverted here. In 1954, Culvert began spouting to 2 feet and rapidly enlarged its crater. Such eruptions have not been seen since the 1988 development of "Persistent Spring" (163), and Culvert has cooled so that its crater is lined with thick mats of cyanobacteria.

163. "PERSISTENT SPRING" is unquestionably another expression of the spring that was buried by the road construction, which earlier led to the development of Culvert Geyser (162). "Persistent" appeared during 1988, when it broke out at the edge of the trail a few feet south of Culvert. The perpetual spouting, a few inches to 2 feet high, rises from a break in a ceramic pipe that is probably an outlet for the buried spring.

164. UNNG-CLC-3 ("VICTORY GEYSER") is across the trail from Culvert Geyser (162) and "Persistent Spring" (163). Its small vent may be another expression of the spring buried beneath the roadway, hence the name. "Victory" broke out during 1998. The fact that it formed in loose gravel rather than solid geyserite indicates that it is a totally new spring where there had never been a thermal feature before. During 1999, some of "Victory's" eruptions sprayed water as high as 7 feet and widely enough to reach the paved trail, and the runoff began killing nearby trees. Unfortunately, it was dormant by 2000 and is now a small hole partly filled with tepid water.

165. LINK GEYSER, so named when it proved to be an important link in the Chain Lakes, is a truly major geyser. While minor activity is very common, major eruptions are rare. Perhaps that is good, because the tremendous eruptions discharge so much water so suddenly that much activity might literally wash away the trail.

Link boils around the edges of its crater at all times, and it has had minor eruptions throughout recorded Park history. Recurring on intervals of 1 to 4 hours and lasting 15 to 30 minutes, these consist of superheated boiling that domes the water 3 to 4 feet high. Major eruptions probably begin during the minor play, although this is by no means certain.

The signs left by a major eruption are unmistakable, making it unlikely that any have gone unnoticed. They have been recorded during only fifteen years of park history.

In 1957–1958 Link showed that it could erupt in series, with several eruptions occurring in the course of one day. This happened again on August 8, 1974, when Link had a series of 8 eruptions. The first interval was 70 minutes, the last 197. Each play lasted less than 1 minute, but so much water was discharged that Firehole River was muddied for a considerable time after each eruption. The play began with a doming of the water within the crater, sending a flood across the surroundings. Then the geyser exploded with jets of water that reached 60 to 100 feet high, carrying rocks with them.

Link's most intense active episode on record was in 1983. Between October 13 and 18, it had 40 major eruptions. These were distributed among 6 individual series that started about 24 hours apart, each consisting of as many as 12 eruptions. Link has had only a few major eruptions since 1983. None have been in series, and some reached no more than 20 feet high. The best season of recent record was 2002, with 12 eruptions. There were 2 eruptions during 2001, 3 more in 2004, and none since then.

Link is directly connected with the other members of the Chain Lakes Complex. On occasion the focus of energy may shift to any one of them, resulting in eruptions. All such action is rare, however. The most commonly cited of these is "Bottomless Pit Geyser." There is a Bottomless Pit among the Chain Lakes, but it is a cool pool, and the geyser referred to was probably either Link or North Chain Lake Geyser (166).

166. NORTH CHAIN LAKE GEYSER was active during 1931–1932 and again in 1953, when it played two or three times per week. A brief series of boiling eruptions 20 feet high followed the Hebgen Lake earthquake, but only one eruption has occurred since 1959. Following Link Geyser's (165) intense series of eruptions in October 1983, the energy within the Chain Lakes shifted to North Chain Lake. On March 25, 1984, it responded with a single eruption, bursting up to 35 feet high.

167. UNNG-CLC-1 ("CLASP GEYSER") is the northernmost of the Chain Lakes, somewhat separated from the others. It was seen to have a single eruption during August 1974, a few days after an active episode by Link Geyser (165). The play was 20 feet high. Although it had a duration of only a few seconds, the eruption seems to have thoroughly upset the system, since overflow, once common, now almost never occurs.

MORNING GLORY GROUP

Lying at the northwestern limit of the Upper Geyser Basin proper is the Morning Glory Group (Map 4.6, Table 4.5, numbers 170 through 179). Named after Morning Glory Pool, the area contains eight known geysers, including what were small mud pots prior to dramatic changes in 1998. The paved Upper Basin trail ends at Morning Glory Pool, but a dirt trail leads downstream along the route of the old highway about half a mile to the next cluster of springs, the Cascade Group. Between the two groups are a narrowing of Firehole River and a hill called Old Faithful View, marking the end of the continuous open valley of the Upper Geyser Basin. The trail continues past the Cascade Group to Biscuit Basin.

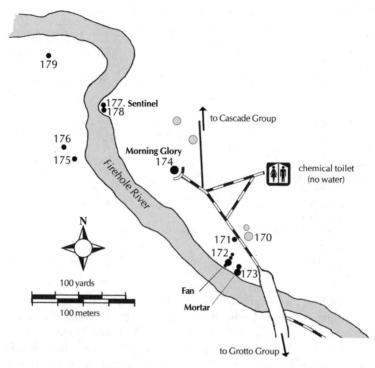

Map 4.6. *Morning Glory Group, Upper Geyser Basin*

170. "NORRIS POOL." Prior to October 1998, two shallow craters across the trail from Spiteful Geyser (171) occasionally exhibited weak mud pot behavior. In 1886 they were described simply as "old craters." Their combined name, "Norris Pools" (plural), arose because of the mud pots' strong, sulfurous odor, similar to that at the Norris Geyser Basin. The southern of the two, the one closest to the footpath where there is a railing, now contains clear water within a deep geyserite crater. It is "Norris Pool" (singular). Clearly, it was a significant hot spring long before recorded Park history, and now it is an important geyser. The other crater has been called "Backwater Spring" because of a weak but long-known response to eruptions by Spiteful Geyser.

The rejuvenation of "Norris Pool" occurred on October 4, 1998. During the initial activity, the geyser had both minor and major eruptions. Typically, there were five to eight minors, one every 20

Table 4.5. Geysers of the Morning Glory Group, Upper Geyser Basin

Name	Map No.	Interval	Duration	Height (ft)
Fan Geyser	172	2½ days–months*	35–45 min	100–125
Green Star Spring	179	[1959?]	unrecorded	unrecorded
Morning Glory Pool	174	[1944]	unknown	40
Mortar Geyser	173	2½ days–months*	35–45 min	40–80
Norris Pool	170	[2002]	2–7 min	20
Sentinel Geyser	177	[2006]	sec–12 min	12–40
Serpent's Tongue	176	steady	steady	2
Spiteful Geyser	171	[1998]	2–8 min	10–30
UNNG-MGG-2 ("Sentinel's Vents")	78	[2005]	sec–3 min	1–3
West Sentinel Geyser	175	steady	steady	2–10

* When active.
[ ] Brackets enclose the year of most recent activity for rare or dormant geysers. See text.

to 60 minutes with each lasting less than 2 minutes, before a major eruption took place. Lasting 6 to 7 minutes, the bursting play of the major sometimes reached over 20 feet high. Immediately after the major, and only then, Spiteful Geyser would erupt. There would then be a few hours of quiet before the minor eruptions resumed. This episode was short-lived and eruptive activity stopped during November 1998. A second, nearly identical active phase took place from August into November 1999, except that Spiteful was not active. Still more activity took place during the summer of 2000, and that time all known eruptions occurred while Fan and Mortar geysers (172 and 173) were also in eruption, almost as if the spring had become a new vent within that amazing complex. Finally, a few solo eruptions happened during 2001 and 2002, but "Norris Pool" has not erupted since then.

171. SPITEFUL GEYSER occupies a deep crater near the trail. In 1884, tour guide G. L. Henderson wrote: "The Spiteful stones unwary heads, her water sources being dry." This is apparently a reference to the steam explosion that formed the jagged crater at what had previously been a cluster of small perpetual spouters. Spiteful lies on a prominent fracture in the sinter that was probably caused by a prehistoric earthquake. Aside from the early eruptions, though, it was dormant until 1964. That activity was highly regular. Eruptions recurred every 15 minutes and splashed water up to 10 feet high from an empty crater. A dormancy led to rejuvenation in the early 1970s. This action was more erratic but also more powerful than before, and some eruptions rocketed thin jets of water as high as 30 feet above a full crater. Dormancy resumed in 1974, and, with the exception of a single eruption seen in 1984, Spiteful remained dormant until the first active phase by Norris Pool (170) in October–November 1998. It has been dormant since then.

172 and **173. FAN GEYSER** and **MORTAR GEYSER** cannot be discussed separately. With only a few modern exceptions, these two major geysers always erupt in concert. Fan's several vents give rise to numerous jets of water. These openings lie along the same fracture that includes Spiteful Geyser (171). Looking downstream toward this complex from the trail, from left to right, these vents are called River, High, Gold, Angle, Main, and East vent. The two cones to the left, next to Firehole River, are Upper and Lower Mortar; Bottom Vent is in front of Lower Mortar, while Frying Pan and Back Vent lie beyond Lower Mortar's massive geyserite shoulder.

The large size of Mortar's cones proves that they were very active in the distant past, but it is also likely that they were dormant, or nearly so, for a long while before the discovery of the Park. The great amount of decayed rock in the area indicates that the geyserite formations had dehydrated, becoming brittle and easily eroded. The modern eruptions of both Mortar and Fan are causing dramatic changes to the geysers, and they and their formations are very different now than they were just thirty years ago. For example, Mortar's Bottom Vent did not exist before the 1980s, but now it is often the dominant geyser during pre-eruption "event cycles," and Fan's separate Angle and Main vents appear to be merging into one. It is not unusual to see large chunks of geyserite tossed high into the air during eruptions.

During most of the first 100 years of Park history, Fan and Mortar played infrequently, and in the 1800s there was considerable confusion about which geyser was which. Mortar was often referred to as Riverside while at the same time other authorities called it Fan, and Fan was sometimes barely mentioned as "Perpetual Spouter." Some references state that Mortar erupted as often as every 8 hours during the 1870s and 1880s, but those records probably referred to Riverside Geyser.

The fan-shaped nature of Fan Geyser's many separate water jets is most obvious when an eruption is viewed backlit by a low sun.

During those early years, it is evident that almost all major activity here was by Mortar *without* Fan. At its best, Fan was rare and weak compared with what it can do today. In essence, Fan and Mortar behaved as two separate geysers during the 1800s. Only one concerted major eruption was recorded during the first fifty years of Yellowstone history, and that was in 1878. Why just one concert probably has to do with the subsurface erosion of the plumbing systems—independent geysers long ago, Fan and Mortar have now become one.

During the intervening years, minor activity occurred often. Even periods considered dormant were punctuated by such play. At these times Fan's High, Gold, and Angle vents would spout about 3 feet high, ordinarily not accompanied by Mortar. On other occasions, especially in 1915, Mortar would have eruptions up to 30 feet high every 2 hours, not accompanied by Fan.

Major eruptions were first seen in 1925 and more probably occurred during the early 1930s, but eruptions by Fan and Mortar remained rare. Records indicate that from then until 1968, only a dozen or so eruptions were recorded. One that was filmed in October 1950 showed the start of Fan and Mortar with an appearance similar to today's, but it ended with Upper Mortar jetting by

*With rare exceptions, major eruptions of Mortar Geyser (*foreground left*) always accompany simultaneous major eruptions by Fan Geyser (*background right*).*

itself without a trace of Fan or Lower Mortar visible in the movie. That is very different from today's activity.

A marked rejuvenation took place in 1969. There have been at least 600 eruptions since that year, all of which have been concerts with the two geysers erupting together. The only years without major action have been 1975, 1978, 1996, 1999, and 2006. There does seem to be a cyclic nature to the major activity—two or three years of frequent eruptions will be followed by one or two years with few or none. Since Fan and Mortar are not known to be connected with any other features except Spiteful Geyser (171) and Norris Pool (170), why such cycles take place is unknown.

Whether active or dormant, Fan and Mortar pass through a continuous series of "hot cycles." By studying these cycles with care, experienced geyser gazers are able to "guesstimate" the time—well, maybe guess the day—of the next eruption. The process is not very accurate. Between major eruptions and throughout periods of dormancy, the action during these cycles is erratic, weak, and (to people familiar with the cycles) disorganized. They are referred to as "garbage cycles." However, when the geysers are undergoing major activity, a series of garbage cycles will occasionally be interrupted by a "strong cycle." During a strong cycle there is a regular, well-organized progression in the activity, as follows: (1) a pause during which neither geyser is splashing; (2) surging within Lower Mortar; (3) splashing and weak jetting in Fan's River, High, Gold, and Angle vents; and (4) a return toward Mortar and another pause. Beginning in 2001, this sequence was altered with the addition of "event cycles." During these, as phase (1) merges into phase (2), splashing can be seen within the main vent of Fan Geyser. Phase (2) is dominated by a series of eruptions by Bottom Vent, continued splashing in Main Vent, and rare minor eruptions by Lower Mortar. By no means does every event cycle result in an "eruption cycle," but many do.

In an eruption cycle, the phase (3) jetting by Fan's High, Gold, and Angle vents becomes steady, reaching 6 to 10 feet high. When this persists for more than a minute or so, the geysers are in "a lock," and a major eruption is almost (almost, but not quite—locks can be unlocked!) certain to begin within a few minutes. Major eruptions most often begin in one of two ways. Usually, it is angled bursting out of the East Vent that triggers the full eruption, but sometimes it is strong surging by Upper Mortar.

The beginning is explosive and virtually simultaneous in both geysers. The main vent of Fan arches up and out to as much as 125 feet high. The spray can reach well across the trail, near and sometimes beyond Backwater Spring, and the horizontal throw of one remarkable eruption was measured as more than 200 feet. Meanwhile, Fan's other vents play as much as 60 feet high. Each jet is at its own angle, so the play does indeed resemble a fan. At the same time, Mortar deserves attention. Both of its vents can play over 50 feet high. Sometimes one of the two temporarily shuts off and the other surges as high as 80 feet. It is truly an indescribable spectacle.

The most impressive part of the eruption takes place during the first 15 minutes. During that time the water jetting is steady, without pause or significant reduction in size. Then the play stops without warning, only to resume a few seconds later. After a few more minutes comes another pause. With each of these pauses, the renewed eruption is weaker. The last weak water jetting is normally seen after a duration of 35 to 45 minutes; periodic steam bursts persist for an additional 45 minutes.

Lesser activity by single vents can still take place, but it usually occurs at times when Fan and Mortar would not be expected to have a major eruption. For example, Upper Mortar had a number of solo eruptions during 2001, each up to 30 feet high and lasting over 3 minutes, and that same summer Fan's East Vent had one independent eruption that jetted water 30 feet outward for a duration of about 1 minute.

When performing at their best, Fan and Mortar's intervals range between 2½ and 5 days (the shortest interval on record is about 22 hours), and the eruptions verge on being predictable. In other years, though, intervals are erratic and can be weeks long. The development of event cycles in 2001 has helped, but until we learn more about the cause and meaning of hot cycle activity, getting to see Fan and Mortar erupt requires either great luck or tremendous patience.

174. MORNING GLORY POOL is one of Yellowstone's most famous hot springs. Before the highway was removed from the area in 1971, Morning Glory was visited by more people than any spring or geyser other than Old Faithful. There are many beautiful pools, but this one, right beside the main entrance to the Upper Geyser Basin, was a natural candidate for popularity.

The fact that Morning Glory once was, and technically still is, a geyser is shown by the runoff channels that lead down to Firehole River. However, only one natural eruption has been recorded, and that occurred in 1944. The duration of the play is unknown, but the height was at least 40 feet, and a tremendous amount of water was thrown out.

Its popularity may have spelled the demise of Morning Glory Pool. In the past it was hot enough to prevent any growth of cyanobacteria within the crater. The color of the pool was a delicate pale blue, unlike that of any other. But so much debris has been thrown into the crater that the vent has been partially plugged. Hot water has smaller egress into the crater, and the temperature has dropped. Cyanobacteria can grow down into the crater, and the color is less beautiful than it used to be.

Because of its known geyser potential, Morning Glory was artificially induced to erupt in October 1950. The purpose was to empty the crater so it could be cleaned. The list of material disgorged included $86.27 in pennies, $8.10 in other coins, tax tokens from nine states, logs, bottles, tin cans, seventy-six handkerchiefs, towels, socks, shirts, and "delicate items of underclothing." Since 1950 several additional attempts to induce eruptions have been made, all without success. Coins and rocks can be seen lining small benches on the crater walls, and cleaning them out is a continuing job—2,785 coins were removed during 2004 alone. The moral is: Morning Glory was not created to be used as a wishing well. If such action continues, Morning Glory could become completely clogged. That would be a tragic loss.

175. WEST SENTINEL GEYSER usually acts as a perpetual spouter. Downstream and barely visible from the boardwalk at Morning Glory Pool (174), the eruption is a surging boiling of the water up to 2 feet high. True intermittent eruptions 4 feet high took place for a brief time following the 1959 earthquake, and a few bursts as high as 10 feet were seen in 1991 and 1992.

176. SERPENT'S TONGUE is a small, somewhat cavernous spring against the hillside behind West Sentinel Geyser (175). It is a perpetual spouter that plays about 2 feet high. The name originated because of eruptive steam bubbles that enter the crater in a darting fashion.

177. SENTINEL GEYSER, also known as East Sentinel, is actually two geysers within one crater. The large cone is adjacent to Firehole River immediately downstream from Cascade Falls. It is an island during much of the year when river water pours into a number of small nearby vents. This apparently stifles any potential eruptions. Even when active, eruptions by either of the geysers are rare.

Both geysers erupt at sharp angles. One, which had its most powerful known activity in the early 1980s, erupts from a vent at the northwest (river) side of the crater. It plays at a 45-degree angle and as high as 40 feet in the upstream direction, toward the riverbank where a shallow gully has been eroded by the spray. Young trees now grow there. The other geyser erupts from the nearer, southeast side of the crater, arching its water 20 feet high and into the river. This is the vent that is more often active, as it sometimes was during the 1990s, with durations as long as 12 minutes, and in 2005–2006, with durations of only a few seconds. July 2002 is the only known time when both vents were active simultaneously. There were several brief eruptions about 20 feet high.

178. UNNG-MGG-2 ("SENTINEL'S VENTS") is composed of several closely spaced holes in the geyserite shield immediately upstream from Sentinel Geyser (177). Only two active episodes have been recorded, in July 2004 and July 2005, each of which lasted about three weeks. So-called minor and major eruptions were about equally common. The minor play splashed to heights of only a few inches for 35 to 60 seconds, whereas the major eruptions could last longer than 3 minutes with some bursts over 3 feet high. The intervals were fairly regular, ranging between 2½ and 4 minutes.

179. GREEN STAR SPRING lies across the river a short distance downstream from Sentinel Geyser. It was named following the 1959 earthquake, when it may have had a few eruptions. No details of the activity are known; it was simply indicated as an "observed geyser" on a 1968 map.

CASCADE GROUP

The Cascade Group (Map 4.7, Table 4.6, numbers 180 through 189) includes Artemisia and Atomizer geysers, both of which are among the most important geysers of the Upper Basin. Because these springs lie over a hill from the Morning Glory Group and are partly hidden by stands of pine trees on the other sides, the rest of the group's geysers have been observed comparatively infrequently because no trail proceeds directly through the Cascade Group. The route of the old highway between Morning Glory Pool and Biscuit Basin is designated as a foot trail, and it affords the best views of most of the area. Across the river, running from the vicinity of Daisy Geyser to Biscuit Basin, is another trail, which provides better views of the geysers near the river as it passes the "Westside Group." It must be emphasized that access to the Cascade Group is restricted to these trails. Wandering off trail among the hot springs is illegal.

180. ARTEMISIA GEYSER was named because of the grayish-green color of some of its geyserite formations, which are similar in color to the foliage of sagebrush (scientific name *Artemisia*; the name is usually pronounced "are-te-MEEZH-ya") Artemisia has one of the largest craters of any hot spring in Yellowstone and is one of the most beautiful blue pools in addition to being a major geyser. There is steady overflow during most of the quiet period. Eruptions begin with little warning. The water level suddenly rises to cause a spectacular flood across the geyserite platform around the pool, and vigorous underground thumping can be felt and heard as steam bubbles form and collapse within the plumbing system. A minute or more passes before Artemisia reaches its full force, which is a massive boiling rather than bursting or jetting. Some of the surges can reach over 30 feet high. The eruptions last between 16 and 33 minutes, and some of the largest surges occur in the last seconds of the activity. Following the eruption the water level drops very slowly, requiring at least 30 minutes to fall about 18 inches. Refilling is extremely slow, taking 3 to 5 hours before overflow is resumed.

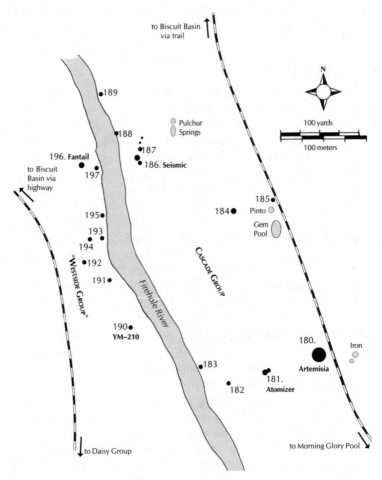

Map 4.7. *Cascade and "Westside" Groups, Upper Geyser Basin*

Table 4.6. Geysers of the Cascade and "Westside" Groups, Upper Geyser Basin

Name	Map No.	Interval	Duration	Height (ft)
"Aftershock Geyser"	187	20–40 min*	4–8 min	6–60
Artemisia Geyser	180	9–34 hrs	16–33 min	30
Atomizer Geyser, major	181	12½–16½ hrs	50 min	40–60
Atomizer Geyser, minor	181	see text	25–90 sec	20–35
Calthos Spring	184	[1959]	1 min	10–15
Fantail Geyser	196	[1990]	10–45 min	40–75
Hillside Geyser	188	[1964]	4 min	3–30
Ouzel Geyser	197	[1989]	seconds	10–50
Restless Geyser	182	steady	steady	1–2
Satellite Geyser	186a	see text	minutes	boil–6
Seismic Geyser	186	20–40 min	4–8 min	4–75
Slide Geyser	183	5–25 min	1 min	horizontal
Sprite Pool	185	3–5 min	1–2 min	inches
UNNG-CDG-1 ("Broken Cone")	189	5 min–hrs	seconds	3–25
UNNG-WSG-1 ("Fracture")	193	seconds	minutes	1
UNNG-WSG-2 ("Bigfoot")	194	1½ min	30–40 sec	6
UNNG-WSG-3 ("Carapace")	195	6 min–6 hrs*	1–5 min	boil–10
UNNG-WSG-4 ("Maelstrom")	191	[1998]	12 minutes	6
UNNG-WSG-5	192	30–40 sec	10 sec	1
"YM-210" ("South Pool")	190	[1992]	10–15 min	sub–15

* When active.
[ ] Brackets enclose the year of most recent activity for rare or dormant geysers. See text.

Relatively little was known about Artemisia's frequency until the mid-1980s, when a study found its intervals to average about 8 hours with little variation. Since then, both the range between short and long intervals and the average interval during any one year have gradually increased. The intervals now vary widely from 9 to 34 hours, with an average near 20 hours.

Minor eruptions were rare until 1985, and although never common, there have been episodes when they occurred as often as once every two or three days. A minor eruption starts and progresses in a fashion identical to a normal eruption, but it ends abruptly after a duration of only 5 minutes. The crater refills within 2 hours, and the following minor interval can be as short as 4½ hours.

The eruption of Artemisia Geyser is not particularly tall, but the huge volume of water discharged ranks it among the most important geysers.

At the north end of the rockwork embankment above Artemisia is a patch of steaming ground. Actually, a ceramic pipe is also there. Usually covered by debris, the pipe is the outlet for Bench Spring, a feature that was buried when the road was built. On the hillside above the old road are two other springs. The larger is Iron Spring, which evidently had a few eruptions up to 4 feet high following the 1959 earthquake. Either it or the other pool was referred to as "Tomato Soup Spring" during the late 1920s.

181. ATOMIZER GEYSER plays from two small cones, both about 3 feet tall, that lie directly beyond Artemisia Geyser (180) near a small stand of trees. The main vent is on the southwestern, flat-topped cone. The northeastern cone is usually active only during major eruptions.

Atomizer shows an interesting progression of activity through the cycle leading to the major eruption. Following a major eruption, several hours will pass before the plumbing system has refilled with water. Then, brief episodes of overflow from the main vent will repeat every few minutes. Accompanied by bubbling, they inspired the geyser's original name, "Restless" (a name since moved to nearby spring #182). It will usually be 8½ to 10 hours after the major and 1 to 4 hours after the first overflow before the first of a series of minor eruptions takes place. Minors then recur every 1 to 2 hours, with three to five (historically as many as eight) occurring prior to the next major eruption.

It is possible to tell approximately where Atomizer is in its minor series by observing the duration and height of a single minor. The first eruption of the series lasts less than 30 seconds and reaches only 20 to 25 feet high. Each subsequent minor is somewhat longer and stronger than the one before, and the last of a series often lasts well over 1 minute and reaches up to 35 feet high.

The intervals between minor eruptions are usually little more than 1 hour long, but there are many exceptions. Sometimes the next-to-the-last ("penultimate") minor interval will approach 2½ hours instead. Then the last interval, that between the last minor and the major, can be as short as 12 minutes; known as "quick comebacks," they are becoming more common with each passing year. Rather rare are "bizarre minors" that splash for 2½ minutes from the northeastern cone, which is normally inactive during the minor activity; these appear to always be followed by a major eruption within 17 minutes. None of these possibilities can be predicted.

The major eruption is indistinguishable from a minor until the end of its first minute or so. At that point, rather than quitting abruptly, the force increases. The height reaches between 40 and 60 feet throughout a water-phase duration of 8 to 10 minutes. The play then gives way to a steam phase that is loud enough to be heard from the trail and lasts as long as 50 minutes. Excepting the rare "bizarre minors," it is only at the time of a major eruption that the northeastern cone joins the other. Nearly sealed in by internal deposits of geyserite, this is the actual "atomizer." During the main vent's steam phase, the atomizer sends steam and fine spray a few feet high while producing a distinct chugging sound.

Atomizer is regular enough to be predicted. The range between successive major eruptions is from 12½ to 16½ hours, with an average near 15 hours. Given the frequency of minor activity and the view of adjacent Artemisia Geyser, a two-hour wait for Atomizer can be wonderfully rewarding.

182. RESTLESS GEYSER (sometimes called "Owl Mask Spring") is a small perpetual spouter whose basin shares some of the sagebrush-gray color with Artemisia Geyser (180). The spring, which splashes just 1 to 2 feet high, is barely visible only from near the Daisy Geyser–Biscuit Basin trail on the west side of Firehole River.

183. SLIDE GEYSER is located on the high bank of Firehole River below Atomizer Geyser (181) and must be viewed from the far side of the river. There is no record of activity by Slide prior to 1974. The vent opens directly onto a precipitous slope, so its erupted water seems to slide rather than flow to the river below. Slide is fairly regular in its performances. Intervals have ranged from 5 to 25 minutes but usually average 18 minutes with little variation. The water bursts out of the cavernous vent so that the play is almost horizontal, with jets sometimes reaching several feet outward. The duration is always about 1 minute. The noisy feature at the edge of the river below Slide is Spitter Spring.

184. CALTHOS SPRING was clearly very active in the distant past. The deep crater sits in the middle of an extensive but highly eroded sinter platform that is drained by a single, wide and deep runoff channel. The only activity on record was triggered by the 1959 earthquake. For a few weeks Calthos underwent irregular eruptions, some of which reached 10 to 15 feet high but lasted less than 1 minute. The water discharged by these eruptions was prodigious, easily filling the runoff channel. This active phase was the only known time when the water level in nearby Gem Pool was below overflow. For the next few years the two springs would alternately ebb and flow. Then Gem resumed the steady discharge that had characterized it since the discovery of the Park. The only further flow from Calthos occurred during 1970 and 1981–1982. It was slight both times and had no observable effect on Gem Pool.

Well-named Gem Pool is one of the bluest of Yellowstone's pools, so it seems unfortunate that its 1872 name, "Great Sky Blue Hot Spring," did not stick.

185. SPRITE POOL supports the runoff channel next to the trail a short distance north of Gem Pool. It behaves mostly as an intermittent spring. During the overflow periods there is steady bubbling over the vent, which can be vigorous enough to cause splashes a few inches high. The quietly bubbling pool a few feet south of Sprite is Pinto Spring.

186. SEISMIC GEYSER is famous for having been the most significant direct geothermal creation of the 1959 Hebgen Lake earthquake. A crack formed in the sinter and became the site of a small steam vent. Activity increased over the next 3½ years until a steam explosion blew out the geyser's crater. The force continued to increase even then, and the newly formed geyser became stronger and more explosive every year. By 1966 the eruptions were reaching anywhere from 50 to 75 feet high. But as the strength of the play grew stronger, the intervals gradually increased. When a small satellite crater was created by another steam explosion in 1971, most of the eruptive energy shifted to the new vent, which was named Satellite Geyser (186a, also known as "Seismic's Satellite"). Eruptions soon became more constant and smaller in size. Ever so slowly, Seismic and Satellite died out. The last large eruptions in Seismic were seen during 1974, and boiling in Satellite has been rare since about 1984.

After 25 years of dormancy, Seismic began having periods of strong, intermittent overflow in May 1999, about a month before "Aftershock Geyser" (187) made its appearance. The pool boils every few minutes, and sometimes it erupts with splashes several feet high.

In addition to Aftershock, Seismic Geyser is unquestionably connected with several other hot springs. The most important of these are Hillside Geyser (188) just downstream and the Pulchur Springs that lie hidden among the trees a short distance up the slope. As Seismic increased in vigor, these other springs declined dramatically. As long as Seismic and Aftershock continue to erupt and pour out heavy flows of water, these other hot springs are unlikely to ever resemble their pre-1959 states. It is interesting that the major-scale development of Fantail Geyser (195) directly across the river in 1986 had no effect on Seismic.

187. "AFTERSHOCK GEYSER" formed in June 1999, clearly as a continuing development of the Seismic Geyser system. The initial eruptions reached 6 to 15 feet high. It was then dormant for several months. With rejuvenation in early 2000, "Aftershock" had its most powerful eruptions that April, with voluminous bursts of water over 40 feet high and 60 feet wide. It rapidly declined, though, and since mid-2000 it has been dormant more often than not. When active, eruptions 6 to 10 feet high occur at the same time as overflow by Seismic Geyser (186), with typical intervals of 20 to 40 minutes and durations of 4 to 8 minutes.

188. HILLSIDE GEYSER was a boiling pool never seen in eruption prior to 1948. It was active with 20-foot eruptions throughout that year but then had little additional activity until a new cycle was initiated by the 1959 earthquake. Major eruptions, up to 30 feet high, occurred with great regularity every 26 minutes and lasted 4 minutes. Such activity lasted into 1961, but then only weaker play was seen. Some of these eruptions were just 3 feet high, and even those stopped in 1964, about when Seismic Geyser (186) began having its first major eruptions. Hillside's water now lies quietly many feet below the surface.

A short distance up the slope from Hillside Geyser, two small 3-foot geysers made their appearance during 1982. After less than a year of action, they ceased to erupt. There is no other record of activity by them, but they appear to lie on the same fracture that passes from Seismic through "Aftershock," so future developments here are possible.

189. UNNG-CDG-1 ("BROKEN CONE GEYSER") plays from a decaying, rust-colored vent adjacent to Firehole River, a short distance downstream from Hillside Geyser (188). The first eruptions of record were seen during the early 1980s. The intervals have always been irregular, ranging from just 5 minutes to several hours long, perhaps on a cyclic pattern. The height is seldom more than 3 feet, although exceptional eruptions during 1983 reached as high as 25 feet. The duration is never longer than a few seconds.

"WESTSIDE GROUP"

The "Westside Group" (Map 4.7, Table 4.6, numbers 190 through 197) contains a number of small hot springs and at least eight geysers. Located directly across Firehole River from the Cascade Group, these springs are probably physical members of that group but are considered separately because of the topographical separation caused by the river. The "Westside Group" is best viewed from the hiking-biking trail that runs through the forest between the Daisy Group and Biscuit Basin.

In addition to the geysers and other hot springs of this area, an area of ground northwest of Fantail Geyser (196) began to heat up during 1984. It is believed that this process was triggered by the Borah Peak earthquake of October 1983 and that it eventually led to the 1986 activity by Fantail. By midsummer 1985 the temperature on the surface of the ground was as high as 205°F (95°C), hot enough to cause a distillation of organic matter in and on the soil. The sickeningly sweet odor of burned sugar could be smelled hundreds of feet away. The ground temperature has gradually cooled since the 1980s, and grass is re-growing in the area.

190. "YM-210," the southernmost feature of the "Westside Group" (sometimes called "South Pool"), served as a numbered water sample site and reference point on the maps produced by the U.S. Geological Survey following the 1959 earthquake. It had a few eruptions at that time, but those records provide no statistics about the geyser's frequency, duration, or height. "YM-210" proved to be a significant geyser during 1989. There were numerous eruptive episodes that spring and summer. In each there was a series of eruptions, starting with a major eruption up to 15 feet high. The water discharge was huge throughout the 10- to 15-minute duration. The major was followed by a series of minor eruptions during the next 5 hours or more. These eruptions rose from a pool level several feet belowground and discharged little water. "YM-210" was active in a similar fashion during the summer of 1991, and a few weak, full-pool eruptions without observed majors took place during early 1992. "YM-210" is now a quiet, pale-green pool with only a trickle of overflow.

191. UNNG-WSG-4 ("MAELSTROM GEYSER") was active for only a few days during May 1998. Although the play reached just 6 feet high, the action was incredibly violent for its size, hence the name. Few eruptions were actually witnessed, and information about the intervals was never obtained; the durations were longer than 12 minutes.

192. UNNG-WSG-5 was first seen as a geyser in July 2006. Playing from a badly weathered crater that is one of those nearest to the constructed bicycle trail, the activity repeated every 30 to 40 seconds for durations of less than 10 seconds. The eruptions reached just 1 foot high.

193. UNNG-WSG-1 ("FRACTURE GEYSER") is a slit-like vent at the top of the Firehole River bank. More forceful than it looks at first glance, the eruption does not seem impressive because its horizontal nature confines it almost entirely to within the slit itself. This geyser is in eruption most of the time. Periods of complete quiet never last more than a few seconds, and the durations can exceed 5 minutes. Because of its off-trail location, Fracture is difficult to see.

194. UNNG-WSG-2 ("BIGFOOT GEYSER") probably began its activity at the time of the 1959 earthquake. It plays from an oblong, foot-shaped crater near several other hot springs, most of which act as small perpetual spouters. "Bigfoot's" vent is at its south end, in the heel of the foot. The intervals between eruptions are highly regular and about 1½ minutes long. Durations range from 30 to 40 seconds. When "Bigfoot" is at its best, the splashing can reach over 6 feet high.

195. UNNG-WSG-3 ("CARAPACE GEYSER") is dormant most years but can be very impressive in its time. Located immediately above the river, the cone's overall shape resembles the carapace (top shell) of a tortoise. The most common form of activity is only a surging, bubbling overflow, but "Carapace" can act as a cyclic geyser. Then, intervals as long as 6 hours separate active episodes when play recurs every 6 to 20 minutes. The eruptions come in two forms. By far more common are minor eruptions, which have durations of about 1 minute and heights of less than 3 feet. Major eruptions are rare but burst water up to 10 feet high for 2½ to 5 minutes. As is the case with "Fracture Geyser" (193), "Carapace" occupies a hidden location and is difficult to observe.

196. FANTAIL GEYSER was *the* geyser story of 1986. It had been a large, superheated pool during all of previous Park history. The water was clear most of the time, but occasionally it was murky because of suspended clay particles. That murkiness might imply that Fantail had infrequent eruptions, but none was positively known until 1985. No play was actually seen even then, but the surrounding sinter platform was washed clean of gravel, and the extent of the washed areas implied heavy overflow perhaps accompanied by bursts a few feet high.

Fantail became a major geyser in April 1986. From its very beginning and persisting through most of that summer, it was a highly regular and imminently predictable geyser, which allowed a great many people to enjoy its unique performances. The intervals were 6 hours long. Following about 2 hours of intermittent boiling of increasing intensity, the eruption began with heavy surging that only gradually built up to the full force. Massive bursts and jets rose from both of the two craters, at their best reaching 75 feet high from one vent and perhaps 50 feet at an angle from the other. The play did not significantly decrease its strength until near the end of the 45-minute duration, when a powerful steam phase began. The show was truly amazing.

Suddenly, in mid-August 1986, Fantail became erratic. The eruptions grew weaker, and the durations of only 10 minutes failed to result in a steam phase. Eruptions ceased completely before the end of October. Fantail had a few additional eruptions between 1987 and 1990, but except for three of full force in August 1988, they were little more than spells of increased boiling months apart. There have been no eruptions of any sort since 1990.

197. OUZEL GEYSER spends much of its time under the water of Firehole River, like its namesake bird. The cold stream usually quenches any potential eruptions, but Ouzel was active along with Fantail Geyser (196) during 1986. Always an irregular and brief performer, some bursts of Ouzel's first eruptions reached over 50 feet high. It quickly died down, however, and was nearly dormant even before the decline in Fantail. Ouzel had a few more, independent eruptions 10 to 15 feet high during 1989 and was seen violently boiling in 2003.

BISCUIT BASIN

The area traditionally included within the Biscuit Basin is more than half a mile long. This span covers two distinct groups of hot springs. One group lies parallel to the route of the old highway through the Upper Basin. This right-of-way does not enter directly among the hot springs, but the trail along it provides reasonably near views of most of them. This group is referred to as the "Old Road Group" of Biscuit Basin. The Sapphire Group of Biscuit Basin, also historically called both the "Main Group" and the "Soda Group," is west of Firehole River.

"Old Road Group" of Biscuit Basin

The "Old Road Group" (Map 4.8, Table 4.7, numbers 200 through 210) occupies an old sinter platform. It is a wide, open area of little relief perforated by many hot springs, but the group contains only about ten, mostly small geysers. It can be reached by following the trail along the old highway route and, like the Cascade Group, is closed to entry.

200. BABY DAISY GEYSER was named because its eruptions are reminiscent of the much larger Daisy Geyser (137). The play is angled sharply toward the east and reaches about 30 feet high. It is the most important geyser in the Old Road Group, even though it has been observed only during brief periods in 1952 and 1959 and more consistently from early 2003 to late 2004.

The 1952 activity began in mid-July and persisted through the end of that summer season; when it ended is unknown. Baby Daisy's eruptions recurred every 105 minutes with remarkable regularity. Along with it, Biscuit Basin Geyser (201) and seven other springs (200b) between the two also erupted, as did nearby "Baby Splendid Geyser" (200a). None of them had ever exhibited any previous animation. The activity of 1959, following the Hebgen

Table 4.7. Geysers of the "Old Road Group" of Biscuit Basin, Upper Geyser Basin

Name	Map No.	Interval	Duration	Height (ft)
Baby Daisy Geyser	200	[2004]	2–2½ min	30
"Baby Splendid Geyser"	200a	see text	unknown	few feet
Biscuit Basin Geyser	201	[1952]	1–1½ min	75
Cauliflower Geyser	202	20–40 min	1–3 min	4–60
Demise Geyser	205	minutes*	minutes	10–30
Dusty Geyser	209	[2004]	3 min	15
Island Geyser	211	near steady	near steady	10
"Mercury Geyser"	207	40 min–2 hrs*	30 sec–min	1–3
Rusty Geyser	208	1–13 min	10–45 sec	10
UNNG-ORG-1	203	5 min*	5 min	1–5
UNNG-ORG-2	204	2–3 min	10–15 sec	3
UNNG-ORG-5	206	[1987]	4–8 min	10–15
UNNG-ORG-6 ("Goldfinger")	210	steady	steady	1–3

* When active.
[ ] Brackets enclose the year of most recent activity for rare or dormant geysers. See text.

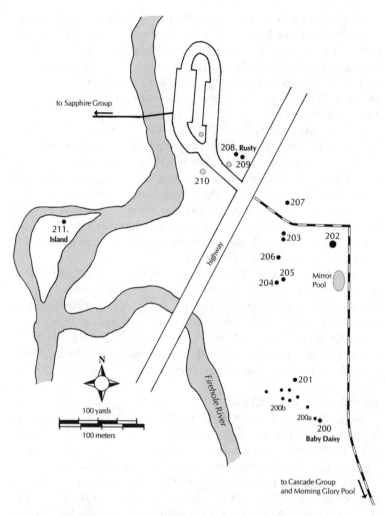

Map 4.8. *"Old Road Group" of Biscuit Basin, Upper Geyser Basin*

Lake earthquake, was similar to that of 1952, except the eruptions were considerably more frequent but less regular, and they were *not* accompanied by activity in any of the nearby geysers. That episode ended sometime during the winter of early 1960. Twenty years later, during 1980, Baby Daisy's runoff channels were washed clean of debris by heavy overflows, but apparently no actual eruptions took place at that time.

The most recent active phase apparently began on February 14, 2003, and was Baby Daisy's best episode on record. Intervals between eruptions ranged between 18 and 85 minutes, with an average of about 50 minutes. The duration was 2 to 2½ minutes, and, as in the previous active phases, the sharply angled water jet reached 30 feet high. This active phase ended without warning on December 8, 2004.

"Baby Splendid Geyser" (200a) is said to have had splashing eruptions in concert with Baby Daisy during 1952, so it is odd that it was not active at all in 1959. Extensively washed areas and deeply scoured runoff channels indicated that "Baby Splendid" played in some fashion at the start of Baby Daisy's reactivation in February 2003, but none of those eruptions was actually witnessed. Later, it was occasionally seen splashing 1 to 2 feet high shortly before the start of some of Baby Daisy eruptions, but these were confined to the crater and produced no runoff.

Northwest of Baby Daisy Geyser toward Biscuit Basin Geyser is open ground that encompasses numerous hot springs. Seven of these (200b) erupted as geysers while Baby Daisy and Biscuit Basin were active during 1952, but which of these openings those seven were was not documented (the indication on Map 4.8 is schematic only). In that area, several small geysers were seen during 2003, and one acted as a vigorous perpetual spouter throughout 2004.

201. BISCUIT BASIN GEYSER was one of the springs that became active at the same time as Baby Daisy Geyser (200) in 1952. It was much larger, with bursting eruptions that reached as high as 75 feet. Its action, however, was very irregular, with several eruptions on some days but none on others. The play lasted only 1 to 1½ minutes. The shallow runoff channels that lead away from the crater were created in 1952, so it is believed that that was Biscuit Basin's first-ever activity. Neither the 1959 earthquake nor the 2003 reactivation of Baby Daisy caused any change in the geyser's appearance of gentle bubbling in water slightly below overflow.

202. CAULIFLOWER GEYSER occupies a large crater surrounded by cauliflower-like nodules of geyserite. Water generally stands several inches below overflow. The eruption begins with a sudden rise in the pool's level, accompanied by boiling that may surge up to 4 feet high. Copious discharge floods the surrounding platform, but much of that water flows back into the crater at the end of the 1- to

3-minute duration. Most intervals fall in the range of 20 to 40 minutes. Between 1980 and 1988, Cauliflower infrequently had large bursts during some eruptions. They reached an estimated 30 feet high, and at least two 1986 eruptions, seen from distant Fantail Geyser (196), probably doubled that. Bursting play of this sort has not been seen since 1988.

203. UNNG-ORG-1 erupts from a small vent at the eastern corner of a shallow crater about 200 feet west of Cauliflower Geyser (202). Its activity began with the 1959 earthquake, and eruptions about 5 feet high recurred every 5 minutes until the late 1970s. It was also active on occasion during the 1980s, but then the play reached only 1 foot. At the northwestern edge of this same crater, another vent was occasionally active after 1983. Its eruptions did little more than cause surface turbulence and slightly heavier overflow. Like ORG-1, the interval was about 5 minutes and the duration a few seconds. Neither of these geysers has been observed in many years.

204. UNNG-ORG-2 is nearly hidden in some trees around 200 feet south of ORG-1 (203). The eruptions, about 3 feet high, are entirely confined within a deep crater and are not visible from the trail. The intermittent steam produced by the play recurs every 2 or 3 minutes and lasts 10 to 15 seconds.

205. DEMISE GEYSER, a few feet north of ORG-2, is erratic and infrequently active. It received its name because the wash of the eruptions seems to be hastening the geyser's demise, eroding the surroundings and filling in the crater. Somehow, though, it manages to keep its vent clear of debris. Demise is probably cyclic, with hourslong active episodes days to weeks apart. Then it erupts every few minutes with jets of steamy spray that can reach over 30 feet high.

206. UNNG-ORG-5 was at its best during 1986 and 1987. Early in that activity a second vent blew out next to the original crater and often took over most of the action. Located at the edge of one of Cauliflower Geyser's (202) main runoff channels, between ORG-1 (203) and Demise (205), some of the bursts reached 10 to 15 feet high. The intervals were as short as 40 minutes. Rocks were occasionally thrown out by the 1987 eruptions. This substantially enlarged the crater and eroded the plumbing system, and many of the last eruptions were thus nothing more than gushing overflow. ORG-5 has not been seen since 1987.

207. "MERCURY GEYSER" is located near the east side of the highway, opposite the entrance to the Biscuit Basin parking lot. Its performances are highly irregular, but when active the intervals are usually 40 to 60 minutes long. The play is 1 to 3 feet high and produces little or no runoff in durations of about 1 minute. "Mercury" is dormant during some seasons.

208. RUSTY GEYSER is one of the most visible geysers in the Upper Geyser Basin. The geyser is located just to the right (north) of the entry road to the Biscuit Basin parking lot. Eruptions recur every 1 to 4 minutes (occasionally as long as 13 minutes), last 10 to 45 seconds, and burst as much as 10 feet high.

209. DUSTY GEYSER erupts from a low sinter cone just a few feet east of Rusty Geyser (208). Its geyserite lacks the iron oxide stain of Rusty, and the duller appearance led to the name. When active during the 1970s and early 1980s, Dusty erupted several times per day. Each eruption lasted about 3 minutes and was a steady jet up to 15 feet high. Dusty and Rusty could often be seen together, forming an impressive duo, but there was no evidence of a connection between the two. Unfortunately, Dusty has been nearly dormant since 1987. A single eruption was witnessed in January 2004, but otherwise it produces only infrequent small splashes.

210. UNNG-ORG-6 ("GOLDFINGER GEYSER") is a perpetual spouter near the southwest side of the road entering the Biscuit Basin parking lot. The name is an allusion to a former horse patrol ranger who was the first to report the activity, in 1980. The gentle splashing is 1 to 3 feet high.

211. ISLAND GEYSER is somewhat separated from the other geysers of the group and may be part of the Sapphire Group. Its crater is on a low, marshy island in Firehole River. After the 1959 earthquake, numerous small springs developed on the island. About a dozen of the springs were geysers. In 1966 one of them took over the major function and began steady spouting to 6 feet. After the 1983 earthquake the play increased in size, with some jets reaching over 10 feet high. Island is a true geyser, albeit barely. Quiet intervals of about 3 minutes are rare, while durations are apparently hours to days long.

Sapphire Group of Biscuit Basin

The fact that "Sapphire Group" is the officially approved name for this area on the west side of Firehole River is not well-known. Other historical names include "Main Group" and "Soda Group." However, the simple term "Biscuit Basin" is more commonly used and is what is shown on nearly all current maps. This name referred to nodules of geyserite that resembled biscuits at several of the springs. The best examples were at Sapphire Pool, but they no longer exist, having been blown away by the powerful eruptions that followed the 1959 earthquake.

The Sapphire Group (Map 4.9, Table 4.8, numbers 215 through 238) is traversed by a boardwalk. Every spring and geyser is easily visible from the trail. At the west side of the basin, another trail leads 1.2 miles up the valley of Little Firehole River to Mystic Falls.

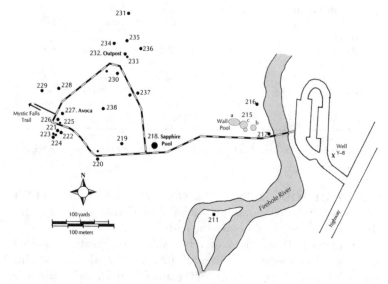

Map 4.9. *Sapphire Group of Biscuit Basin, Upper Geyser Basin*

Table 4.8. Geysers of the Sapphire Group of Biscuit Basin, Upper Geyser Basin

Name	Map No.	Interval	Duration	Height (ft)
Avoca Spring	227	1 min*	sec–1 min	4–25
Black Diamond Pool	215c	[2007]	10–40 sec	15–80
Black Opal Pool	215b	[1953]	seconds	20–50
Black Pearl Geyser	237	[1967]	sec–min	4–30
Coral Geyser	237	near steady	near steady	subterranean
Fumarole Geyser	238	frequent	seconds	sub–10
Jewel Geyser	219	4–12 min	1–3 min	5–20
Mustard Spring, East	230	2–3 min	5 min	6–10
North Geyser	231	extinct?	minutes	15
Sapphire Pool	218	[1991]	sec–5 min	6–125
Sea Weed Spring	228	[2003]	seconds	2
Shell Spring	220	min–hrs	min–hrs	6
Silver Globe Geyser	221	minutes*	seconds	1–25
Silver Globe Cave Geyser	222	minutes*	seconds	see text
"Silver Globe Pair Geyser"	223	irregular*	1 min	1–12
Silver Globe	226	seconds*	seconds	1–4
Slit Geyser	224	11 min–1½ hrs	1½–2½ min	10
UNNG-BBG-2 ("Red Mist")	236	seconds	seconds	2–10
UNNG-BBG-3	225	frequent	minutes	1–2
UNNG-BBG-4 ("Outpost")	232	6–12 min*	2–3½ min	2–8
UNNG-BBG-5 ("Sentry")	233	8–28 min*	1–4½ min	1–2
UNNG-BBG-6 ("Green Bubbler")	234	6 min*	6 min	1–2
UNNG-BBG-7 ("Yellow Bubbler")	235	min–2 hrs*	20–26 min	1
UNNG-BBG-8	216	near steady	long	1–6
UNNG-BBG-9	217	erratic	seconds	1–3
Wall Pool	215a	[1959?]	seconds	unknown
West Geyser	229	[1988?]	unknown	5–30

* When active.
[ ] Brackets enclose the year of most recent activity for rare or dormant geysers. See text.

215a. WALL POOL; 215b. BLACK OPAL POOL; and 215c. BLACK DIAMOND POOL occupy large hydrothermal explosion craters that definitely did not exist before the early 1900s.

Wall Pool (215a) was created by one or more steam explosions that took place sometime between 1902 and 1912, when it was photographed and described as a "new feature." Additional explosions in 1918, 1925, and 1931 probably formed Black Diamond Pool

(215c) and an unnamed crater as new features east of Wall Pool without affecting Wall Pool itself. Excepting small activity in some secondary vents, these springs had been quiet since 1931, until 2006 (see discussion later in this section; Wall Pool might have been weakly active for a brief time following the 1959 earthquake).

The development of Black Opal Pool (215b), the spring closest to Firehole River, took place in a series of titanic steam explosions that occurred from January to March 1934. The large, angular boulders of sandstone and conglomerate scattered around this area were blown out by those blasts and led to the original name "Black Boulder Geyser." Small chunks of sandstone were found as far as 1,000 feet to the north. Follow-up episodes of geyser eruptions, typically single bursts 20 to 30 feet high that did not expel additional rocks, were seen in 1937 and in 1947–1948. Black Opal's last known activity occurred in 1953, when rocks and sand were thrown as far as 50 feet.

The area's explosive nature had largely been forgotten until August 2005, when the entire Wall Pool–Black Opal Pool complex was slightly energized. At least six vents underwent small eruptions. The two largest jetted as high as 2 feet, one from a crack at the northwest edge of Black Opal and another from a small crater near Wall. These proved to be a prelude to developments in 2006.

February 2006 saw the formation of UNNG-BBG-8 (216), along with hot ground and small geysers between there and Black Opal. Then, on July 13, 2006, Black Diamond Pool underwent a powerful eruption of black, muddy water. A great deal of mud and rock was thrown out by bursts as high as 50 feet. Numerous additional explosive eruptions took place in the following days. Some were no more than 15 feet high, but at least two exceeded 80 feet and showered the boardwalk with slurries of mud and sand. Continuing activity of this sort was observed in March 2008.

216. UNNG-BBG-8 made its appearance as a new area of steaming ground in February 2006. By April 2006 it was erupting from a small crater near Firehole River. Developments continued, and by midsummer splashing was visible from at least four different vents, two of which reached as high as 6 feet. Continuing activity is essentially perpetual.

Numerous other steaming vents and one small geyser formed within an extensive area of hot ground between Black Opal Pool (215b) and BBG-8 at the same time as BBG-8 formed. Probably marking a developing fracture zone, this area is being carefully monitored because of the explosive histories of nearby Wall Pool, Black Opal Pool, and Black Diamond Pool (215).

217. UNNG-BBG-9 began to form during early February 2007 from two of the many tiny vents that developed in 2006 at the same time as the hot ground between Black Opal Pool (215b) and BBG-8 (216) developed. Eruptions were only 1 to 3 feet high, but they quickly enlarged the vent into a distinct hourglass-shaped crater. Eruptions stopped within a few weeks, but because BBG-9 is within a few feet of the boardwalk, it is being closely watched.

218. SAPPHIRE POOL gained early notoriety as one of the most beautiful pools in Yellowstone. The crater is of great depth, giving the water an incredibly rich blue color. Since discovery, Sapphire has been known as a geyser. Minor eruptions occurred every few minutes, doming the water about 6 feet high, resulting in heavy overflow.

No spring in Yellowstone was more greatly affected by the 1959 earthquake than Sapphire. The day after the shocks the crater was filled with muddy water, constantly boiling with vigor. Four weeks later Sapphire began having tremendous eruptions. Fully 125 feet high and almost equally wide, these eruptions were among the most powerful ever known in Yellowstone. At first, the intervals were as short as 2 hours, and each play lasted 5 minutes.

The huge eruptions were short-lived. As time passed, Sapphire began to have brief dormancies. As these grew longer, the active periods grew shorter, as did the intervals between their eruptions. The force of the play decreased, too, and by 1964 no eruptions were more than 20 feet high. Still, it was not until 1971 that Sapphire finally cleared all muddiness from its water, at about the same time all eruptive activity stopped.

Only time will tell whether Sapphire will ever again undergo major eruptions. There is plenty of evidence that the geyser damaged its plumbing system. Prior to 1959 the crater was circular and 15 feet in diameter; today it is oval and measures 18 by 30 feet. The explosive activity that enlarged the crater also eroded substantial amounts of material from deep within the plumbing system. The frequent minor eruptions seen before 1959 no longer occur, and the only trace of periodicity is a slight variation in the rate of boiling. Nevertheless, on August 9, 1991, Sapphire Pool had at least two eruptions about 20 feet high. Clearly, some potential still exists.

219. JEWEL GEYSER is appropriately named, whether the term came from the beads of pearly sinter about the vent or from the sparkling droplets of water of its eruption. It was originally called "Soda Geyser" by the Hayden Survey, who considered this the most important geyser of their "Soda Group."

Jewel's eruptions are very regular. Although intervals are known to range from 4 to 12 minutes, they usually average near 8 minutes, depending on the number of bursts during an eruption—the more bursts there are, the longer the subsequent interval. Each eruption consists of 1 to 10 bursts (usually 3 to 5) separated from one another by several seconds. The largest bursts are up to 20 feet high. Jewel was altered temporarily by both the 1959 and 1983 earthquakes. Over the next few years, some of the bursts reached as high as 40 feet and were angled sharply enough to reach the boardwalk (which has since been moved farther away).

In early November 1992 the casing of the old "Y-8" research drill hole adjacent to the Biscuit Basin parking lot blew out (see Map 4.9). Interestingly, hot springs near the well such as Rusty Geyser (208) were entirely unaffected, but Jewel Geyser nearly 1,000 feet away was seriously impacted. Although the well was plugged within days of the blowout, Jewel's water level dropped dramatically and has not recovered. The eruptions are still beautiful, but the water level is so low that there is little discharge during even the biggest bursts.

220. SHELL SPRING lies within a yellowish crater that someone imagined looked like a clamshell. Shell is a cyclic geyser whose total period is many hours to perhaps a day long. When it is in an active phase, each eruption tends to last a little longer and raise the water level a little higher. Near the end of a cycle the crater begins to overflow, and one last eruption, with bursts up to 6 feet high and a duration as long as an hour, fully drains the system. Shell then requires several hours to recover and begin the first weak eruptions of a new cycle.

Silver Globe Complex (numbers 221 through 227)

Few geysers were more dramatically affected by the 1983 Borah Peak earthquake than this set of intimately related springs. There was a great increase in eruptive activity, and observations indicate that probably no other group of springs anywhere shows such extensive cyclic behavior compounded by frequent exchanges of function among its members. The descriptions here are confined to the most spectacular activity; to try to explain all the relationships would be impossible. As of 2007, vigorous activity by Avoca Spring (227) had made the complex all but dormant, but often a few minutes spent here is highly rewarding.

221. SILVER GLOBE GEYSER is the blue pool immediately below the boardwalk. This spring contains two vents. When active at its best, the vent farthest from the walk plays the highest, spraying massive bursts as high as 25 feet. The nearer vent may accompany the other by erupting up to 15 feet. Intervals can be as short as 5 minutes. More common are frequent minor eruptions 1 to 4 feet high that rise only from the vent nearer the boardwalk. Whether major eruption or minor, the duration is never longer than a few seconds.

222. SILVER GLOBE CAVE GEYSER plays from a cavernous opening in the cliff face immediately to the left of Silver Globe Geyser (221), and their eruptions are closely related. Constantly churning within the cavern, this geyser is likely to spray outward only in concert with or immediately following action by Silver Globe Geyser. The most typical eruptions are largely confined to the visible catch basin. So-called super eruptions are rare, but in 1986 they sent powerful jets of water as far as 50 feet outward, away from the boardwalk. The geyser's cavern creates a dangerous overhang here, so please stay on the boardwalk.

223. "SILVER GLOBE PAIR GEYSER" erupts from two small craters between Silver Globe Geyser (221) and Slit Geyser (224). It usually undergoes only small, brief eruptions every few minutes. "Pair" was the last of the Silver Globe geysers to show major activity, as no bursting eruptions of large scale were reported until 1989. During major activity, which is rare, some bursts reach as much as 12 feet high. The spray is angled onto the slope near Silver Globe Geyser, where erosion implies that such activity had never previously occurred. When "Pair" is the focus of energy, eruptions can recur as frequently as every 5 minutes. Most of the large bursts take place at the very beginning of an eruption, and the duration seldom exceeds 1 minute.

224. SLIT GEYSER plays from a narrow rift in the sinter next to a circular geyserite basin, which is probably a sealed-in pool. Slit is the most distant of the Silver Globe Complex from the boardwalk. It is most active when the other Silver Globe geysers are not. Intervals vary from 11 minutes to 1½ hours but tend to be quite regular at any given time. The duration is 1½ to 2½ minutes, during which a fan-shaped spray of water is jetted about 10 feet high. Slit often pauses for 10 to 30 seconds about halfway through the play.

225. UNNG-BBG-3 is a small crater across the boardwalk from the Silver Globe geysers. It has frequent eruptions at those times when Silver Globe Geyser (221) and Silver Globe Cave Geyser (222) are vigorously active. At other times the activity is less frequent and somewhat weaker. Even at its best, only a little water is sprayed above ground level. In early 2006, when the thermal energy shifted from Silver Globe (226) to Avoca Spring (227), BBG-3 fell totally dormant, and its crater quickly decayed into rubble.

226. SILVER GLOBE is the pool within a deep crater immediately next to Avoca Spring (227). It was named by G. L. Henderson in 1888, and his description would lead one to believe this was the most fascinating and gorgeous of all Yellowstone hot springs. Instead, it is a rather prosaic feature. The water level always lies well below overflow. The vent is at the far back side of the crater beneath a thick overhanging shelf of geyserite Henderson called the "Zygomatic Arch." When Silver Globe is active, Avoca Spring is dormant, and the other Silver Globe geysers across the boardwalk exhibit minimal activity. Silver Globe's intervals and durations are both only seconds long. The play might reach several feet high if it wasn't for the underside of the overhang, which blocks the more vigorous splashes. On the other hand, when Silver Globe is dormant, the water level stands much lower, and bubbling in the vent is slight.

227. AVOCA SPRING played as a small geyser during part of 1934, but otherwise it was always a steadily boiling and overflowing, but non-erupting, spring until the 1959 Hebgen Lake earthquake. After those shocks, it first became a powerful steam vent, and by the end of the year it had become a geyser. Generally cyclic, there was a series of minor eruptions. Recurring at intervals of 1 minute, the eruptions were 4 to 6 feet high and lasted around 5 seconds each. After a dozen or more minors, Avoca would have a major eruption. Up to 25 feet high, they had durations of 20 to 25 seconds and were followed by 6 to 25 minutes of quiet before the next series of minors began. These performances continued until after the 1983 Borah Peak earthquake.

At first, the Borah Peak earthquake seemed to have had little effect on Avoca Spring, but when the Silver Globe geysers had their first major eruptions in 1985, Avoca fell nearly dormant. And when adjacent Silver Globe (226) was active, Avoca fell completely silent, with vigorous boiling in some small side vents but none within the main crater. This relationship among the springs continues. As the activity shifts unpredictably within the Silver Globe Complex, Avoca irregularly turns on and off. It has acted as a noisy, steamy perpetual spouter since early 2006.

228. SEA WEED SPRING was named in 1887 because of the thick, stringy cyanobacteria that grew within the crater. Vague references say it boiled and possibly erupted during 1897. In 1988, at the same time West Geyser (229) had unseen eruptions, the bacteria of Sea Weed was disturbed, and a large runoff channel was carved into the surrounding gravel. The actual nature of the action that caused those changes is uncertain, but it may have been heavy overflow rather than true eruptions. Nevertheless, in May 1996 and August 2003, Sea Weed Spring did erupt, with brief bursts 2 feet high seen over the course of several days.

229. WEST GEYSER is the pool beyond Sea Weed Spring (228). It was active as a geyser within the first two weeks after the 1959 earthquake. No eruptions were witnessed by any reporting observer, but splashed zones surrounding the crater indicated that the play might have reached 30 feet high. In similar fashion, West had some eruptions in both 1986 and 1988. Again, none was seen, and this time the washed areas implied lesser heights. In August 2003, at the same time Sea Weed was active, West underwent intermittent overflow and disrupted its mats of cyanobacteria, but whether any actual eruptions took place is unknown.

230. MUSTARD SPRINGS, EAST and **WEST,** are separated by 50 feet. The two were similar, quiet pools until 1983. In the early part of that year, *before* the Borah Peak earthquake, the water level rose in

East Mustard and fell in West Mustard. Now East Mustard is active as a geyser. Bursts reach 6 to 10 feet high and begin when the pool is several feet below overflow. The duration of 5 minutes is longer than most intervals. West Mustard lies cool and quiet at a low level.

231. NORTH GEYSER was named because it was the northernmost of all the hot springs in the Upper Geyser Basin. Its site is invisible, over the rise about 300 feet beyond the boardwalk. Unless it was active under the name "Mustard Spring" in the 1880s, North's only known eruptions followed the 1959 earthquake. The play was vigorous splashing up to 15 feet high. North has been dormant since 1963, and the vent has disappeared so completely that in 1998 Park Service researchers could not positively locate the site.

The "Outpost Complex"

The "Outpost Complex" (numbers 232 through 236) lies on the flat area north of the boardwalk where there are several springs. Most commonly active only as small spouters, if at all, they have largely been ignored. However, they became significant geysers in 2001 and 2002, were nearly dormant in 2003, and erupted again from 2004 to 2007. The fact that all five geysers are affected by simultaneous changes indicates that they comprise a distinct complex of related springs.

232. UNNG-BBG-4 ("OUTPOST GEYSER") is the largest member of this cluster of springs. It erupts from a fairly obvious crater about 75 feet from the boardwalk. Its intervals range from 6 to 12 minutes, and most of the play lasts 2 to 3½ minutes. "Outpost" usually splashes 2 to 4 feet high, but it can have bursts as high as 8 feet during those eruptions *not* accompanied by "Sentry Geyser" (233).

233. UNNG-BBG-5 ("SENTRY GEYSER") is immediately to the front-right of "Outpost Geyser" (232), where its vent is within the geyserite shoulder of "Outpost's" crater. "Sentry" was named because most of its eruptions occur as precursors to those by "Outpost," typically preceding the other geyser by only a few seconds. However, some eruptions of "Outpost" are not accompanied by "Sentry," which in turn has occasional independent eruptions. Intervals therefore vary considerably, ranging from 8 to 28 minutes. The durations are equally variable, ranging from less than 1 minute to over 4½ minutes. Smaller than "Outpost," "Sentry's" play is 1 to 2 feet high.

234. UNNG-BBG-6 ("GREEN BUBBLER GEYSER") is northwest of "Outpost Geyser" (232). The crater is hidden by a low geyserite mound, so only the top of the 1- to 2-foot eruptions can be seen. "Green Bubbler's" action tends to be quite regular. The intervals and durations both average near 6 minutes when the geyser is active, but dormant periods are common.

235. UNNG-BBG-7 ("YELLOW BUBBLER GEYSER") lies almost directly beyond "Outpost Geyser" (232) near an area densely covered with stonecrop plants, a succulent that bears brilliant yellow flowers in midsummer. "Yellow Bubbler" is the least active member of the "Outpost Complex," with recorded intervals as long as 2 hours. Durations are also long, ranging between 20 and 26 minutes. The height of the eruption is only about 1 foot, and, since "Yellow Bubbler" is the most distant of these geysers from the boardwalk, it is difficult to see.

236. UNNG-BBG-2 ("RED MIST GEYSER") is visible to the right of "Outpost Geyser" (232), perhaps 100 feet from the boardwalk. The most active member of the "Outpost Complex," its brief but frequent play is often little more than fine spray rising above a mound of geyserite that is often coated with red-orange cyanobacteria. At times, however, Red Mist jets considerable streams of water as high as 10 feet. The intervals and durations are both measured in seconds.

237. BLACK PEARL GEYSER and **CORAL GEYSER** lie on opposite sides of the boardwalk. Which is which is uncertain, but Black Pearl is probably the one west of the walk, Coral to the east. Both geysers have been dormant since 1967.

Black Pearl's best activity was in 1946, when eruptions up to 30 feet high were seen. Although the geyser is dormant in most years, the 1959 earthquake stimulated an active episode that lasted until 1967. That play was 4 feet high at intervals and durations both seconds to minutes long.

Coral probably had episodes of small eruptions during the early years of the Park, but the only activity known in detail followed the 1959 earthquake. At intervals of 8 to 15 minutes, the play reached up to 10 feet high and lasted 5 to 8 minutes. In 2004, Coral began nearly constant splashing at depth, the sound of which implied subterranean eruptions of considerable force.

238. FUMAROLE GEYSER was one of the great many features throughout Yellowstone that had a brief episode of activity following the 1959 earthquake, first as a steam vent and then as a 10-foot geyser. No further action was seen until the mid-1980s. Most of Fumarole's activity is subterranean, but spray occasionally rises several feet above the ground.

BLACK SAND BASIN

Black Sand Basin (Map 4.10, Table 4.9, numbers 239 through 254) is named for the obsidian gravel found in many parts of this group. Originally, the name "Black Sand Basin" referred only to today's

Table 4.9. Geysers of Black Sand Basin, Upper Geyser Basin

Name	Map No.	Interval	Duration	Height (ft)
Cliff Geyser	245	minutes	sec–min	20–40
Cucumber Spring	241	steady	steady	2
Green Spring	246	[2005]	3–5 min	3–12
Grumbler, The	243	with Spouter	2–10 hrs	subter-ranean
Handkerchief Geyser	250	steady	steady	4–8
Handkerchief Pool	248	frequent	hours?	2–3
"Jagged Spring"	244	irregular	seconds	15
Ragged Spring	244	frequent	seconds	1–3
Rainbow Pool	252	[1996]	sec–min	15–100
Spouter Geyser	243	2–5 hrs	2–12 hrs	6–8
Sunset Lake	254	seconds*	seconds	1–35
UNNG-BSB-1	242	infrequent	seconds	3–10
UNNG-BSB-2	247	5 min*	5 min	1–5
UNNG-BSB-3 ("Cinnamon Spouter")	251	steady	steady	2–6
UNNG-BSB-4	253	extinct?	–	–
UNNG-BSB-5	249	frequent	minutes	3–6
UNNG-BSB-6	240	frequent	sec–min	1–20
Whistle Geyser	239	[1991]	1–3 hrs	30–70

* When active.
[ ] Brackets enclose the year of most recent activity for rare or dormant geysers. See text.

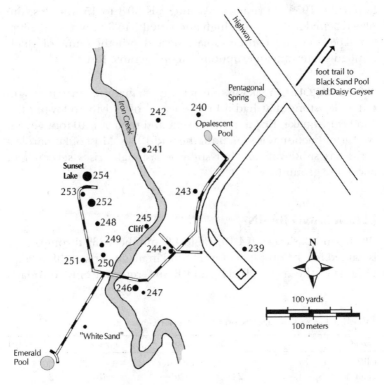

Map 4.10. *Black Sand Basin, Upper Geyser Basin*

Black Sand Pool (146), whereas this area was called "Sunlight Basin." Black Sand Basin is a relatively small cluster of springs, well to the west of the rest of the Upper Geyser Basin. It is easy to reach and explore. A parking lot gives access to a boardwalk that approaches all the important springs.

The best-known feature in the group is Emerald Pool. Not a geyser, Emerald is a deep green color as a result of the combination of the blue of its water with the yellow-orange of cyanobacteria lining the crater walls. Along the boardwalk spur leading toward Emerald Pool is a small spring, sometimes called "White Sand Spring," that occasionally acts as a small geyser up to 2 feet high but most often is an empty hole. Near the road entrance to Black Sand Basin is five-sided Pentagonal Spring, which apparently has always been a perfectly quiet pool. Nearer the parking lot, Opalescent Pool probably had an explosive origin but now contains no hot spring and is only a catch basin for runoff from Spouter Geyser. The milky, opal-like color is caused by suspended particles of colloidal silica that form in the cooling water.

The stream that flows through Black Sand Basin is commonly called Iron Creek, but its full official name is Iron Spring Creek. The name was given when early explorers thought the orange cyanobacteria mats in the runoff channels were deposits of iron oxide.

239. WHISTLE GEYSER is very rarely active. The grand total of known eruptions since the first was inferred in 1878 is only thirty-six. The best year on record was 1957, when seven eruptions were recorded. There was one in 1968, one in 1990, and two in 1991 (May and July). There is little doubt that Whistle is an old geyser that has so nearly sealed itself in with internal deposits that it has lost most of its water supply.

Whistle is capable of two kinds of eruptions. Most of those observed have been steam-phase eruptions, where initial 30-foot jets of water last less than 30 seconds before a powerful steam phase that persists for 2 to 3 hours sets in. The eruptions of 1990 and May 1991 were this type. Water-phase eruptions have perhaps happened only twice, in 1931 and July 1991. At first splashing water only a few feet high, Whistle does not begin the steam phase until at least 4 minutes into the eruption. Then, as the water gives way to steam, the jetting can reach at least 70 feet high. This steam phase is actually more powerful than that of the steam-phase type of eruption, but it is briefer, decreasing rapidly within an hour of the start.

240. UNNG-BSB-6 evidently acted as a geyser during the 1960s, when it was mapped as such by the U.S. Geological Survey, but no additional information is available. On May 3, 2007, it was observed jetting at an angle to as high as 20 feet, and similar but smaller jets were seen over the next few days. Thereafter, the action died down to intermittent splashes just 1 to 6 feet high. Angular chunks of broken geyserite litter the area around the vent, proof of an explosive origin.

241. CUCUMBER SPRING was once a quietly flowing spring next to Iron Creek. The 1959 earthquake created a small steam vent on

Spouter Geyser is one of those features whose eruption durations are considerably longer than the intervals, so much so that it was long thought to be a perpetual spouter (hence the name) rather than a geyser. Credit, NPS photo by Jeremy Schmidt.

one shoulder of the crater, and it soon became a perpetual spouter a few feet high. Another explosion in 1969 enlarged the crater and merged it with Cucumber's. The combination now discharges a heavy stream of water. The spouter continues to play up to 2 feet high. It is best viewed from the high, bridge-like boardwalk next to Sunset Lake (254).

242. UNNG-BSB-1 is within a cluster of springs known as the Brown Spouters, beyond Cucumber Spring (241). At least one of these springs is a geyser. Activity is infrequent and brief, with splashes 3 to 10 feet high. Its location far down the stream makes BSB-1 very difficult to see.

243. SPOUTER GEYSER was first thought to be a perpetual spouter—hence the name—but it was recognized as periodic by 1887. Eruptions last anywhere from 40 minutes to 3½ hours, and these durations strongly control the length of the following intervals, which range between 2½ and 8½ hours. Overall, Spouter is in eruption about one-third of the time. The splashing play is 6 to 8 feet high.

Two additional vents erupt in concert with Spouter. One is a small vent on Spouter's shoulder that splashes 1 to 2 feet high. The other, in a deep hole immediately next to the concrete sidewalk, is The Grumbler, whose noisy play can reach up to 4 feet above its subterranean pool.

244. "JAGGED SPRING" and RAGGED SPRING. At the start of the boardwalk is a large, irregular crater that probably did not exist before the mid-1930s, when the surface geyserite collapsed inward. This is "Jagged Spring," whose first recorded eruptions took place during 1941. A few feet beyond "Jagged" is Ragged Spring, which erupts from a much smaller and more symmetrical vent. Both geysers play in unison. Small splashing is nearly constant in both, but every few minutes Ragged increases its action to splash as high as 3 feet. It is only then that brief bursts sometimes send violent jets of water as high as 15 feet in "Jagged." The duration is never longer than a few seconds.

245. CLIFF GEYSER plays from a wide crater just across Iron Creek from a boardwalk platform. The geyserite on the stream side of the crater forms the cliff of its name. Cliff's performance has gone through wide variations through the years. Dormant at times, during the 1960s it commonly had long eruptions separated by intervals of 12 hours. Cliff became progressively more frequent during the 1970s and 1980s, and by 1985 the intervals were as short as 30 minutes. These short intervals continue, interrupted by frequent minor eruptions that last only a few seconds. All eruptions begin from an empty crater. As the bursting continues, the crater slowly fills, and the deepening pool tends to focus the water into jets that reach as high as 40 feet. Often an eruption ends at about the time the crater reaches overflow, but exceptionally long durations can continue for several minutes more. Given that the boardwalk spur ends only a few feet across the creek, Cliff is a most impressive geyser.

246. GREEN SPRING, called "Lesser Emerald Pool" during the 1870s, was noted as "a bulger" during the 1880s, a term that implies that eruptions of some sort were seen at that time. However, the first clear descriptions of intermittent activity did not come until 1934, when Green played a few times up to 20 feet high. Since then it has had activity during only a few years, the most recent being 2004 and 2005. A typical active phase consists of a series of eruptions, each sending bursts 8 to 12 feet high every 30 minutes for durations of 3 to 5 minutes. The geyser usually continues this action for a day or less. The behavior of 2004–2005 was different, in that solo eruptions recurred at intervals of hours to several days; similar action occurred during February–March 2008. Green is also known to sometimes flood its surroundings with heavy overflow but no actual eruption; one of these episodes in 2006 was so voluminous that the high temperature killed fish in Iron Creek.

247. UNNG-BSB-2 is a small vent just to the east (left) of Green Spring (246). During 1980, it was active as a small geyser as much as 5 feet high. Inactive by 1981, it may have had a few eruptions during the winter of 1992–1993. It reactivated as a small geyser during 2007, erupting 1 to 3 feet high every few minutes.

248. HANDKERCHIEF POOL is the famous spring that once attracted nearly as many visitors as any other single feature in Yellowstone. It was possible to place a handkerchief at one end of the spring and have convection pull it down a vent and out of sight. A few moments later it would reappear in another vent. Unfortunately, in 1926 somebody jammed logs into one of the openings. Later, eruptions by Rainbow Pool (252) washed gravel into the crater, completely obliterating the site. In 1950 Handkerchief began to reappear as a bubbling spot in the gravel. It was shoveled out, and the logs were removed. Handkerchief Pool now performs as either a perpetual spouter or a geyser. The geyser activity, most recently seen in 2000, commonly has durations far longer than the intervals. The play reaches no more than 2 or 3 feet high.

249. UNNG-BSB-5 is a geyser that plays from a small vent located between Handkerchief Pool (248) and Handkerchief Geyser (250). It was first reported in May 1996. Eruptions are frequent and up to 6 feet high.

250. HANDKERCHIEF GEYSER was named in allusion to Handkerchief Pool, even though it bears no resemblance to the pool and is 150 feet away. It was first described in the 1930s, when the play was 3 feet high. It was not reported again until 1986–1987, when there were eruptions as high as 15 feet. It then declined into another dormant period. Since rejuvenating in 1996, Handkerchief Geyser has generally acted as a perpetual spouter between 4 and 8 feet high. There are several other small geysers and spouters in this general area.

251. UNNG-BSB-3 ("CINNAMON SPOUTER") was first noted as a "sizzling fracture" in 1933. Nothing more was seen of it until it broke out as a geyser directly underneath the boardwalk in April 1988 (the walk has since been moved). The geyser has gradually developed an ever-larger crater stained with a cinnamon-red deposit, and some splashes reach more than 3 feet high. The eruption is nearly constant, with rare intervals only seconds to minutes long.

252. RAINBOW POOL has a crater nearly 100 feet across. It is ordinarily a quiet spring with light overflow. A few minor eruptions during the early 1930s were a portent of the future. In 1938 it proved to be a powerful geyser, erupting throughout that summer up to 80 feet high several times per day. By 1939 it was much weaker. Rainbow was then an erratic performer through 1948, when eruptions were seldom more than 15 feet high. There was a single eruption in 1973. Finally, Rainbow had several series of eruptions in late May 1996. The most powerful yet known, some bursts were clearly more than 100 feet high. Rainbow Pool has not erupted since 1996.

253. UNNG-BSB-4 played from a small vent between the boardwalk and Rainbow Pool (252). It has disappeared, but in 2004 a small perpetual spouter appeared beneath the boardwalk a few feet away. It remains active as a curious source of noise.

254. SUNSET LAKE, even larger than Rainbow Pool (252), has undergone small eruptions for many years. Because the high temperature of the spring causes dense steam clouds to form, it is difficult to observe the play except on hot days. The largest known eruptions occurred during 1981 and 1984, when some bursts were 35 feet high. More typical are eruptions with bursts recurring every few seconds and reaching 10 to 20 feet high. Sunset fell nearly dormant during 2003. It continues to overflow heavily, but eruptions are small and infrequent.

MYRIAD GROUP

The Myriad Group (Map 4.11, Table 4.10, numbers 255 through 268) is the thermal area behind and to the west of Old Faithful Inn. Only the Three Sisters Springs lie close to the road, and no trails penetrate the area. True to its name, the area contains more than a thousand hot spring vents (a count done during post-earthquake mapping in 1959 came up with 1,113 features). The area is extremely dangerous to explore, and unauthorized entry is illegal.

The Myriad Group is the site of several important geysers, and some have been major in scale. In addition to the geysers described here, the group contains dozens of other small geysers and perpetual spouters. Most are too small to see from a distance, but often a geyser of size will make a temporary appearance. For example, in May 2004 a spring known only as "Myriad #35" underwent daily eruptions up to 12 feet high; its activity had ceased before the end of June. Also located within the group are the largest and best mud pots in the Upper Geyser Basin, one of which is named Pink Cistern.

Every now and then an entire group of hot springs will undergo a sudden and dramatic increase in activity without any observable cause. Known as "energy surges," an especially strong example

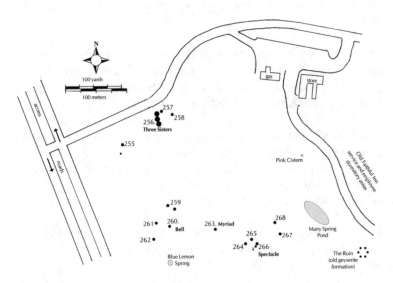

Map 4.11. *Myriad Group, Upper Geyser Basin*

Table 4.10. Geysers of the Myriad Group, Upper Geyser Basin

Name	Map No.	Interval	Duration	Height (ft)
Abuse Spring	265	[1990]	1 min	5–125
Basin Spring	255	minutes*	3–10 min	2–6
Bell Geyser	260	30 min–5 hrs*	2 min	1–4
Blue Lemon Spring	262	[1959]	unknown	unknown
Cousin Geyser	258	[1985]	12 min	25
Lactose Pool	268	infrequent	10 min	20–30
Little Brother Geyser	257	sec–min*	seconds	6–25
"Middle Sister"	256	[2005]	seconds	5–30
Mugwump Geyser	256	5–20 min*	seconds	5–30
Myriad Geyser	263	[1955]	5 min	80–100
Pit Geyser	261	[1959]	unrecorded	unre-corded
Round Geyser	264	[1990]	1 min	10–150
"South Sister"	256	[1959?]	unrecorded	few feet
Spectacle Geyser	266	rare	1–3 min	2–75
Strata Geyser	262	8–21 min	seconds	
subterranean				
Three Crater Geyser	256	minutes*	seconds	3–10
Trail Geyser	259	[1988]	1–2 min	3–50
UNNG-MYR-1	256	rare	seconds	3–6
UNNG-MYR-2	266a	erratic	sec–min	1–6
("Squirtgun")				
West Trail Geyser	259	steady	steady	1–20
White Geyser	267	2–15 min*	20–40 sec	12

* When active.
[ ] Brackets enclose the year of most recent activity for rare or dormant geysers. See text.

struck the Myriad Group during the winter of 1987–1988. Some of the action resulting from those changes continues even now.

255. BASIN SPRING was found to be active as a geyser during early 1984, undoubtedly as a result of the 1983 Borah Peak earthquake. The activity was somewhat erratic and perhaps cyclic. Most intervals were only a few minutes long, but others extended for hours. There were also several brief dormant periods in which the pool would fill and quietly overflow, as it had during all of its previous history. When active, eruptions would last 3 to 10 minutes, during which violent boiling domed the water as high as 6 feet. Basin Spring stopped playing in late 1984 and was not seen again until 2004. From then into 2007 it had occasional eruptions, most of which were no more than 2 feet high. When not erupting, Basin's crater serves as a drain for runoff from the Three Sisters Springs (256).

A few feet south of Basin Spring, a small vent played as a geyser during 1987. During its brief active phase the intervals were 3 minutes, durations 2 minutes, and heights 1 to 3 feet. Few eruptions have been seen since 1987.

256. THREE SISTERS SPRINGS sit next to Three Sisters Road, the access route to the front of Old Faithful Inn. The three craters, all filled with pale-blue water, are connected both above and below-ground and incorporate at least four geysers. Only the vents within the North Sister, closest to the road, had any eruptive history prior to the 1959 earthquake. Most of this activity was referred to simply as "North Sister," but two of the three vents have their own names.

Mugwump Geyser is a large vent at the near-right side of North Sister. It was named because of brief but noisy eruptions in 1884. The Mugwumps were Republican politicians who refused to support the party's presidential nominee, James G. Blaine, and voted instead for Grover Cleveland. Many people felt the Mugwumps complained loudly for little or no reason, accomplishing nothing. The geyser seemed similar, but, in fact, Mugwump's eruptions can reach up to 30 feet high, and the concussions of steam explosions can be heard and felt at the service station more than 500 feet away. Play 5 to 10 feet high is more typical, however. There have been several active episodes in recent years, most recently in 2003–2006 when intervals were about 18 minutes long. Each eruption lasts only a few seconds.

Three Crater Geyser is the back-left vent within North Sister. It has been very active in recent years. Like Mugwump, the usual height is about 10 feet, and durations are just a few seconds long. Also like Mugwump, Three Crater sometimes makes a loud popping sound during its eruptions.

The third vent within North Sister is UNNG-MYR-1. Near the front rim of North Sister, it is rarely active, with bursts 3 to 6 feet high.

"Middle Sister" was active for a short time following the 1959 earthquake. Its play was frequent and up to 15 feet high, but it stopped erupting when Three Crater Geyser activated a short time later. Either "Middle Sister" or "South Sister" had a single 30-foot eruption observed in July 1993. "Middle Sister" was also briefly active in January 2005, when eruptions consisted of single bursts 5 feet high that repeated every 12 to 20 minutes.

"South Sister" has rarely been active, if at all. There might have been a few small eruptions following the 1959 earthquake, and the July 1993 eruption might have been here. It is possible, though, that all reports of "South Sister" actually refer to "Middle Sister."

257. LITTLE BROTHER GEYSER is closely related to the Three Sisters (256). Its first known eruptions were during 1926, but it gained little attention until 1950 when it briefly splashed 3 feet high every 5 minutes. The next year of activity was 1958, with eruptions to 12 feet. Following the 1959 earthquake, Little Brother reactivated along with the Three Sisters. Those erratic eruptions reached as high as 30 feet. There have been few active phases since then. One took place during May–June 1999, when Little Brother was a frequent performer; most intervals varied between 9 and 17 minutes in length, lasted 50 to 60 seconds, and were 15 to 25 feet high. Very different was activity in early May 2005 and in winter 2006, when intervals as short as 40 seconds led to equally brief durations of only 4 to 6 seconds and heights seldom greater than 6 feet.

258. COUSIN GEYSER is another relative of the Three Sisters Springs (256). It underwent its first known active spell during July–August 1980. Lasting as long as 12 minutes, the eruptions were a steady stream of murky water jetted as high as 25 feet at an angle toward the east. The play recurred every 1½ to 3 hours for a few days before Cousin began a slow decline into dormancy. Cousin reactivated slightly during 1985, when a few weak eruptions were followed by several days of intermittent overflow.

259. TRAIL GEYSER and adjacent **WEST TRAIL GEYSER** are small, shallow pools activated by the 1959 earthquake. For several weeks they underwent powerful concerted eruptions. Playing hourly, Trail reached 50 feet and West Trail about 20 feet high. The play lasted 1 to 2 minutes. Dormant by December 1959, they were quiet, cool pools until the energy surge of 1987–1988. This time West Trail was the stronger of the two, with some bursts of 15 feet, and Trail often failed to join its companion. West Trail continues to play as a small perpetual spouter.

260. BELL GEYSER, like the next two geysers, is virtually impossible to see from the roadways. It is a bell-shaped pool that contains two vents. Most activity rises from a vent within the wide, deep "body" of the bell. Eruptive episodes are uncommon, but the 2-minute play that splashes up to 4 feet high then recurs irregularly every 30 minutes to 5 hours. A smaller vent near the "top" of the bell rarely erupts to about 1 foot.

261. PIT GEYSER is normally a quiet pool. The massive geyserite formation that surrounds the crater implies a significant amount of activity in its past. Historically, though, Pit played only for a short time following the 1959 Hebgen Lake earthquake. No details about those eruptions were recorded, except that they apparently were several feet high.

262. STRATA GEYSER is frequently active, but virtually all eruptions take place deep underground within a jagged vent in layered deposits of old geyserite. Intervals range between 8 and 21 minutes, while the durations are only a few seconds long. The height of the subterranean play is unknown but probably small.

At the edge of the forest south of Strata Geyser is the small pool known as Blue Lemon Spring. It apparently had a few small eruptions following the 1959 earthquake, but no details were recorded.

263. MYRIAD GEYSER was named because in its time it was the largest geyser ever seen in the Myriad Group. The only recorded eruptions were during 1954 and 1955. In those summers it erupted daily, with intervals ranging from 5 to 13 hours. The eruptions were between 80 and 100 feet high. Shot out at an angle, they strongly resembled those of Daisy Geyser (137) during the 5 minutes of play.

264. ROUND GEYSER, not to be confused with Round Spring (127) elsewhere in the Upper Basin, ranks as the largest geyser of the Myriad Group and one of the largest in Yellowstone. It erupts from an impressively deep crater, 4 feet in diameter and almost perfectly round.

Round's eruptions resemble those of Old Faithful. Except for some possible action in 1933, no modern activity took place until just after the 1959 earthquake. Those eruptions were just 10 feet high. Following this brief period and another, even briefer one in

Round Geyser is the largest ever observed in the Myriad Group, with some erup-tions over 150 feet high.

1961, Round was dormant until 1966. During the renewed activ-ity the eruptions were more powerful, sometimes reaching 50 feet high, and they gradually gained strength as they continued. By the mid-1970s some were 150 feet high. The intervals were quite regular, averaging around 14 hours within a 9- to 18-hour range. During the quiet period the pool periodically boiled around the edges. During one of these "hot periods" the water level suddenly rose, and heavy overflow accompanied vigorous boiling by the pool. There were *always* three such surges, separated by about 10 seconds, before the eruption. A fourth surge suddenly rocketed the water, needing only a few seconds to reach maximum height. The eruption itself lasted less than 1 minute but was followed by a series of equally short but impressively noisy steam-phase eruptions.

Round returned to dormancy in 1981. A few additional eruptions took place during the energy surge of 1987–1988, and three or four were seen in late summer 1989. Minor eruptions, only 30 feet high and without the steam phases, were seen during the winter of 1989–1990. Round has not played since then, but it continues to boil and overflow, so more active phases probably lie in its future.

265. ABUSE SPRING was named because of the vast amount of debris thrown into the crater by early Park concession employees. As early as the 1890s it as well as Round Geyser (264), Spectacle Geyser (266), and other springs served as sources of hot water for a vegetable greenhouse. Further damage occurred in the 1920s when a laundry facility was actually constructed over the pool, and it was used by a contractor's kitchen in 1928. The crater still shows the abuse it received over the years. The first known eruption by Abuse was a result of the 1959 earthquake. Although it was not observed, the amount of debris thrown out indicated that it was very powerful.

No further play occurred until 1974, when nearby Spectacle Geyser also activated. At first, there was frequent exchange of function between these two springs. Abuse would have one or two eruptions, then Spectacle would play repeatedly over the course of several days. The eruptions by Abuse were massive domes of water 15 feet high. Both geysers returned to dormancy before the end of 1974. In May 1976 unprecedented eruptions took place as both Abuse and Spectacle reactivated as truly major geysers. Intervals were as short as 90 minutes, and many of the eruptions were in concert. Abuse always reached at least 90 feet high and sometimes may have exceeded 125 feet, while Spectacle jetted simultaneously as high as 75 feet. For a one-week period, Abuse and Spectacle were among the most powerful geysers in the world. But that was nearly the end of the show. During the next few weeks Abuse had only a few more small eruptions, and it was dormant by the end of June 1976. The only action since then consisted of a few eruptions just 5 feet high during the winter of 1989–1990, at the same time nearby Round Geyser (264) also had some minor eruptions.

266. SPECTACLE GEYSER has had an uncertain history. During an episode of construction at Old Faithful Inn during 1928, Abuse Spring (265) was used as a hot water source by a laundry and by a contractor's camp cook. Perhaps because of this use, a nearby hole broke out and began erupting beneath the laundry. In an attempt to stop the eruptions, the crater was filled with sand. That effort failed, the building was removed, and the hole is now called Spectacle Geyser. Its crater has grown through the years, as every episode of eruptions has enlarged the crater to a measurable degree. It is now several feet across.

The first modern action by Spectacle occurred during 1968, when it had some small eruptions that gained little notice. Then in 1974, at the same time nearby Abuse Geyser began erupting, Spectacle joined in. Of the two, Spectacle was usually the larger and more active. An active cycle was initiated by one or two eruptions by Abuse. Then, over the next several days Spectacle erupted about every 20 minutes. Each play lasted 1 to 3 minutes and was 25 feet high. The eruptions vigorously jetted at an angle, beautiful and strongly resembling a small-scale Daisy Geyser (137). After several months of activity, both Spectacle and Abuse were dormant until May 1976. Over a period of about a week, both Spectacle and Abuse became tremendous geysers. Erupting as often as every 1½ hours, Spectacle would play at least 75 feet high, and Abuse joined it in concert at over 90 feet. Following these major eruptions, which ended on May 31, Spectacle continued minor activity through most of that summer even after Abuse had fallen dormant. Since then, Spectacle has been dormant most of the time. Between 1978 and 1981, it sometimes erupted shortly after eruptions by Round Geyser (264), but those plays ended with Round's dormancy in 1981. A few 30-foot eruptions took place during 1983 and again in 1986. Occasional very small eruptions still take place.

266a. UNNG-MYR-2 ("SQUIRTGUN GEYSER") erupts up to 6 feet high from a small hole in the artificially enlarged runoff channel that leads away from Spectacle Geyser (266). Such action is most common when Spectacle is also active. Otherwise, "Squirtgun" tends to play as a 1-foot perpetual spouter.

267. WHITE GEYSER is the most visible geyser in the Myriad Group. It is active almost all the time. Eruptions recur at intervals of 2 to 15 minutes and last 30 seconds. The height is about 12 feet. Dormant periods have been recorded, during which a small spring nearby raises its water level and undergoes eruptions too small to be seen from the roads. This was the case throughout 2007. Such dormancies tend to be brief, and White is the only geyser in the Myriad Group likely to be seen by a casual viewer.

268. LACTOSE POOL is a muddy, milky white spring near White Geyser (267) and is undoubtedly the original "White Geyser" of the 1880s. It spends most of its time bubbling gently from a water level deep within its crater. On infrequent occasions, seemingly always in late summer, it enters active episodes that may persist for several days. Some bursts of the muddy water reach 20 to 30 feet high over durations as long as 10 minutes.

PIPELINE MEADOWS GROUP

No established trail leads into the Pipeline Meadows, the open area across Firehole River from the Old Faithful Lodge cabins. The area is located upstream from the main portion of the Upper Geyser Basin. The springs of the Pipeline Meadows Group (Map 4.12, Table 4.11, numbers 270 through 275), also known simply as the "Meadows Group," are scattered throughout the area. Only a small number of hot springs are located here, but six of them are geysers.

270. DILAPIDATED GEYSER is the first Pipeline Meadows spring encountered if one walks upstream along the river. It plays from a badly weathered cone next to a deep crater with a considerable overhang. The size of the cone and depth of the runoff channels indicate a great deal of activity in its past. However, although it was listed as a geyser in the early days of the Park, no eruptions of Dilapidated were ever recorded until the late winter of 1980. It was then active into 1981 and more briefly in both 1987 and 1988. Cyclic in its activity, Dilapidated would experience a series of eruptions followed by a day or more of quiet. Intervals were about 2 hours long. Throughout the 2- to 5-minute duration the geyser would burst from the cone, some spray reaching 30 feet high. The only active phases since 1988 occurred in late May 2002, when at least five eruptions took place during a 2-day period; in early May 2005, when only one eruption was seen; and occasionally through-

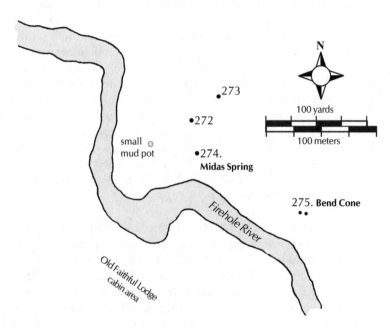

Map 4.12. *Pipeline Meadows Group, Upper Geyser Basin*

Table 4.11. Geysers of the Pipeline Meadows Group and Upper Springs, Upper Geyser Basin

Name	Map No.	Interval	Duration	Height (ft)
Bend Cone, "Northwest"	275	near steady	near steady	1
Dilapidated Geyser	270	[2006]	sec–5 min	10–30
Midas Spring	274	[1996]	30 sec	2
UNNG-PMG-2 ("Stiletto")	272	2½–14 hrs*	10–12 min	2–15
UNNG-PMG-4 ("Secluded")	271	2 min–hrs*	sec–2 min	3–10
UNNG-PMG-5	273	steady	steady	1–6
UNNG-UPG-1	none	frequent*	sec–min	1–3

* When active.
[ ] Brackets enclose the year of most recent activity for rare or dormant geysers. See text.

out 2006 and early 2007. The eruptions that were witnessed were smaller and briefer than those of the 1980s, usually reaching only 10 feet high for less than 1 minute.

271. UNNG-PMG-4 ("SECLUDED GEYSER") was first observed in 1985 when its steam cloud was noticed on a cold morning. There is no known report about this spring even existing prior to then, yet the crater and its surroundings prove it to have been active for a long time. Perhaps it was not noticed before because of its location well up the hillside, where it is behind trees and isolated from the other members of the Pipeline Meadows Group. "Secluded" is known to have both minor and major eruptions. Only minor eruptions occur in most years, and they are always the more common type, lasting a few seconds and bursting about 3 feet high. Major eruptions, when they take place at all, are usually hours apart. They have durations of more than 2 minutes and reach 7 to 10 feet high. Months-long dormant periods are known.

272. UNNG-PMG-2 ("STILETTO GEYSER") erupts from a small vent surrounded by a round geyserite platform. Leading from the platform are several deep runoff channels that owe their existence to eruptions since 1981. There is no certain record of activity for any time before then. The first action was erratic. With intervals of minutes to many hours, "Stiletto" would squirt water about 10 feet high for a duration of just 1 minute. Dormant in 1986, it renewed activity in 1987 on a very different pattern that continued through 1998. The intervals ranged from 9 to 14 hours. Small splashing pre-play, which started with little variation 4 hours before the eruption, served as an indicator. In 1998 the activity dramatically increased and produced intervals as short as 2½ hours. The play was steady jets that reached as high as 15 feet, and the duration of 10 to 12 minutes ended with a weak steam phase. Unfortunately, the development of nearby PMG-5 (273) in early 2004 produced near dormancy in "Stiletto," but small eruptions still infrequently take place.

273. UNNG-PMG-5 was a steam vent prior to February 2004, when eruptions 6 feet high blew out enough rock to reveal a jagged crater. Now PMG-5 is a perpetual spouter. The height is only a foot or so, and most of the action is confined to the crater, but this activity caused nearby "Stiletto Geyser" (272) to fall nearly dormant.

PMG-5's vigor decreased slightly during 2005, and "Stiletto" then resumed weak eruptions.

274. MIDAS SPRING was named in the 1920s when the crater was lined with golden-yellow cyanobacteria. It had a long history as the most active geyser in Pipeline Meadows, playing every 3 to 8 minutes until it entered a dormant period in 1996. The 30-second eruptions were not impressive, never more than 2 feet high, but think about this—a geyser that erupts every 5 minutes, as was typical of Midas Spring, will have more than 100,000 eruptions in one year. In other words, even the smallest geysers are very significant features. Since its current dormancy began, Midas has cooled, and it again contains a mat of cyanobacteria.

275. BEND CONE is actually a pair of large geyserite cones merged together into a single feature at the upstream end of the Pipeline Meadows. At the top of each cone is a small spring. Active as perpetual spouters, the one to the northwest occasionally plays as a true geyser, with intervals seconds to a few minutes long. Both springs splash just 1 foot high.

OTHER UPPER GEYSER BASIN GEYSERS

Geysers are known to exist in at least four other hot spring groups within the Upper Geyser Basin.

Near the Pipeline Meadows, about ¼ mile along the trail to Mallard Lake, is the Pipeline Creek Group. Most of its springs are mud pots and acid pools, but one large pool within a sinter-lined basin often acts as an intermittent spring and rarely has splashing eruptions as much as 4 feet high.

Along Firehole River upstream from the Pipeline Meadows area and near the highway are the Upper Springs (also called the "Upstream Group"). Among the several springs here is one deep blue pool (UNNG-UPG-1) that generally acts as a perpetual spouter but has occasional episodes of truly periodic activity. Some of its bursts reach 3 feet high.

In the woods across the highway east of Black Sand Basin are the Pine Springs, which consist of two clusters of hot springs. At the high point of this area, visible from the highway, is a jumble of ancient geyserite that appears to be the remnant of a large cone.

Known as "The Old Cone," it is probably what is left of an extensive mound or platform rather than a cone. South of The Old Cone and visible from the highway near the Old Faithful highway interchange bridge is the "Mud Spring Group." Mud Spring itself is a small pool that has had a few eruptions since 1986, most of which are known only on the basis of washed areas and fresh runoff channels. Not far from there is UNNG-PIN-1, the most active geyser at Pine Springs. It lies so deep within a crater that its activity is entirely subterranean, even though some bursts can reach 10 feet above the pool level. The "Fracture Group" is the northern cluster of the Pine Springs, located east of The Old Cone. Numerous craters open along a series of fractures that cut a sinter platform so old it is mostly covered with forest. A number of these vents were active as geysers following the 1959 earthquake, but only one presently undergoes small eruptions. UNNG-PIN-2 plays every minute or so, splashing 2 feet high for a few seconds.

Finally, the Hillside Springs lie northwest of Black Sand Basin, where they are visible as a series of terraces colored orange by cyanobacteria on the steep slope below the cliffs. On the valley floor below the Hillside Springs is Sinter Flat, a rarely visited area where there are some old spring craters. One of these contained a small geyser in the 1980s.

CHAPTER FIVE

Midway Geyser Basin

The Midway Geyser Basin (Map 5.1, Table 5.1) is a relatively small area. The hot springs are mostly confined to a narrow band of ground that parallels a 1-mile stretch of Firehole River. Additional springs extend up the Rabbit Creek drainage to the east, at the head of which are assortments of mud pots, small geysers, and one exceptionally large pool.

Topographically, the Midway Geyser Basin is part of the Lower Geyser Basin (Chapter 6), but it has always held separate status because it is separated from the Lower Basin by a forest of lodgepole pines. First known as the Halfway Group and Hell's Half-Acre, then as Egeria Springs, Midway contains only a few geysers of note. Possibly dozens of other hot springs are geysers, too, but they are small and little is known about them.

Despite its small size, Midway is the location of some of the largest single hot springs in the world. Grand Prismatic Spring is more than 270 feet across and 120 feet deep; possibly the only larger bona

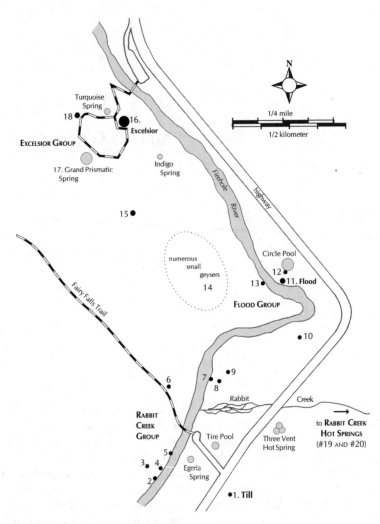

Map 5.1. *Midway Geyser Basin*

fide *single* hot spring anywhere is the lake in the Inferno Crater of New Zealand's Waimangu Valley, but that is a distinctly volcanic feature. Excelsior Geyser discharges a steady stream of more than 4,000 gallons of water every minute. In fact, a large proportion of the springs at Midway are of extraordinary size.

The Midway Geyser Basin includes four named hot spring groups. The Rabbit Creek Group includes the springs in the lower

Table 5.1. Geysers of Midway Geyser Basin

Name	Map No.	Interval	Duration	Height (ft)
Catfish Geyser	10	15 min	5 min	2–3
Excelsior Geyser, historic	16	[1890]	–	300
Excelsior Geyser, major	16	[1985]	2 min	30–80
Excelsior Geyser, boiling	16	[2000]	seconds	5–10
Flood Geyser	11	1½–45 min	20 sec–8 min	10–25
Grand Prismatic Spring	17	[1880s]	unknown	few feet
Opal Pool	18	[2003]	seconds	5–80
Pebble Spring	8	1–2/year	1 hr	10
Rabbit Creek Geyser	19	15–25 min	1 min	sub–10
River Spouter	7	see text	steady*	10
Silent Pool	6	[1992]	minutes	3
Tangent Geyser	12	infrequent	1–5 min	1–10
Till Geyser, major	1	5½–11 hrs	30 min	10–20
Till Geyser, minor	1	see text	1–2½ min	10–20
Tromp Spring	15	[1960s]	unrecorded	1–3
UNNG-MGB-1	2	rare	min–hrs	6–15
UNNG-MGB-2	3	steady*	steady	1
UNNG-MGB-3	4	1–3 min*	1–2 min	1–10
UNNG-MGB-4	9	1–5 min	seconds	3–6
UNNG-MGB-5	5	rare	30 min	2–3
UNNG-MGB-7	14	see text	–	–
UNNG-MGB-8	20	minutes	seconds	1–10
West Flood Geyser	13	[1993]	1½–6 min	10–40

* When active.
[ ] Brackets enclose the year of most recent activity for rare or dormant geysers. See text.

Rabbit Creek drainage and along Firehole River in that vicinity, at the southern end of Midway. Along the highway and river in the middle portion of the Midway Basin is the Flood Group, dominated by Flood Geyser. The most accessible group, served by the parking lot and boardwalk at the "Midway Geyser Basin" signs on the highway, is also the best-known area, properly called the Excelsior Group. One of the best views of this area is from Midway Bluff, the hill across the highway from the parking lot, where one is high enough to get a good view of the pools and their coloration. The hike is well worth the few minutes it takes. Finally, the Rabbit Creek Hot Springs are those scattered along the valley that extends eastward toward the source of Rabbit Creek. These springs are spectacular, but much of the area is dangerous. Explorers should always assure that others know about any planned hike into Rabbit Creek *and* check back in after the trip.

RABBIT CREEK GROUP

The Rabbit Creek Group encompasses nine geysers, several large pools, and some mud pots. Although named for Rabbit Creek, which meets Firehole River among these springs, the majority of the features hug the banks of the river rather than being scattered along the creek. None of the geysers is particularly large.

1. TILL GEYSER is named for the glacial gravel that composes the ridge it rises from. Active during the 1880s when it was named Rabbit Geyser (also an officially approved name), it seems to have been something of a forgotten geyser until the time of the 1959 earthquake. Because of its location against the hillside, Till is easy to see by people traveling southbound on the highway but nearly invisible to those driving in the opposite direction.

Till erupts from a complex of vents. The two main craters are situated at the top and bottom of the cluster; between them are several smaller apertures that also jet water. The system fills slowly after an eruption and does not usually reach the first overflow until 30 to 50 minutes before the next play. Moments before the eruption begins, the overflow becomes periodic and then gushes as the jetting starts. The greatest height comes from the upper main vent, where some jets reach 20 feet high. At the same time, the lower main vent sprays horizontally outward as far as 10 feet or more. The smaller vents sputter a few inches to a few feet high. The entire operation lasts 30 minutes, with a few brief pauses near the end of the activity. Till is a highly regular geyser, and its intervals vary little from 10¾ hours.

Till's major eruption is always followed by a long series of minor eruptions. These recur every 7 to 20 minutes and continue for as long as 3½ hours after the major. Each minor has a duration of 2 to 5 minutes and jets water nearly as strongly as the major does.

"Steel Bridge Springs" (numbers 2 through 5)

The "Steel Bridge Springs" comprise a small cluster of unnamed features south (upstream) of the old steel road bridge that now serves as the trailhead to Fairy Falls and Imperial Geyser (see Chapter 6). All of these springs are on the west side of Firehole River, and all except MGB-5 erupt from small cones perched at the top of the steep riverbank. Eruptive activity is generally uncommon in any of them, and the cluster as a whole has been largely inactive since the late 1980s—indeed, weathering has now produced enough soil to allow vegetation to cover much of the once-bare geyserite.

2. UNNG-MGB-1 is the farthest upstream among the "Steel Bridge Springs." Little is known about this geyser. During most seasons it is dormant, or nearly so, but in others it is reported several times per month. Most eruptions splash 6 to 8 feet high, either with extremely short intervals or single durations as long as several hours. Jetting eruptions that last a few minutes and reach at least 15 feet high have been reported just two or three times.

3. UNNG-MGB-2 is located on the hillside above MGB-1 (2) and MGB-3 (4). When it is active, MGB-2 often acts as a perpetual spouter just 1 foot high, but a brief eruption 2 to 3 feet high was seen in July 2006. Because the low cone is covered with brown cyanobacteria and has grass growing on its sides, it is not an obvious feature.

4. UNNG-MGB-3 erupts from the northernmost of the cones perched above the river. The eruptions are usually less than 1 foot high. During active periods, the play recurs every 1 to 3 minutes and lasts 1 to 2 minutes. Brief eruptions 10 feet high were seen in 1988 and 2003.

5. UNNG-MGB-5, a small pool slightly elevated above the river, is the closest of the "Steel Bridge Springs" to the bridge itself. Several times in the mid-1980s it was seen splashing 2 to 3 feet high. Nothing is known of its intervals; durations were at least 30 minutes long.

6. SILENT POOL was always known as a quiet spring until 1989. For the next three years it erupted in cycles with considerable regularity. Several days to weeks would pass between active episodes. Then eruptions would recur every 2 to 3 hours for a day or more before Silent lapsed into another inactive phase. Durations were several minutes, with splashes up to 3 feet high. Silent Pool has been dormant since 1992.

7. RIVER SPOUTER was apparently created by the 1959 earthquake when a crack formed in the sinter at the base of an old cone. (This cone is probably the original Catfish Geyser of 1871. That name was later transferred to a geyser [#10] in the Flood Group, and it is also used for a geyser in the Upper Basin. The reason for the name is unknown in each case.) Because the vent is often under the water of Firehole River, River Spouter is evident only when the river is low. Then the steamy jets—steady as a perpetual spouter at times, frequent and intermittent as a geyser during a few years, and dormant in other seasons—will sometimes reach 10 feet high. During these eruptions the old cone raises a commotion within its vent but seldom splashes any water onto its sides. River Spouter has been mostly dormant since the mid-1980s.

8. PEBBLE SPRING is a small geyser located on the flat above River Spouter (7). Its vent is centered in a round sinter bowl tinted rich orange-brown by iron oxide minerals. Pebble undergoes perpetual splashing no more than 2 feet high, but on infrequent occasions much larger bursting eruptions occur. Generally seen only once or twice per year, these spray 8 to 10 feet high and persist for about 1 hour.

9. UNNG-MGB-4 is a complex of related perpetual spouters. Some level of eruptive activity is always going on here, but the number and strength of erupting vents are always changing. Most often several vents are active at once, some of the play reaching 3 feet high, with occasional surges to 6 feet. The largest pool among these sometimes acts as a geyser, with intervals of 1 to 5 minutes and durations of a few seconds.

FLOOD GROUP

The Flood Group contains just three geysers of size, and only Flood Geyser itself is frequently active. There are also several pools and smaller springs. Best included as part of this group are the clusters of numerous small springs on the open flat to the west and northwest of the river, collectively described here as #14.

10. CATFISH GEYSER is different not only from the Catfish in the Upper Geyser Basin but probably also from the spring originally given the name in the Midway Geyser Basin (see #7). This Catfish is a large pool that has shown little variation in its activity over the

Flood Geyser is the largest reliably active geyser in the Midway Geyser Basin, with doming bursts that sometimes reach as high as 25 feet.

years. Boiling is nearly constant, but about every 15 minutes the water rises. Heavy discharge is then accompanied by more vigorous boiling and minor bursting that throws the water 2 to 3 feet into the air. The duration is near 5 minutes.

11. FLOOD GEYSER is the largest active geyser in the Midway Geyser Basin, being frequent and discharging considerable water. Even so, it was largely ignored for much of its history—for years, the highway to Old Faithful crossed the river downstream from Flood, leaving it isolated. Now the road crosses the hillside above Flood, where a large turnout provides a wonderful view. Still, no systematic observations of Flood's performances were conducted until the early 1970s, and the full extent of its complex behavior was not revealed until 1985.

Historically, Flood's activity has consisted of minor, intermediate, and major eruptions, classified according to the duration. Minor eruptions have durations of just 20 to 60 seconds and are followed by intervals of 3 to 5½ minutes. The durations of the intermediates are 2 to 5 minutes and commonly result in intervals of 12 to 25 minutes. Major eruptions last 6 to 8 minutes and produce intervals as long as 26 to 50 minutes. The aspect that has varied most has been the relative proportions between the different kinds of eruptions—some years had few or no minors, while on other occasions it was the intermediates that were uncommon; it appears that majors always take place. A big change took place in 2007, when Flood was dominated by major eruptions at intervals as long as 3 hours; only a few minor eruptions took place during these intervals.

Regardless of the variety, all eruptions of Flood look about the same. Water is bulged upward by expanding steam bubbles rising into the crater, which burst large globular splashes. The height ranges between 10 and 25 feet. There is a tendency for the bigger splashes to occur during major eruptions, but this is not an absolute rule. All eruptions have a heavy discharge of water, the total during major play amounting to several thousand gallons. This may slosh well beyond the crater. Signs admonish visitors to stay away from Flood. Some people who failed to do so were seriously burned during 1994, as was a careless fisherman in 2002.

12. TANGENT GEYSER plays from a jumble-filled fracture that runs along the southwest edge of Circle Pool, the large round spring just north of Flood Geyser (11). Its best performances were during the early 1980s, when water could be jetted in a fan-shaped spray as high as 10 feet. Durations were 1 to 5 minutes. Tangent's intervals were always erratic, ranging from a few minutes to several hours. Only small, infrequent eruptions have been seen since 1985.

13. WEST FLOOD GEYSER was named because of both its location near Flood Geyser (11) and its resemblance to Flood when in eruption. Although the two geysers are directly across Firehole River from one another and thus not far apart, there seemed to be no connection between them until 2007, when Flood experienced much longer intervals at the same time West Flood began episodes of frequent, heavy overflows. West Flood was originally described as a quiet pool. No eruption was recorded until 1940, when there were bursts up to 40 feet high. During its more recent active phases, eruptions recurred every 45 minutes to 4 hours. The bursts were 10 to 12 feet high and lasted 1½ to 6 minutes. West Flood was dormant from 1993 until a brief episode of minor activity during April 2007, when frequent periods of heavy overflow produced occasional bursts up to 8 feet high. Boiling overflow periods without true eruptions continued throughout 2007.

14. UNNG-MGB-7 is this book's designation for the numerous hot springs in the grassy areas west of Firehole River. At least fourteen of these vents have been known as geysers, but their small sizes and inaccessible locations make any detailed study difficult. Only one has been given an informal name. "Tentacle Geyser" has a series of fractures that radiate away from the vent, so it somewhat resembles a jellyfish, a fact not visible from the road. Its play can reach 10 feet high, as can that of two or three other geysers among this cluster of springs.

15. TROMP SPRING lies far out on the open flat about halfway between the Flood Group and the Excelsior Group. It evidently has had just two known episodes of geyser activity: in 1887, when it was described in general terms as a geyser, and in the early 1960s (probably induced by the 1959 earthquake), when eruptions 1 to 3 feet high recurred every 1 to 2 hours.

EXCELSIOR GROUP

The Excelsior Group is served by a parking lot and a boardwalk loop where there are "Midway Geyser Basin" signs along the highway. There are only three geysers, and two of them have not erupted for more than 100 years. The only other features here are Turquoise Spring and Indigo Spring.

16. EXCELSIOR GEYSER is one of the brightest stars in the world of geysers—when it is active. The last of its truly stupendous eruptions was in 1890 (there may have been some activity in 1901, too). During the ten years before then, Excelsior underwent several active episodes of vigorous activity. Although most eruptions reached "only" 100 feet, some were 300 feet high and as wide as they were high. Considering the size of the geyser, the amount of activity was amazing. For example, during the eleven days between September 27 and October 7, 1881, Excelsior played 63 times, giving an average interval of only a little over 4 hours. The duration was about 2 minutes. At other times, some intervals were as short as 1 hour with durations as long as 3½ minutes. The water level within the crater between eruptions stood about 2 feet below overflow, an important point when the geyser of the 1800s is compared with today's steadily overflowing spring.

Excelsior Geyser, the only geyser of truly major size in the Midway Geyser Basin, last had a full, major eruption in 1890 (possibly 1901), when it played as high as 300 feet. Lesser activity, some of it up to 75 feet high, was seen in 1946 and 1985. Credit, NPS photo by F. J. Haynes, 1888.

The present Excelsior is quite different, although still impressive. The crater measures more than 200 by 300 feet. This entire volume was blasted out by eruptions, and the crater was relatively small prior to the eruptions of the 1880s. In fact, although it is impossible to know for sure, it might not have existed at all in 1839—trapper Osborne Russell made no mention of Excelsior yet clearly described Grand Prismatic Spring (17). It is certain that the crater was greatly enlarged by the eruptions of the 1880s that threw out vast quantities of rock. The huge, azure pool boils at numerous points, proving an abundant source of heat. The discharge is tremendous, too, amounting to a measured 4,050 gallons per minute, or more than 5.8 million gallons per day. Old Faithful Geyser needs nearly two months to discharge as much water as Excelsior does in a single day.

This flow is constant, and that tells us a lot. Excelsior was a geyser, a periodic hot spring. So the flow of the water might be expected to be intermittent or at least variable, even when no eruptions are occurring. The fact that Excelsior boils from many places other than the main vent indicates that the eruptions of the 1880s tore some of the crater and the plumbing system apart. In effect, Excelsior has been leaking, unable to generate the pressure needed for eruptions. Most observers felt Excelsior could not possibly erupt until it healed its wounds.

It was therefore a surprise when Excelsior had true bursting eruptions during September 14–16, 1985. Eruptions were frequent during that 46-hour period. Most were minor in size, with the biggest bursts reaching perhaps 30 feet high. There were also a few "major" eruptions. Although not nearly of the scale of earlier times, the largest major sent jets of muddy gray water up to 80 feet high and 100 feet wide, and spray reached across the river into the parking lot. All of the eruptions (major and minor) lasted nearly 2 minutes, with intervals that ranged between 5 and 66 minutes. During the active episode, the water discharge was several times greater than at any known time since 1890.

Unfortunately, the 1985 eruptions do not seem to have been a prelude to renewed activity on a major scale. Excelsior did have several brief episodes of intermittent boiling violent enough to be construed as eruptions during June and July 2000, but the geyser now looks the same as it did before the 1985 activity. The evidence is that Excelsior's entire history of perhaps less than 170 years has been one of short series of sudden, explosive events. It may be its nature to experience brief periods of powerful eruptions followed by decades of relative quiet.

17. GRAND PRISMATIC SPRING, the Park's largest hot spring, was the first Yellowstone spring to have been individually described in an identifiable way, having been seen by Osborne Russell in 1839. It merits inclusion here because Russell called it a "boiling lake," the geological report of 1878 said it "rises and falls in a series of wavelike pulsations," and in 1883 it "sometimes splashes." However, it has apparently been a perfectly quiet pool since 1883.

18. OPAL POOL is a significant geyser, but its eruptions are rare and exceedingly brief. No activity was known until 1947, when Opal played several times to as high as 50 feet. Similar action occurred in 1949 and 1954, but then nothing further was seen until 1979. The only seasons without known eruptions since then have been 1982, 1983, 1990, and 2004 through 2007. Most modern eruptions are less than 30 feet high, although some estimated at 70 to 80

Opal Pool is an extremely rare performer, but its eruptions can reach up to 80 feet high, making it the largest geyser other than Excelsior in the Midway Geyser Basin. Credit, Photo by Genean Dunn.

feet have occasionally been reported. No matter what the size, the play is often only a single, virtually instantaneous burst of water. Sometimes successive bursts a few seconds apart stretch the total duration to 1 minute. During the minutes to hours before an eruption, strong convection is visible over the vent, but this can also happen without resulting in an eruption. Sometime in early 2005, Opal completely drained, but it refilled as a beautiful green pool in early 2008.

RABBIT CREEK HOT SPRINGS

There is no trail into the Rabbit Creek Hot Springs. The distance from the highway to the main hot spring area is about 1 mile. Springs of minor importance are scattered all along the way, especially along a fracture zone that includes little-known "Tuba Geyser" and "Two Hole Geyser." Most people hiking into this area park at a roadside turnout and follow Rabbit Creek toward its source at the head of the valley. Most of the flow of Rabbit Creek originates in a single large pool that rivals any other for size and beauty. Interestingly, it has no name. Indeed, only one feature here has been named, and it is the only consistently active geyser in the group.

This is a very dangerous area—many of the springs have wide overhangs around their edges, superheated water is common, and the mud pot areas can look solid but are actually only a dried crust on top of boiling mud. Always notify others of your planned departure *and* return if you go here. And be careful—any thermal burn is a serious injury, especially when it happens away from a developed area.

19. RABBIT CREEK GEYSER is located on the hillside well to the south of the pool at the head of the stream, where it lies within a deep, jagged crater. Most of its eruptions consist of violent boiling 2 to 3 feet high that is confined to the subsurface. This play recurs every 15 to 27 minutes and normally lasts about 1 minute. Occasional major eruptions include bursts that can jet several feet aboveground, producing a total height of 10 feet or more.

Upslope above Rabbit Creek Geyser is "Scaffold Spring," named because of the haphazard arrangement of downed logs within the crater. Still farther uphill are the "Rabbit Highland Hot Springs," where several pools look very much like cream of tomato soup because of iron oxide suspensions in their waters; one of these was named Brick Pool in 1960.

20. UNNG-MGB-8 is a pool at the base of the slope a short distance north of Rabbit Creek Geyser (19). The vent is beneath an overhanging, cliff-like formation near the eastern corner of the crater. The activity is highly variable and usually consists of random splashes just 1 to 2 feet high. Vigorously jetting true eruptions as much as 10 feet high happen at undetermined intervals.

Still farther north a short distance beyond MGB-8 is another deep pool that undergoes intermittent boiling and probably has true, bursting eruptions on irregular intervals.

CHAPTER SIX

Lower Geyser Basin

The Lower Geyser Basin (Map 6.1) is Yellowstone's largest by area. The hot springs occur as scattered groups that cover 5 square miles within a valley twice as large. Most of these groups include geysers, several of which are as large and famous as those of the Upper Basin. Great Fountain especially deserves its reputation as one of the most magnificent geysers in the world.

Most of the individual clusters of geysers in the Lower Basin are readily accessible. Roads approach them in several areas. Boardwalks allow the visitor to further explore the more important groups, and most of the remote springs can be reached by dirt trails. A few groups, however, remain visible only at a distance from the highway or boardwalks.

The majority of the springs here are of the same clear-water type as those of the Upper and Midway basins, but the Lower Basin is notable in that there are also some areas of muddy, acid activity. These are often closely spaced with the geysers, an unusual

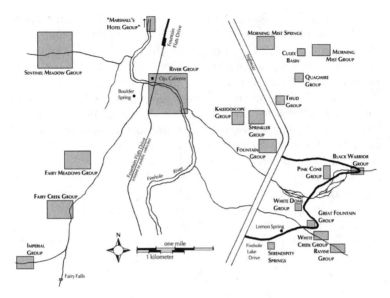

Map 6.1. *Index Map of the Lower Geyser Basin*

occurrence. The Fountain Paint Pots, within the Fountain Group of geysers, are the best-known examples. The Pocket Basin Mud Pots in the River Group require a hike to see but are larger and more varied.

FIREHOLE LAKE DRIVE

Firehole Lake Drive is a one-way loop road that leaves the main highway at the southern end of the Lower Geyser Basin and passes through several important hot springs groups before rejoining the highway opposite the Fountain Paint Pot parking lot. To drive the road nonstop is worthwhile, but to stop and explore the thermal areas is far better. One can easily spend days here without seeing the same thing twice.

SERENDIPITY SPRINGS

The Serendipity Springs are located in the small Serendipity Meadow, visible through a gap in the forest to the right (southeast) of Firehole Lake Drive, approximately $^1/_3$ of a mile from the road's entrance at the highway and where tiny Erythemis Creek (named for the genus of a common dragonfly) passes under the road (see Map 6.1). The springs are all small. Some act as weak perpetual spouters, and there is one geyser. UNNG-SDP-1 was first seen to erupt during 2004. The eruptions, up to 2 feet high, rise from one of two vents in an hourglass-shaped pool about 3 feet long. Intervals are less than 1 minute long, and the play lasts 20 to 35 seconds. The spring that is farthest from the road, and just beyond the edge of the meadow, is a perpetual spouter that might occasionally act as a true geyser.

A short distance farther along Firehole Lake Drive, on the left side of the road, is Lemon Spring. It was named because its water temperature is often just right to support brilliant yellow cyanobacteria. However, during 2006 Lemon underwent three, day-long episodes of boiling and heavy overflow. Although it never actually erupted, it may someday prove to be a true geyser.

GREAT FOUNTAIN GROUP

Great Fountain Geyser is the namesake of a group consisting only of itself and three other geysers (Map 6.2, Table 6.1, numbers 1 through 4). All lie on the west side of Firehole Lake Drive; the springs east of the road comprise the White Creek and Ravine groups. Using the road as the separation is a matter of convenience, as the two groups might comprise a single system at depth. The Great Fountain Group also includes Surprise Pool, which is superheated and intermittent in its boiling overflow but has never been known to erupt.

1. FIREHOLE SPRING is immediately below the road at a pullout just south of White Creek (and is rightfully a member of the White Creek Group). A pretty, rich blue pool, it has erupted throughout Park history. This was one of the very few hot springs to completely escape earthquake effects, and the bursting play reaches as much as 6 feet high. Firehole Spring is essentially a perpetual spouter, although occasional seconds-long pauses in the activity lead some to call this a geyser. The name does not directly relate to the Firehole River or Valley; this "firehole" is so called because of the flashing flame-like appearance of steam bubbles as they enter the bottom of the pool. The same effect can be seen in several

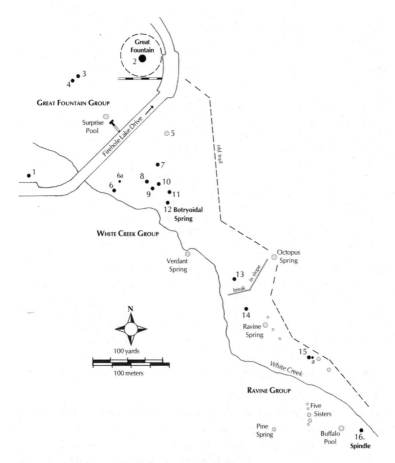

Map 6.2. *Great Fountain, White Creek, and Ravine Groups, Lower Geyser Basin*

of the springs in the Black Warrior Group, farther along Firehole Lake Drive.

2. GREAT FOUNTAIN GEYSER was observed and accurately reported on by the Cook-Folsom-Peterson party in 1869. It was their first experience with a large geyser, and they were fortunate to arrive on October 1, 1869, apparently just as Great Fountain began an eruption (or perhaps one of its late bursts). Duly impressed, Cook later

Table 6.1. Geysers of the Serendipity Springs, Great Fountain, White Creek, and Ravine Groups, Lower Geyser Basin

Name	Map No.	Interval	Duration	Height (ft)
A-1 Geyser	8	[1993]	5–10 min	6
A-2 Geyser	9	[1996]	6–8 min	10
A-Zero Geyser	7	25–29 min	25–40 sec	10–20
Botryoidal Spring	12	2½–5½ min	10–20 sec	12–20
Diamond Spring	13	[1987]	seconds	3
Firehole Spring	1	near steady	near steady	4–6
Great Fountain Geyser	2	8–17 hrs	45 min–2 hrs	10–250
Logbridge Geyser	6	16–90 min*	25–35 sec	10–12
"Prawn Geyser"	3	28–50 min*	minutes	1–6
Spindle Geyser	16	1–3 min	30 sec	1–3
Tuft Geyser	15	50–100 min	5–30 min	2
UNNG-GFG-2	4	irregular*	minutes	2–5
UNNG-SDP-1	—	30 sec–1 min	10–12 sec	1
UNNG-WCG-3a ("Eclipse")	15a	[1984]	2 min	8
UNNG-WCG-4	14	40–50 min	30 sec	subterranean
UNNG-WCG-5	10	days?	minutes	3–5
UNNG-WCG-6	11	min–hrs	minutes	inches–4
UNNG-WCG-7	5	near steady	near steady	1–3
UNNG-WCG-8	6a	hours	1–2 min	1–3

* When active.
[ ] Brackets enclose the year of most recent activity for rare or dormant geysers. See text.

wrote, "We could not contain our enthusiasm; with one accord we all took off our hats and yelled with all our might."

Great Fountain has always attracted much attention. The large crater is set in the middle of a broad, raised sinter platform decorated with exquisite catch basins, rims, and geyserite beadwork that lent it the alternate early name "Architectural Fountain." The vent itself is about 16 feet across and filled with clear, boiling water. The setting is impressive even when the geyser is not in eruption.

Great Fountain's eruptions are regular enough to be predicted. The superheated water surges and boils during a long preplay period. It takes close looking, but the most significant time for a prediction is the start of overflow from the crater onto the platform. This usually happens 70 to 110 minutes (average 85 minutes) before the beginning of an eruption. With this, gentle superheated boiling begins around the edge of the pool. The boiling becomes periodic and progressively stronger as the over-

Great Fountain Geyser is the largest in the Lower Geyser Basin. Although most of its play reaches 100 feet or less, eruptions estimated as tall as 230 feet have been observed.

flow continues. Eventually, a heavy surge known as the "one-meter boil" starts the eruption. This boiling is often (but not always) followed by a quiet pause a few minutes long. This in turn leads to heavy surging, which becomes violent and domes the water several feet high. Then, Great Fountain bursts into the sky. Many of the eruptions are "only" 100 feet high, but even these are spectacular since the bursts are very wide and frequently jetted at an angle. Sometimes the bursting will reach 150 feet and splash onto the roadway. "Superbursts" are rather rare. They usually happen near the beginning of an eruption, sometimes following a massive, quiet doming of the entire crater volume in an amazing "blue bubble." Superbursts have been measured to reach as high as 230 feet. (Not all blue bubbles result in superbursts, and surperbursts without blue bubbles are more common.) The activity continues with bursts of widely varying heights for several minutes. Then, Great Fountain pauses for a few minutes while leading up to a second eruptive period. This process is repeated four to seven times, so the whole eruption lasts 45 minutes to 2 hours.

Every geyser, even one as intensely observed as Great Fountain, has something new to tell us. Only during 1993 was it realized that there is a regular relationship between the duration of an eruption and the length of the following interval. Perhaps this had not been known before because the calculation demands that the duration be timed very accurately, requiring the observer to wait out the very last, weak bursts. The resulting mathematical relationship (approximate and varying from year to year) says that the next interval (in hours) will be about equal to 7 times the duration (in hours) plus 6; that is, $I = 7D + 6$.

During all the pre–1959 earthquake years, Great Fountain's intervals averaged near 12 hours. The tremors caused some sort of underground changes, for while Great Fountain continued to be regular, the intervals were cut to only 8 hours. Very slowly, the average increased until it again achieved about 12 hours in 2001. In 2007, the exact average of 483 recorded eruptions was 11 hours 59 minutes within a range of 8½ to 17 hours; one extraordinarily long interval of 22 hours was also recorded in 2007.

On rare occasions, Great Fountain enters what is called "wild-phase" activity. For many hours to several days it erupts almost continuously. Water is thrown 10 to 50 feet high several times per minute. Although not nearly as high as normal eruptions, this play is impressive. What causes wild-phase action is unclear, but it might be related to changes in the non-geothermal groundwater level, since most occurrences have happened in the late summer and early fall. The most recent wild phase took place in late October 2006. There have also been occasions of excessively long overflow before eruptions, as well as a few in which Great Fountain failed to refill normally after an eruption. In both of these cases, the resulting interval can be as long as 3 days.

Overall, Great Fountain is an extremely reliable performer. Because it is predictable and the only such major geyser in the Lower Basin, it performs for thousands of people every year. Few are disappointed by the display.

3. "PRAWN GEYSER" made its appearance during 1985. The name is an allusion to the geyser's small ("shrimpy") size as compared with nearby Great Fountain. Located within Great Fountain's southernmost runoff channel, this geyser shows some relationship to Great Fountain's activity. Although intervals are known to range between 28 and 50 minutes, most are on the short side unless Great Fountain has just erupted; then the next one or two intervals will be substantially longer. Lasting several minutes, the eruptions reach between 1 and 6 feet high. Long dormancies, as in 2006–2007, have been known.

4. UNNG-GFG-2 was first seen during 2004. It plays from a vent a few feet from that of "Prawn Geyser" (3) and might represent the same plumbing system, since GFG-2 shows the same relationship to Great Fountain (2) as does "Prawn"—short intervals near the time of Great Fountain but longer intervals or inactive at other times. Steamy spray is jetted as high as 5 feet.

WHITE CREEK GROUP AND RAVINE GROUP

Scattered along White Creek east of Firehole Lake Drive are many springs. Generally known as the "White Creek Group," there is a natural separation between the upper and lower areas that historically were separated into two groups. Together they include at least twelve geysers (Map 6.2, Table 6.1, numbers 5 through 16). Both groups can be seen by following an old trail that winds upstream from the Great Fountain parking lot.

The cluster of springs nearest the road is the White Creek Group proper (sometimes called the "Lower White Creek Group").

This group's geysers are all visible from the road as well as the trail. At the edge of the stream near Botryoidal Spring (12) is the remnant of Verdant Spring; once described as the loveliest pool in Yellowstone, its crater was completely filled with muddy debris by a flash flood following the forest fires of 1988, and it is now a patch of steaming holes.

Upstream and perched on top of geyserite terraces is the Ravine Group, so named in 1911 but also called the "Upper White Creek Group." There are only three small geysers there but also several pools of special note. Octopus Pool is a clear, pale-blue intermittent spring; Ravine Spring is the largest among a series of shallow pools; the runoff from the small connected pools of the Five Sisters Springs supports a profuse growth of brilliantly colored cyanobacteria; the crater of Pine Spring contains a jumble of fallen logs; and Buffalo Pool boasts wide, dangerous overhanging ledges of geyserite around its bone-littered crater.

Much of the pioneering work on *thermophilic* ("temperature-loving") life has been conducted along White Creek. The bacteria *Thermus aquaticus* was discovered near here. An enzyme called "Taq polymerase," used to amplify and copy DNA for forensic "fingerprinting" and medical diagnostic procedures, was initially recovered from this bacteria. Research based on Taq polymerase has become a multibillion-dollar industry, and researchers are presently studying other *thermophilic* organisms for similar enzymes that may have unique uses. If you see experiments in progress, please look but do not touch. As always, remember to stay on the trail, as casual wandering among the hot springs is illegal.

5. UNNG-WCG-7 is the spring closest to the road and to the Great Fountain Geyser parking area. It was a quiet pool until May 2005, when it was found splashing from a lowered water level. The play is nearly perpetual and usually less than 1 foot high, but occasional surges can spray as high as 3 feet.

6. LOGBRIDGE GEYSER is so called because it lies near some cut logs placed across White Creek so long ago that they have nearly rotted away. The spring was known only as a small perpetual spouter until 1985, when it began having frequent eruptions far larger than any seen before, sometimes reaching as high as 15 to 20 feet. The runoff cleared gravel out of old runoff channels, showing that similar activity had taken place long before. After 1985, Logbridge continued to play as a geyser, but the intervals were hours long and the activity much weaker—until 2006. The action is apparently cyclic. Intervals of 16 to 45 minutes are occasionally interrupted by quiet episodes as long as 90 minutes, with two to seven eruptions in a series. The initial eruption of a new series is followed infrequently within 3 minutes by a second brief but full-force eruption. The 30-second play is as high as 12 feet.

As seen from the road, there is a small bubbling pool a few feet to the left of Logbridge. A short distance still farther left is a geyser (6a. UNNG-WCG-8) apparently never seen before 2006. Its intervals are hours long, while the 1- to 2-minute eruptions reach up to 3 feet high.

7. A-Ø (or **A-ZERO**) **GEYSER** was named in allusion to its proximity to the somewhat more officially named A-1 (8) and A-2 (9) geysers. A-Ø is quite regular in its performances. Intervals vary from 25 to 29 minutes, and it is easy to miss the eruption since the duration is as short as 25 seconds. The play is 10 to 12 feet high. Rare major eruptions, seen only a few times during the early 1980s, lasted several minutes and reached at least as high as 20 feet.

8. A-1 GEYSER shows a clear relationship to nearby A-2 Geyser (9). If A-1 is active, then A-2 is nearly dormant; if A-2 is frequent, A-1 is completely dormant. The crater of A-1 is an irregular oval, largely filled with geyserite rubble. During active phases, A-1 plays every 30 to 40 minutes, bursting up to 6 feet high for 5 to 10 minutes.

A different pattern of activity was seen during 1993. A-1 was active, even though A-2 was also playing. The intervals were long and erratic, and the play reached no higher than 2 feet. Perhaps this presaged the 1996 activation of Botryoidal Spring (12), which caused an immediate and complete dormancy in both A-1 and A-2.

9. A-2 GEYSER lies 25 feet east of A-1 Geyser (8). It plays from a shallow basin containing three large and several small vents. The bulk of the eruption comes from just two of these openings. Except in 1970 and 1971, when A-1 was the dominant member of this group, A-2 was the most important geyser along White Creek until Botryoidal Spring (12) began its powerful eruptions in 1996. There were both major and minor eruptions. Most common were the majors, in which intervals of 1 to 2 hours separated plays lasting

6 to 8 minutes. After an eruption the crater drained and then slowly refilled, and the next play started at about the time of first overflow. Minor activity recurred on intervals as short as 5 minutes with durations of less than 1 minute, and the crater did not drain during those episodes. All eruptions were about 10 feet high. A-2 has been completely dormant since Botryoidal Spring started its major activity in 1996, and grasses are beginning to grow in its decaying basin.

10. UNNG-WCG-5 started erupting in 2004, when it jetted steamy, fan-shaped spray as high as 5 feet from a small crack just northwest of the old crater rim of A-2 Geyser (9). The few eruptions seen in 2005 implied that the intervals were days long. The duration was not accurately determined, but the play lasted several minutes. This might be the same geyser as one that was briefly active but never fully described during 1996.

11. UNNG-WCG-6 is a compact cluster of vents within a depression a few feet northwest of Botryoidal Spring (12). The activity is probably cyclic, as the intervals seem to be either minutes or hours long. The play lasts a few minutes and consists of splashes inches to 4 feet high.

12. BOTRYOIDAL SPRING erupts from a pool about 100 feet southeast of A-2 Geyser (9). The crater is surrounded by massive geyserite shoulders covered with botryoidal (grapelike) globules of sinter beadwork. Historically, Botryoidal behaved nearly as a perpetual spouter, its activity infrequently interrupted by a few seconds of semi-quiet. In late 1996 it began truly intermittent activity. Eruptions recur every 2½ to 5½ minutes. Most begin when a huge bubble of steam bulges the pool upward as a "blue bubble" that bursts to throw water 12 to 20 feet high. The remainder of the eruption, only 10 to 20 seconds in duration, involves much smaller splashes.

13. DIAMOND SPRING is well separated from the rest of the White Creek Group but is still below the break in slope. It was named because of the shape of its crater. Diamond was a quiet spring until the 1959 earthquake. It apparently erupted powerfully that night, but no reasonably frequent eruptions were seen until 1973. These were brief and no more than 3 feet high. Washed areas showed that further strong action took place after the 1983 earthquake, but otherwise only a few small eruptions were seen in 1987. Diamond remains active as an intermittent spring.

14. UNNG-WCG-4, the nearest of the Ravine Group geysers to the road, occupies a deep crater on the sinter platform just beyond the break in slope above Diamond Spring (13). Its 4-foot eruptions are confined to the subsurface, and the fact that it is a geyser was only discovered around 1980. Eruptions are fairly regular, recurring on intervals of 40 to 50 minutes whenever checked. The duration is about 30 seconds.

15. TUFT GEYSER and **15a. UNNG-WCG-3a ("ECLIPSE GEYSER")** are small geysers of which nothing was known prior to the 1970s. The more active is Tuft. It plays from a number of small openings next to a crescent-shaped deposit of geyserite only a few inches high. The eruption is noisy sputtering out of the openings, with some spray reaching perhaps 2 feet high. Tuft's activity is such that it can be difficult to distinguish the interval from the duration. The sputtering begins before there is any overflow and continues after the last discharge, and sometimes it never completely quits before the next eruption begins. Taken on balance, most intervals are between 50 and 100 minutes long, with overflow durations of 5 to 30 minutes.

"Eclipse Geyser" (15a) is rarely active. It erupts from a round hole, about 8 inches in diameter, between Tuft and a small blue pool. When it is active it completely "eclipses" Tuft, rendering it dormant. During the most recent active phase, in 1984, the play recurred every 35 to 40 minutes, lasted 2 minutes, and sent a steady jet of water 8 feet high.

16. SPINDLE GEYSER is the most upstream geyser of the Ravine Group, lying in an orange-brown crater beyond the Five Sisters Springs and Buffalo Pool. (Walk wide around Buffalo. It has extremely dangerous overhangs around its edges, making it easy to see why skeletons lie on the bottom of the crater.) Few geysers illustrate the typical cycle of activity of a fountain-type geyser as well as Spindle does. Eruptions recur every 1 to 3 minutes. Steam bubbles can be seen rising into the crater, expanding as they go before throwing the water 1 to 3 feet high. Following the eruption the water level drops several inches, far enough to stop all over-flow. The pool then begins to rise almost immediately, the flow becoming heavy just before the next eruption. Thus, the alternating buildup and loss of pressure within the system is readily seen. In 1985, during an erratic episode of less frequent eruptions, some play 10 feet high was seen.

Beyond Spindle Geyser, hot springs continue far up the canyon, but none are known to be geysers. Black Spring, named because of black mineral deposits in its crater, is the only feature to have been named. The stream's headwater areas were severely burned during the forest fires of 1988. Summer thunderstorms in 1989 produced several muddy flash floods, and one partially filled Spindle's crater with mud. The geyser did not clear its crater and fully recover activity until 1993.

WHITE DOME GROUP

The White Dome Group (Map 6.3, Table 6.3, numbers 20 through 26) is a tiny cluster of hot springs bisected by Firehole Lake Drive. It encompasses an area little more than 300 feet square, yet it includes six geysers and one perpetual spouter. The most important of these is White Dome Geyser, which based on the size of its cone is one of the oldest hot springs in Yellowstone.

Visible in the fringe of forest several hundred feet across the grassy flat to the east is an unnamed cluster of springs that includes Mushroom Pool, the actual discovery site of the *Thermus aquaticus* bacteria. Immediately adjacent to Mushroom is "Toadstool Geyser,"

Table 6.2. Geysers of the White Dome Group and The Tangled Geysers, Lower Geyser Basin

Name	Map No.	Interval	Duration	Height (ft)
Cave Spring	25	steady	steady	subterranean
Crack Geyser	22	[2007]	3–7 min	10
Gemini Geyser	23	5–25 min*	sec–3 min	10
Pebble Geyser	21	[2003]	seconds	1–20
UNNG-TTG-1	28	2½–13 min	seconds	10
UNNG-WDG-1	26	near steady	near steady	sub–5
UNNG-WDG-2	24	[1993]	seconds	1
White Dome Geyser	20	9 min–3 hrs	2½ min	30

* When active.
[ ] Brackets enclose the year of most recent activity for rare or dormant geysers. See text.

which underwent eruptions as high as 10 feet in 1968 and 1969 and by doing so destroyed some biological experiments.

20. WHITE DOME GEYSER has a very long history. The massive geyserite cone built by the spray of its eruptions is over 12 feet high. Yellowstone's only larger cone structures are at Castle Geyser and the nearly extinct White Pyramid Geyser Cone, both in the Upper Geyser Basin. The cone in turn sits atop a mound of sinter formed by an even older hot spring. Traces of an old crater can be seen atop this mound, and White Dome might have been a fountain-type geyser of considerable power centuries ago. The modern White Dome has nearly sealed itself with deposits of sinter inside

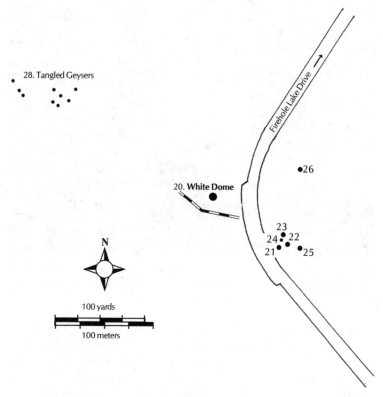

Map 6.3. *White Dome Group, Lower Geyser Basin*

The thin water jet of White Dome Geyser reaches 20 to 30 feet above the top of its 12-foot geyserite cone. On average, the eruptions repeat about every half hour.

the vent, for the remaining opening is only 5 by 7 inches. White Dome is nonetheless a fascinating geyser, so beautiful and symmetrical that it used to be a symbol of the old Yellowstone Library and Museum Association (now the Yellowstone Association).

While the size of the cone seems to hold promise of big eruptions, the resultant display is disappointing to some viewers. An eruption begins after a few moments of splashing pre-play. At first the action consists mostly of steam, but a steady jet of water soon makes up the bulk of the discharge. The maximum height of about 30 feet is maintained for the first half of the 2½-minute eruption, which then declines into weak, steamy spray. White Dome's intervals show wide variation. Although most are in the range of 20 to 35 minutes, some are as short as 9 minutes and others are as long as 3 hours. No way has been found to predict what White Dome's next interval will be. In addition, intervals as long as 6 hours took place during 1992 and 1993, years in which there were also at least two episodes of minor eruption series with intervals just 8 minutes long, durations of a few seconds, and heights of only around 10 feet. Such action was never recorded before or since, although a few minor eruptions not in series were seen during 2007. White Dome was apparently dormant much of the time between 1878 and 1906, based on the fact that only two brief tour descriptions and no geological reports mention it in those 28 years. It has no known connections with other hot springs, so why White Dome is so erratic is a mystery.

21. PEBBLE GEYSER is the small pool close to the road just across from the White Dome parking area. This spring was not known to erupt until August 1968, when it underwent several eruptions. Because it is located so close to the edge of the road, much gravel gets pushed and thrown into the crater. The first of the 1968 eruptions scattered those pebbles across the road, giving the geyser its name. Those eruptions were brief but over 20 feet high. Pebble continued occasional activity into 1969, when nearby Crack Geyser (22) became active. Since then, Pebble has had only infrequent eruptions, 1 to 4 feet high, most occurring a few moments preceding or immediately following those by Crack or Gemini (23). These and a few still smaller, independent eruptions seen during 2003 rose from a partly drained pool. In 2006 and 2007, Pebble infrequently filled to overflow, then drained (but did not erupt) during eruptions by Crack.

22. CRACK GEYSER developed along a fissure in the sinter platform. Formed by the 1959 earthquake, this fracture was first the site of a fumarole. By 1960 it was erupting and became known as Crack Geyser. After a few months of activity it went dormant with the rejuvenation of Gemini Geyser (23). Additional activity took place in

1969 and 1983. In 1987 and 1988 an interesting pattern developed in which Crack and Gemini alternated frequent eruptions. Crack would erupt for 3 to 4 minutes, sending a fan-shaped spray 10 feet high. About 20 minutes later would be an eruption by Gemini, and then around 40 minutes after that would be another by Crack. This action was highly regular and persisted for about 10 months. Crack fell dormant in late 1988 and was not seen again until it reactivated for only a few days in July 2006 and again in June 2007. Some of these eruptions were preceded and/or followed by eruptions of Gemini. The intervals were as long as 8 hours, with the play lasting as long as 7 minutes.

23. GEMINI GEYSER plays from two small vents, one perched at the top of a tiny cone and the other within an adjacent small pool. At the beginning of an eruption, bubbling water begins to well from the vents, and this progressively builds into the eruption. The full height of 10 feet is reached within moments. Both of the twin water jets are angled slightly away from the road. Except for 1987–1988, when Gemini and Crack Geyser (22) alternated eruptions, Gemini has acted as a cyclic geyser. During active phases, eruptions recur every 5 to 25 minutes and last a few seconds to 3 minutes. An active phase can last several hours, but then a day or more is required before activity resumes. However, Gemini has been dormant much more than active. Gemini was essentially dormant from 1994 until a few eruptions days apart were seen in 2004. Short episodes of activity occurred during 2005, infrequent individual eruptions were seen in 2006 during the few days when nearby Crack was active. Finally, in June–July 2007, Gemini resumed vigorous activity when eruptive series several hours long included intervals as short as 9 minutes. Unfortunately, this action ended after only a few weeks.

24. UNNG-WDG-2 is a small hole in the sinter between Gemini (23) and Pebble (21) geysers. During 1987–1988 it played briefly between the alternating eruptions of Gemini and Crack (22). It was also weakly active during 1993, occasionally sputtering about 1 foot high.

25. CAVE SPRING lies deep within the large cavern to the northeast of Crack Geyser (22). The spring is eruptive, probably as a perpetual spouter. Even to people illegally off the roadway, it is almost impossible to see.

26. UNNG-WDG-1, about 150 feet north of Gemini Geyser (23), is a small pool within a sinter-lined bowl. It was first observed during 1971. Intervals were 2 to 3 hours long, but durations of 20 minutes and splashes 5 feet high made it fairly obvious. Activity continued through 1986, when the intervals and durations were both only seconds to minutes long. WDG-1 is now a weak perpetual spouter, with nearly all of its splashing invisible below ground level.

THE TANGLED GEYSERS

Tangled Geysers (Map 6.3, Table 6.2) is the official name for the group of springs that lie across the marshy flat west of White Dome Geyser (20). Little is known about the Tangled Geysers except that at least eight of the two dozen springs have histories as small geysers. Such action is uncommon because the springs' vents are usually drowned by the water of Tangled Creek.

28. UNNG-TGG-1 designates the only member of the Tangled Geysers to have been consistently active in recent years. Eruptions recurred about every 13 minutes when first seen during 2002, and the geyser gradually grew more frequent and regular. Most measured intervals fell between 3½ and 6 minutes long. Steady jets up to 10 feet high lasted 20 to 25 seconds. Following a brief dormancy in May–August 2007, eruptions resumed with intervals of only 2½ minutes, but these durations were as short as 10 seconds, with jetting weaker than before.

PINK CONE GROUP

Pink Cone, for which this cluster of springs is named (Map 6.4, Table 6.3, numbers 30 through 37), is by far the best-known geyser of this group, but all its other members are geysers, too, except Shelf Spring. The eruptions tend to be of considerable size, frequency, and regularity. This is interesting because the group was considered unimportant during the geological surveys of the 1870s and 1880s, when geyser activity apparently was infrequent and weak. Pink Cone rarely performed before 1937, Bead was infrequent into 1938, it was unusual to see Narcissus erupt prior to the 1950s, Pink only became active after the 1959 earthquake, and Labial and Box Spring grew frequent only following the 1983 quake. In fact, all of

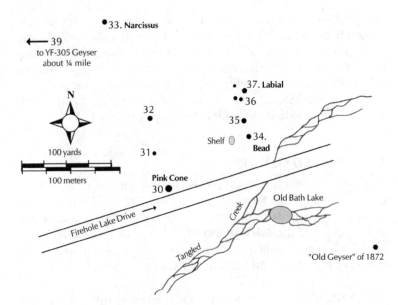

Map 6.4. *Pink Cone Group, Lower Geyser Basin*

the geysers increased their activity following those tremors, some nearly doubling their frequency, and most of the increased action has persisted with little change.

Wandering around within the Pink Cone Group is illegal. Signs warn you to stay on the roadway, and by positioning yourself properly it is possible to see all the geysers from there. The area is perforated by many spring holes with overhanging rims. Shelf Spring is deep and highly superheated; its thin sinter rim projects far out over the water, making any approach very dangerous.

To the southeast is Old Bath Lake (also known as "The Tank," "Tank Spring," and "Ranger Pool"). It apparently is a prehistoric, manmade dam on Tangled Creek and is officially closed to visitation because of its archaeological importance. On the hillside beyond is an unnamed group of springs. One of them was called "Old Geyser" in 1872, perhaps because of its appearance rather than any geyser action. Small eruptions may have taken place there in August 2007, but this could not be confirmed because of the archaeological restrictions.

Table 6.3. Geysers of the Pink Cone Group and Underhill Springs, Lower Geyser Basin

Name	Map No.	Interval	Duration	Height (ft)
Bead Geyser	34	21–38 min	2½ min	25
Box Spring	35	15–70 min	sec–2½ min	3–8
Dilemma Geyser	31	minutes*	seconds	inches–3
Labial Geyser	36	5–6½ hrs	1 min	25
Labial's Satellite, East	37	days–weeks*	5–15 min	6
Labial's Satellite, West	37	infrequent*	minutes	10
Narcissus Geyser (short mode)	33	2–3 hrs	5–8 min	15–30
Narcissus Geyser (long mode)	33	3–6 hrs	13–20 min	15–50
Pink Cone	30	18–23 hrs	1½–2 hrs	30
Pink Geyser	32	5–11 hrs	11–17 min	15–20
"YF-305"	39	4–6 min	10 min	2–6

* When active.

30. PINK CONE, with its brownish-pink cone, is colored by a trace of manganese oxide; if a bit more were present, the color would be black. This same coloration shows up in Pink (32) and Narcissus (33) geysers. Because of this similarity and the fact that the three lie along a line, they are probably connected along some sort of subsurface lineation—a long crack or perhaps a fault. The connection cannot be too direct, however, as none of these geysers' performances are affected by the others.

Pink Cone was named by members of the 1878 geological survey of Yellowstone. They recognized that it must have been a geyser because of its shape, but they saw no eruptions (which is why "Geyser" is not part of the official name). It was active during 1887, but on a pattern different from today's. The frequency of that play was not recorded, but the duration was about 30 minutes and the height only 5 to 10 feet. Interestingly, a secondary vent on the platform near the cone jetted vigorously as high as 10 feet before producing a noisy steam phase; this vent is now a rare performer that seldom reaches higher than 2 feet.

Not a single eruption of Pink Cone was recorded between 1889 and 1936, but in 1937 it was active, with intervals as long as 50 hours. Pink Cone maintained such intervals until the 1959 earthquake markedly increased its performance. For a while, the intervals were less than 50 minutes long, with durations of the same

Pink Cone erupts steady jets as much as 30 feet high from a small cone next to Firehole Lake Drive.

length. Through time since 1959, the intervals have very slowly increased. The present range is from 18½ to 23 hours, but Pink Cone tends to vary little from its average, which was near 20½ hours in 2007. With occasional splashing pre-play, the eruptions begin abruptly and almost instantly reach the maximum height of 30 feet. The duration is about 100 minutes. The steady water jet pulsates some, and as the play continues these pulses become more extreme. Near the end of an eruption they cause brief, total pauses that in turn become progressively more dominant and merge into the end of the action. Some final puffs of steam may occur more than 2 hours after the start of the eruption.

Pink Cone lies immediately next to the road. When it was built, the route was cut right through the broad mound on which the geyser sits. Some minor parts of the plumbing system were tapped into by the road cut, resulting in a series of tiny vents, the "Roadside Bubblers." These vents often begin to flow small streams of water before Pink Cone's eruption, but this lead time varies from near zero to over 5 hours. They also sputter a little water while Pink Cone is playing. It is amazing that Pink Cone was not seriously altered by the roadwork, but now the cut provides a cross-section of a geyserite mound, built up layer after layer by uncountable eruptions.

31. DILEMMA GEYSER plays from a small double vent about midway between Pink Cone (30) and Pink Geyser (32). Tiny, very brief eruptions during 1984 were hardly worth noting, except that observations of the surroundings revealed old runoff channels leading from the vents. The "dilemma" was about the nature of activity that had produced the channels, since that of 1984 was not capable of such erosion. Action that started during 1989 solved the problem. Although Dilemma's more vigorous eruptions are only 2 to 3 feet high, they are preceded and accompanied by runoff that scours the channels. The seconds-long eruptions continue to take place every few minutes. but sometimes, as throughout 2007, they produce little or no visible splashing at the surface.

32. PINK GEYSER lies 235 feet beyond Pink Cone (30). The rose-colored sinter basin of this spring is 7 feet in diameter; the vent itself is only inches across and comes to the surface at an angle. Pink apparently was not active before the 1959 earthquake and was then erratic until the 1983 earthquake. Those tremors stimulated Pink into its present cycle. The intervals average near 6 hours, usually

Narcissus Geyser bursts as high as 30 or more feet every 2½ to 6 hours.

with little variation except that intervals longer than 11 hours are occasionally recorded. The play, which lasts from 11 to 17 minutes, is jetted 15 to 20 feet high at an angle in the uphill direction. It's a very pretty eruption concluded by a brief steam phase.

33. NARCISSUS GEYSER is located farthest from the road, nearly hidden from view behind a stand of lodgepole pines. Therefore, it is seldom seen except via its distant steam clouds. The geyser erupts from a soft-pink crater filled with pale-green water in a lovely setting.

Possibly an infrequent performer until the 1950s, Narcissus is now highly regular in its activity. If it were in a more accessible location, it would be one of Yellowstone's most predictable geysers. It has short mode (minor) and long mode (major) eruptions that alternate with one another on a regular basis. The short mode eruptions take place after intervals that average near 2½ hours. They start when the crater is only about half full. These eruptions last 5 to 8 minutes, and some bursts reach 30 feet high. The long mode action begins after longer intervals of 3 to 6 (usually about 4½) hours, when Narcissus has completely filled its crater and then undergone several episodes of brief, pulsing overflows that recur every few minutes. The major eruptions last longer than the minors, in the range of 13 to 20 minutes, and some of these bursts can approach 50 feet high. In most geysers long durations lead to long intervals, but Narcissus is one of the few in which the long mode eruptions are almost invariably followed by the short mode. The alternating long-short-long-short pattern fails only a few times per year, when there are two consecutive major eruptions or, less often, two consecutive minor eruptions. After any eruption, the crater drains completely while forming a whirlpool over the vent.

34. BEAD GEYSER is known for its extreme regularity. It also used to be known for its fine collection of "geyser eggs"—small, loose, ¼- to 2-inch spheres of geyserite that slowly form in the splash basins of most geysers. The geyser eggs that gave Bead its name are long gone, having been removed by Park visitors almost as soon as their existence was known.

Bead is one of the most regular geysers in Yellowstone, a bit surprising, perhaps, since it was rare at best before 1938. The intervals vary some as time passes, and currently they range between 21 and 38 minutes. The eruption begins suddenly after water has slowly risen within its vent. Just as it reaches the overflow level, the geyser suddenly surges. Within a few seconds, water jets are reaching 25 feet high. The duration is about 2½ minutes. The bursting is nonstop until it ends as suddenly as it begins.

35. BOX SPRING was named during post-1959 earthquake studies, and in the successive years it had rare eruptions up to 5 feet high. The first consistent activity took place during 1984, probably as a result of the 1983 quake. Intervals were around 7 hours. The play was vigorous bursting up to 10 feet high that decreased to minor bubbling by the end of the 5-minute duration. Starting about 2000, Box became more frequently active but somewhat weaker, with intervals between 15 and 70 minutes. Bursts 3 to 8 feet high are continuous throughout the duration. Most eruptions last only a few seconds, but occasional "major" eruptions can persist for as long as 2½ minutes.

36. LABIAL GEYSER was an infrequent performer prior to the 1983 earthquake, the intervals being 12 or more hours long. The tremors doubled its frequency, so it now plays every 5 to 6½ hours. Labial's cone is hidden from view by a geyserite mound. A viewer on the road can see nothing of Labial except the eruption itself and so, unfortunately, cannot observe the interesting preliminary activity that takes place within Labial and some nearby vents. During the quiet interval the water level rises and falls every few minutes. The times of high water result in boiling and sloshing within Labial. Each cycle brings the water level a bit higher, and the related vent begins to overflow about an hour before the eruption. During the last few minutes Labial's own surging becomes violent, and some splashes may be seen from the road. One of these triggers the eruption. The play lasts only about 1 minute, but the sharply angled water jet is fairly wide and as much as 25 feet high. A second vent bursts to about 6 feet, and the related spring splashes 1 to 2 feet high. Following the main eruption, Labial continues to have bursts that spray from the vent; these rarely lead to a second and even a third briefer but full-force eruption during the next half hour or so.

37. LABIAL'S SATELLITE GEYSERS, EAST and **WEST,** play from craters at the top of the geyserite mound that prevents a direct view of Labial Geyser (36). The eastern of the two has an ornate crater and may have been the original Bead Geyser of 1878, rather than the spring (#34) that now bears the name. Until 1989 these geysers appeared to be related to Labial, with more frequent activity shortly before Labial played, but since 1989 eruptions have been less frequent and apparently independent of Labial. The geysers are seen only one or two times per year. Action usually starts in the East Satellite, where bursting play 6 feet high may last as long as 15 minutes. The West Satellite often, but not always, joins its neighbor after a few minutes. It is the taller of the two, jetting steamy spray as high as 10 feet. West Satellite has rare independent eruptions, too.

UNDERHILL SPRINGS

A number of nearly extinct springs dot the ground about ¼ mile west of the Pink Cone Group. Without explanation, one of these—which one is unclear, but it may be the intermittent steam vent described later—was given the name "Underhill Spring" on a map published in 1904. The U.S. Geological Survey adopted that name for the entire group during the mapping project following the 1959 Hebgen Lake earthquake. Some observers have continued to use that name for the one geyser in the area, but that is officially considered an unnecessary and unacceptable duplication of the name. The simplest access to the area is by walking through the woods from near the end of Firehole Lake Drive (Map 6.4, Table 6.3, number 39).

39. "YF-305," a water sample site for the U.S. Geological Survey, is the one geyser among the Underhill Springs. The eruptive activity was "discovered" in 1999; how long it might have been active before then is unknown, as previously there had been little reason to visit the area. "YF-305" is one of those uncommon geysers in which the durations are longer than the intervals. Most play lasts 6 to 10 minutes, while the intervals of complete quiet are as short as 2 minutes. Overall, "YF-305" is in eruption about 60 percent of the time. The action is a vigorous series of bursts that reach 2 to 6 feet high.

A short distance east of "YF-305" is an old geyserite formation topped by debris-filled vents that act as intermittent fumaroles. The formation lies immediately next to a hillside, and some observers feel this is the original "Underhill Spring."

BLACK WARRIOR, OR FIREHOLE LAKE, GROUP

The Black Warrior Group (Map 6.5, Table 6.4, numbers 40 through 44) is an area of very high water output. Total discharge is about 3,500 gallons per minute, which flows from the area through Tangled Creek. Most of the hot springs are large, quiet pools. The biggest is Firehole Lake, east of the road. Supplied with water through several vents, its temperature is about 160°F (70°C). The water flows from there through Black Warrior Spring, into Black Warrior Lake, and then to Hot Lake. Only a collecting basin along the runoff, Hot Lake has a few hot springs around its eastern perimeter but is not actually a spring itself.

The reason for the odd name "Black Warrior" was never explained but surely has to do with the mineral deposits within the group. The springs around Firehole Lake have produced some unique formations. Many are dark in color, with craters coated by heavy, powdery deposits of black manganese oxide minerals. This is not so unusual by itself, but nowhere else in Yellowstone is so much

Table 6.4. Geysers of the Black Warrior (Firehole Lake) Group, Lower Geyser Basin

Name	Map No.	Interval	Duration	Height (ft)
Artesia Geyser	42	steady	steady	3–10
Gray Bulger Geyser	41a	near steady	near steady	1–6
Primrose Springs	44	frequent*	seconds	1–10
Steady Geyser	40	near steady	near steady	1–30
Sulfosel Spring	43	[1959]	unrecorded	unre-corded
Young Hopeful Geyser	41	near steady	near steady	1–6

* When active.
[ ] Brackets enclose the year of most recent activity for rare or dormant geysers. See text.

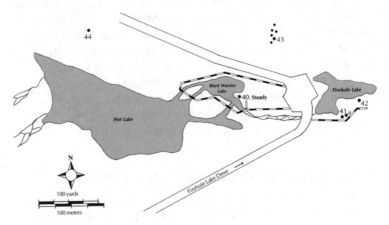

Map 6.5. *Black Warrior (Firehole Lake) Group, Lower Geyser Basin*

deposited as here. Also, this is the only geyser group in Yellowstone where travertine, a form of calcium carbonate, is deposited along with siliceous sinter. Young Hopeful Geyser deposits crystalline calcite within its vent, Steady Geyser forms travertine "geyser eggs," and part of Firehole Lake is bordered by low travertine terraces.

The Black Warrior Group contains five geysers plus well-named Steady Geyser, a large perpetual spouter that proved to be a periodic geyser during 2004.

40. STEADY GEYSER, also called **BLACK WARRIOR GEYSER,** used to be the largest perpetual spouter in Yellowstone. In recent years it has declined to a much smaller size, and in June 2004 it was seen completely quiet for the first known time. Steady has two vents that alternate between eruptive and dormant periods. The shifts of energy between these closely spaced openings are very slow, sometimes taking several years to complete. The top vent spouts straight up, in good times reaching 30 feet high. The other vent plays at an angle and may be 12 feet high. It is the lower vent that has been active for most of the past twenty years, with only a bit of splashing seen in the upper vent. The current activity, playing from both vents, is less than 3 feet high.

41. YOUNG HOPEFUL GEYSER, named in 1872 for uncertain reasons, used to erupt through eleven separate vents. Most of the time the play was only 2 to 6 feet high, but infrequently Young Hopeful erupted with considerably more force, reaching 15 feet in 1878 and as high as 20 feet during 1939. It also had frequent dormancies lasting days to months. About 1975, a steam explosion converted the vents into two larger craters. Most of the splashing, which is nearly perpetual, is just 1 to 2 feet high.

41a. GRAY BULGER GEYSER. There has been considerable confusion as to which feature should bear what name along the south shore of Firehole Lake. A careful consideration of the historical records leads to these conclusions. The name "Young Hopeful Geyser" originally applied to only some of its eleven original vents. Those nearer the break in slope and closest to the boardwalk separately comprise Gray Bulger Geyser, which in 1872 was called "Minute Spouter." Given this conclusion, Gray Bulger has been active throughout most of Park history, playing sometimes as a geyser and otherwise as a perpetual spouter 1 to 3 feet high—until 1975. In that year some small steam explosions enlarged the vents of it and its immediate neighbor, Young Hopeful (41). Gray Bulger began having explosive eruptions to heights as great as 25 feet. That kind of action continued through 1977. Gray Bulger has been generally weak since then, but play 6 feet high is still seen on occasion.

42. ARTESIA GEYSER was named following the 1959 earthquake, when it began playing as a perpetual spouter. It has sometimes incorrectly been called Gray Bulger, but there is no record of it erupting before 1959. In 1975, at the same time explosions and increased activity began in Young Hopeful (41) and Gray Bulger (41a), Artesia also began to have powerful eruptions. They recurred every few seconds and lasted only seconds, but the water was jetted as high as 25 to 30 feet. A second vent, barely noticed before, began playing almost horizontally, squirting outward as far as 20 feet. After a short dormancy in 1977, Artesia resumed activity. The present play is perpetual and usually less than 5 feet high.

See the Gray Bulger entry, #41a, for more about the confusion among the names here. In 1975, a visitor walking on the boardwalk was severely burned by a 25-foot eruption. The official reports cite Gray Bulger, but the event actually happened at Artesia. The boardwalk is now located farther from the spring.

43. SULFOSEL SPRING is one of several small, nearly insignificant springs on the slope north of the road known collectively as the Sulfosel Springs. Although no details were recorded, Sulfosel was listed as an active geyser for a short time following the 1959 earthquake.

44. PRIMROSE SPRINGS (a single pool whose name is official in the plural form) had its only known vigorous activity during short periods following the earthquakes of 1959 and 1983. In both cases the infrequent and brief eruptions reached as high as 10 feet. In 1988 and since 1994, Primrose has marginally acted as a very small geyser, with intermittent bubbling that sometimes splashes up to a foot high. Closer to the road near Primrose is Fissure Spring, a pool with wide shelves colored by cyanobacteria. The long narrow spring a short distance farther along the road is Dart Spring.

FOUNTAIN GROUP

The Fountain Group (Map 6.6, Table 6.5, numbers 50 through 73) is one of the largest single collections of geysers in the Lower Geyser Basin. Several are large, and most are connected as members of the Fountain Complex. Geysers notwithstanding, the most popular attraction here is the Fountain Paint Pots, the largest easily accessible mud pots in the Park. Since the 1959 earthquake increased their activity, the basin in which they lie has been considerably enlarged. At one time the expansion threatened to engulf walkways and roads, and concrete remnants of the old construc-

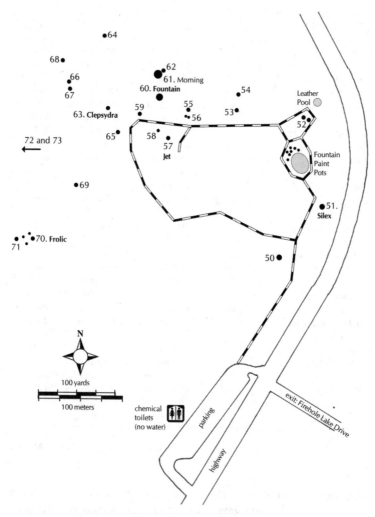

Map 6.6. *Fountain Group, Lower Geyser Basin*

tions can be seen overhanging the mud at several points near the modern boardwalk.

The Fountain Group is traversed by a boardwalk, with access from a large parking lot. Leaflets are available for self-guided tours of the half-mile loop trail, and they point out many of the fascinating aspects of the area in addition to the geysers and mud pots.

Table 6.5. Geysers of the Fountain Group, Lower Geyser Basin

Name	Map No.	Interval	Duration	Height (ft)
Bearclaw Geyser	56	min–hrs	sec–min	inches–1
Celestine Pool	50	frequent	sec–min	1–4
Clepsydra Geyser	63	see text	near steady	25–45
Fitful Geyser	67	seconds	seconds	5–7
Fountain Geyser	60	3½–11 hrs	22–69 min	10–100
Frolic Geyser	70	5 min–hrs	1–2 min	8–50
Jelly Geyser	65	rare	seconds	3–12
Jet Geyser	57	see text	1 min	20
Mask Geyser	72	10–50 min*	sec–min	6
Morning Geyser	61	[1994]	10–32 min	150–200
"Morning's Thief"	62	w/ Fountain	seconds	3–50
New Bellefontaine Geyser	66	seconds	seconds	10–20
Old Bellefontaine Geyser	69	10–60 min*	seconds	2–6
Old Cone Geyser	54	rare	hours	1–5
Red Spouter	52	steady	steady	4–8
Silex Spring	51	[2006]	minutes	10–35
Spasm Geyser	59	hours	min–hrs	1–25
Sub Geyser	64	rare	seconds	sub–10
Super Frying Pan ("Sizzler")	58	3–6 hrs	10–17 min	3–15
Twig Geyser	55	2–3 hrs	see text	2–10
UNNG-FTN-1	53	[1986]	2 min	1
UNNG-FTN-4	71	infrequent	seconds	10
UNNG-FTN-6 ("Stalactite")	68	1–8 min	seconds	12–15
UNNG-FTN-7	73	11–23 min	5 min	10

* When active.
[ ] Brackets enclose the year of most recent activity for rare or dormant geysers. See text.

50. CELESTINE POOL is the closest spring to the parking area. A relatively large pool, it is a geyser whose eruptions consist of frequent superheated boiling. Celestine appears to have an underground connection with Silex Spring (51) because when that geyser is active, Celestine can burst as high as 4 feet.

51. SILEX SPRING is one of Yellowstone's prettiest pools. It fills a deep crater below the boardwalk. The overflow runs across a wide area and supports a profuse growth of multicolored cyanobacteria. Silex has always been observed to undergo boiling periods with consequent heavier overflow, but true eruptions have been uncommon. They were first seen during 1946 and 1947 and then for a while following the 1959 earthquake, when several days of steady surging tossed water 10 feet into the air. No further eruptions took place until 1973, when Silex began an active phase that persisted into 1979. Silex was completely dormant from 1979 until August 2000. It then erupted erratically until vigorous action occurred in 2005 and 2006. Those eruptions were cyclic, with active episodes repeating as often as every 1½ days. An initial eruption would begin from a full pool with little warning. Although heights of 10 to 20 feet were most typical, some bursts could reach higher than 35 feet throughout durations of 6 to 10 minutes. This major eruption would be followed by a series of brief minor plays as high as 10 feet, sometimes interrupted by as many as three additional majors over the course of about 3 hours. Following any eruption, the crater drained with an impressive whirlpool. Unfortunately, the last eruption series was on December 28, 2006, and Silex is apt to lie quiet for years before another episode of eruptions begins.

52. RED SPOUTER (actually two adjacent craters) is a direct product of the 1959 earthquake—this was a flat, grassy area until steam vents formed along a fracture created by the shocks. The vents soon developed into mud pots. Now, during seasons when the water table is high or following heavy rains, the two vents act as perpetual spouters, throwing muddy, red-brown water up to 8 feet high. At low water, as is typical of midsummer, Red Spouter reverts to noisy steam vents. In reality, then, Red Spouter is a pair of fumaroles occasionally drowned by surface water; they are not true geysers.

The large pool across the boardwalk north of Red Spouter is Leather Pool, which apparently had a few "geyser-like" eruptions during the first week after the 1959 earthquake. The activity ceased as soon as Red Spouter developed into a distinct spring.

53. UNNG-FTN-1 plays from an old rift. Until eruptions began in 1985, the crack was filled with soil and grew lush grasses and wildflowers aided by a bit of steam heat and moisture. The activity revealed a series of vents that eventually enlarged into a string of small craters. The play was a series of splashes, few of which reached more than 1 foot high. The best activity was in 1986, when intervals were 10 to 15 minutes and durations about 2 minutes long. FTN-1 has been inactive since that year, and the craters are again largely filled with soil.

54. OLD CONE GEYSER rises from a jagged vent on the near (south) flank of Old Cone Spring, a nearly extinct geyserite mound named in 1878. One of Old Cone Geyser's best summers on record was 2005, when it was seen several times, but it often goes through an entire year without any reported activity. The splashing play 1 to 5 feet high lasts several hours when it does take place.

Fountain Complex (numbers 55 through 68)

The Fountain Complex probably merits the term "complex" more than any other group of springs. Every one of its fourteen members is a geyser, and the far-reaching subterranean relationships produce extensive exchanges of function. The activity of everything within the complex is dependent on the current behavior of all the other geysers. Three—Fountain, Morning, and Clepsydra—are included in any list of Yellowstone's most important geysers.

On the hill above the Fountain Complex, a short boardwalk spur serves as the "Fountain Overlook." The entire complex can be seen from there, as can the distant Kaleidoscope and Sprinkler groups.

There is often a well-established pattern of behavior among these springs. The patterns vary over the years, but here is a typical scenario: late in Fountain's eruption or just after it quits, the entire complex will be quiet. Even normally constant Clepsydra may pause for a short time, but it will restart within a few minutes and will then be the only active geyser for much of Fountain's interval. As the system begins to recover, Twig may have brief eruptions. One to 2 hours before the next Fountain, Spasm sometimes startles people when it begins to play, with 20-foot jets immediately next to the boardwalk. Less than an hour before Fountain, eruptive frequency will pick up. Twig may play with only brief pauses and is frequently joined by Bearclaw. Jet will begin erupting with intervals that grow progressively shorter. A few minutes before Fountain, Super Frying Pan often starts, and even distant Sub may play higher than ground level. Then Fountain begins to play. Jet erupts every 2 to 4 minutes. Clepsydra loses some of its water, so one of its vents becomes a roaring steam jet. Twig, too, may lose water so as to noisily send steamy jets over 20 feet high. Then Fountain quits and the cycle begins anew.

This scenario should be taken only as a guide to the kinds of relationships that exist here. There can be considerable variation from year to year. For example, in 2003 Spasm served as a reliable indicator for Fountain, beginning to play 30 to 40 minutes before Fountain; in 2004 that relationship failed, as Spasm sometimes erupted nonstop throughout Fountain's interval, and in 2005 it sometimes failed to play at all during the interval. The Fountain Complex is an extremely dynamic set of geysers. Dramatic changes can happen overnight, and the reader should understand that the descriptions here relate only to the activity typical of the past few years.

55. TWIG GEYSER lies near the foot of the stairway leading down from the overlook. To the right of the boardwalk, the shallow crater is about 4 feet in diameter. The play consists of splashes a few feet high, with occasional jets that reach up to 10 feet. Twig's intervals depend strongly on the activity of the rest of the Fountain Complex, especially on whether it is Fountain (60) or Morning (61) Geyser that is active. Twig is generally quiet for the first part of Fountain's interval. It then begins to play erratically but increasingly during the last 1 or 2 hours before Fountain. Then Twig can be in eruption 50 percent of the time. It continues to erupt throughout Fountain's activity and frequently jets as high as 20 feet as it nearly goes into steam phase near the end of some eruptions by Fountain.

Twig's behavior was quite different during the active phases by Morning in 1991, when it was known to go as long as 14 hours between short series of brief eruptions.

56. BEARCLAW GEYSER consists of the small vents between the boardwalk and Twig Geyser (55). Gurgling sounds are more likely to attract attention than are Bearclaw's eruptions, most of which reach only a few inches high. In general, it is most active when Fountain is approaching or is in eruption, and therefore when Twig is also active. During these times the play is erratic, turning on and off every few seconds to minutes.

57. JET GEYSER erupts from an elongated cone that appears to have developed along an old fracture in the sinter. This same break may extend through Super Frying Pan (58), Spasm (59), Clepsydra (62), New Bellefontaine (66), and Fitful (67) geysers. Jet's cone proves it has had a long history, but it apparently was virtually dormant during the first fifty-plus years of Park history, when only one eruption (in 1886) was described prior to 1927. A few dormant periods have been recorded since then, too, but it is normally active under the control of the nearby geysers.

During the first few hours following an eruption by Fountain (60), Jet is quiet and often remains so until 1 or 2 hours before the time of the next Fountain. Then Jet begins an active series in which eruptions recur every several minutes. In general, the intervals grow shorter as the time for Fountain approaches. Jet is most vigorous while Fountain is in eruption, with intervals as short as 2 minutes. Most durations are shorter than 1 minute. There are several vents within the cone. One of the spouts reaches vertically as much as 20 feet high, while some of the others are nearly horizontal.

On the rare occasions when Morning Geyser (61), rather than Fountain, is active, Jet behaves in a very different fashion. There is no extended quiet period. Instead, Jet remains active throughout Morning's interval, but with longer intervals and durations.

58. SUPER FRYING PAN is a poor (but unfortunately official) name, because a true frying pan is a type of acid hot spring vastly different from a geyser. It has also been called "Sizzler Geyser." Super Frying Pan plays from several vents and cracks in the geyserite that probably formed at the time of the 1959 earthquake. The first slight activity was seen in 1961 and more in 1964, but the geyser did not break out in distinct form until 1975. The various openings are gradually merging to form a single distinct crater, and consequently the eruptions are developing a bursting and jetting action. The height is as great as 15 feet for durations of 10 to 17 minutes.

Super Frying Pan's relationship to the other geysers of the Fountain Complex varies. Typically, though, it has intervals of 3 to 6 hours, so it plays at least once and often twice during Fountain's intervals.

59. SPASM GEYSER's ragged, jerky pattern of play fits its name fairly well—except that the name was originally applied to today's Jelly Spring (65) and this was the original "Jet Geyser." During its early history, Spasm sometimes reached as high as 40 feet, with an appropriately jet-like column of water. In 1963 a steam explosion occurred during an eruption. The old crater was enlarged, a new vent was added to the system, and the nature of the play changed. As a member of the Fountain Complex, Spasm is variable in its performances, which seem to be somewhat different every year. The play always begins abruptly, with initial bursts sometimes over 20 feet high—rather startling to people standing at the boardwalk railing immediately next to the crater—but it then rapidly dies down to surging that is only 1 to 5 feet high. The eruptions, once started, used to continue until about the time Fountain (60) quit its next eruption, which means the duration could be as long as several hours. However, play in 2006 and 2007 often was as short as 10 minutes and could recur several times during Fountain's interval.

60. FOUNTAIN GEYSER has long been considered the major geyser of the Fountain Group. Morning Geyser (61) can be larger, but it is seldom active. Fountain's pool is the nearer of the two beyond the name sign. Broad and deep, it is a rich azure-blue color. The high sinter shoulders about the crater suggest that Fountain has been active for a very long time.

The pool is calm throughout the long quiet period, and an eruption begins with little warning other than a sudden rise in the water level. Huge bursts propel water from the crater, sometimes appearing as gigantic "blue bubbles" that explode to throw water in all directions. The play is often as wide as it is high. Most bursts are 10 to 20 feet high, but some taller than 40 or 50 feet are common, and a few over 80 feet high are seen during many eruptions. Fountain was at its very best in 1991 when it erupted in concert with Morning. Some of that play exceeded 100 feet high.

Interesting patterns often show up in Fountain's activity. In many years it is regular enough that it could be predicted, except for one problem. Since 2000 the intervals have made abrupt switches, alternating between short and long modes. The short mode intervals average 3½ to 5 hours and the long mode 6 to 7½ hours. (Historically, similar modes tended to fall at 5½ hours and 11 hours, but the 11-hour mode has not been seen for several years.) The switches between modes take place without transition, and there is no warning that a switch is about to occur. Whichever mode is in force, the typical duration is 25 to 32 minutes.

"Wild-phase" activity, in which Fountain plays only 4 to 10 feet high nonstop for as long as two weeks, is quite rare. It was observed twice during the spring of 2000. Similar action but with durations

Fountain Geyser is the dominant member of the Fountain Complex, but its eruption frequency can change abruptly because of the many subsurface connections with other members of its complex.

of only a few hours occurred in the summers of 2001 and 2004 and three times during 2005. Wild phases are apparently triggered when a normal-looking eruption begins after an exceptionally short interval of only about 2½ hours and has a duration as short as 3½ minutes.

61. MORNING GEYSER is one of the most powerful geysers in Yellowstone, and its eruptions are often more spectacular than anything seen in the Park since the days of Excelsior Geyser (see #16 of the Midway Geyser Basin). Tremendously explosive bursts can reach 150 to 200 feet high and may spread 60 to 100 feet wide. Unfortunately, though, Morning is seldom active.

The first recorded eruptions of Morning were in 1899, when it was called "New Fountain Geyser." That active phase lasted about three months. The second period of action was even shorter, spanning two months in 1909. There was 1 eruption during 1921 and another in 1922. Nothing more was seen until 1945, when renamed "Fountain Pool" began an active phase that ultimately produced 62 observed eruptions through 1949. Morning proved to be one of those geysers where strong wind (common in Yellowstone's after-

Concerted (that is, simultaneous) eruptions by Fountain Geyser (right) and Morning Geyser are extremely rare, having been seen only during the first two days following the 1959 earthquake and five times during 1991. Credit, Photo by Tom and Genean Dunn.

noons) blowing across a large surface pool will delay or eliminate eruptions. The majority of those eruptions occurred during the morning hours, so "Fountain Pool" became Morning Geyser.

After a dormancy, Morning was irregularly active from 1952 until the time of the 1959 earthquake. That jolt caused a shift in the energy flow of the Fountain Complex. Clepsydra Geyser (63) grew more vigorous, and Morning's activity became weaker. Morning was dormant within three weeks of the earthquake. It did not erupt again until 1973, when there were several eruptions, some of which were only about 50 feet high. Additional active episodes that approached the power of old occurred during 1978, 1981, 1982, and 1983.

None of this matched the unprecedented action seen in 1991. After eight years of dormancy, Morning had 5 eruptions in early May. Then, on July 4 and 5, there were 2 eruptions. Both of these were in concert with Fountain Geyser (60), an event previously seen only immediately following the 1959 earthquake. After another brief dormancy, Morning rejuvenated on August 9, and during the next twenty days there were at least 118 eruptions. Morning was regular enough that informal predictions based on average intervals of 3¾ hours were posted at the parking lot. The durations were as long as 32 minutes. The first 1 and last 2 of these eruptions were in concert with Fountain, as had happened in July.

That active phase ended on August 29, 1991, and Morning was dormant until late March 1994. On a few occasions during those years there were "Fountain stalls" in which Fountain Geyser had extraordinarily long intervals and the water level rose in Morning. The effect was that the system was "trying" to shift energy to Morning, but it might have been years before it finally did so had it not been for an earthquake. A tremor of magnitude 4.9 on March 26 was centered a few miles northwest of Madison Junction. Morning rejuvenated on either March 30 or 31, 1994. The activity resembled that of August 1991, with some intervals shorter than 6 hours. That active episode lasted only a week, and, with a single exception, Morning has been dormant since 1994. Morning did have a "minor" eruption just 30 feet high for a duration of 1½ minutes on July 6, 2006; this took place as the first event within the Fountain Complex after an exceptionally long, 69-minute-duration eruption by Fountain. (Morning was said to erupt on March 2, 2006, and May 13, 2007, when play 100 feet high was reported, but it is likely that those eruptions were actually extraordinarily powerful play by "Morning's Thief" [62].) As a major member of the Fountain Complex, Morning is severely affected by any other activity in the group. On most occasions, apparently too much water and energy are lost from the system through other hot springs, and Morning is a rare sight.

62. "MORNING'S THIEF" erupts from a round opening at the far right (northeast) corner of Morning Geyser's (61) crater. It was named at a time when small, long-duration eruptions 3 to 8 feet high seemed to rob Morning of the energy needed for an eruption. That conjecture has never been proven. Much larger action has been seen during the 2000s, when "Morning's Thief" has often played during eruptions by Fountain Geyser (60). These eruptions seem to be growing stronger with each passing year, and in 2007 some bursts reached over 50 feet high for durations of a few seconds.

63. CLEPSYDRA GEYSER was named after a mythical Greek water clock because of its supposed regularity. At times it did erupt for a few seconds at intervals of almost exactly 3 minutes, but its history shows that there were many periods of erratic behavior and one of complete dormancy. Any regularity ended permanently at the time of the 1959 earthquake. In the few weeks following the tremors, Clepsydra gradually grew stronger and steadier and entered what is called "wild-phase" activity. This continued with few pauses into 1963. At that time, Fountain (60) reactivated from a dormancy, and Clepsydra would stop playing for a short time following each eruption by Fountain. When Fountain returned to dormancy in 1964, Clepsydra immediately resumed its nonstop wild phase. This kind of activity has characterized Clepsydra ever since—if Fountain is dormant, then Clepsydra rarely stops playing; if Fountain is active, Clepsydra may pause for a few minutes near the end of Fountain's eruption but seldom at any other time.

Clepsydra plays from several vents. The two largest open within a cone of geyserite stained a distinct yellow color. One of the vents jets as high as 45 feet, while the other reaches about 25 feet at a slight angle. A number of other openings splash a few feet high. The forceful wild-phase activity shows no signs of abating nearly fifty years after the earthquake. It must now be considered Clepsydra's normal activity.

64. SUB GEYSER plays within a deep crater out across the sinter flat northwest of Fountain Geyser (60). Eruptions are actually quite common, but nearly all are confined to the invisible depths within the crater, hence the name (abbreviated from "subterranean"). Exceptional bursts can reach more than 10 feet aboveground and therefore as much as 20 feet above the subsurface pool level. Very rare, such jets appear to take place only during the last few minutes before Fountain begins playing. In 1956, Sub briefly took over Clepsydra Geyser's (63) role within the Fountain Complex, playing as high as 20 feet aboveground while Clepsydra was dormant for the first known time. This gave Sub a second name, "Clepsydra's Thief." Sub still shows the relationship to Clepsydra by occasionally having brief eruptions large enough to be seen from the boardwalk at about the time Clepsydra quits during Fountain's play.

65. JELLY GEYSER erupts from a crater that measures 16 by 30 feet but is mostly shallow, so the water is a pale greenish-blue. Prior to the 1959 earthquake Jelly was often active, with intervals as short as 10 minutes. Most of the play lasted only a few seconds but reached 3 to 12 feet high. For the last several years, though, Jelly has usually been dormant. Showing its relationship to Morning Geyser (61), Jelly played a number of times while Morning was active in 1991, had two additional eruptions during "Fountain stalls" in 1992, and was seen a handful of times during 1993. Most of these eruptions consisted of only two or three quick bursts less than 5 feet high. Similar eruptions were seen during an eruption of Fountain (60) in June 2004 and following the questionable eruption of Morning on May 13, 2007.

66. NEW BELLEFONTAINE GEYSER is located directly beyond Clepsydra Geyser (63), at the visible end of the rift that starts at Jet Geyser (57). A very active geyser, it repeats its fountain-like play at intervals seldom longer than a few seconds. The durations are equally short. The height is as great as 20 feet.

67. FITFUL GEYSER lies on the near side of New Bellefontaine Geyser (66). It usually passes unnoticed or is mistaken as part of New Bellefontaine. Both the intervals and durations are seconds long. The height is 5 to 7 feet. Fitful and New Bellefontaine are the most visible members of the nine geysers and spouters among the Gore Springs, a separately named part of the Fountain Complex mostly hidden from view down the slope to the west.

68. UNNG-FTN-6 ("STALACTITE GEYSER") is the only named member of the Fissure Springs, a fracture-controlled group that lies hidden down the west-facing slope below the similar Gore Springs (see #67). Every few minutes "Stalactite" sends a few jets high enough to be visible from the boardwalk. Even then, only the top of the eruption is actually seen, and the full height is as great as 12 to 15 feet. The brief durations are separated by intervals of 1 to 8 minutes.

69. (OLD) BELLEFONTAINE GEYSER is not fitting of the name, "beautiful fountain," which was originally applied to another member of the Fountain Group (perhaps Mask Geyser, #72). Active periods are rare, and years sometimes pass between them. Intervals then range from 10 to 60 minutes during episodes that may last a few days. However, most eruptions are solo and can be so brief as to consist of a single splash of water less than 6 feet high. Many such eruptions might therefore go unnoticed.

70. FROLIC GEYSER was first recorded in 1964. It has probably been active in every year since then, but its performances are variable. Only a few eruptions are recorded in some seasons, while at other times the intervals can be as short as 5 minutes. The play typically lasts less than 40 seconds and reaches 8 to 10 feet high. During extraordinary action seen only during the early 1980s, steady jets sometimes exceeded 50 feet high in eruptions that lasted several minutes.

71. UNNG-FTN-4 plays out of a jagged crater a few feet southwest of Frolic Geyser (70). Rather uncommon, the eruptions last only seconds while spraying up to 10 feet high. This area apparently includes at least two additional geysers among its complex of jagged vents, but their activity is rare and little is known about them except that they play only a few feet high.

72. MASK GEYSER is a pool near the western base of the geyserite mound of the Fountain Group. The largest of the Pithole Springs, it is actually one of Yellowstone's most beautiful pools and may be the original "Bellefontaine Geyser" and, before that, the original "Jelly Spring." Unfortunately, it is only marginally visible from the distant boardwalk. Mask's eruptions are quite variable. The intervals commonly range between 10 and 50 minutes. The play lasts a few seconds to a minute or two and reaches 10 feet high. Dormancies are known.

73. UNNG-FTN-7 is still farther out across the sinter flats west of the Fountain Group and beyond Mask Geyser (72), where there are several additional springs. At least two of these have been seen as geysers. The more active is an unnamed pool (FTN-7) that is visible from the distant boardwalk only when it is in eruption; otherwise, the crater appears empty. It plays every 11 to 23 minutes. The duration is about 5 minutes, and the eruption reaches as high as 10 feet. The other geyser in the area is less regular and perhaps 4 feet high.

KALEIDOSCOPE GROUP

The Kaleidoscope Group (Map 6.7, Table 6.6, numbers 75 through 85) is a relatively compact set of hot springs. Its most obvious feature

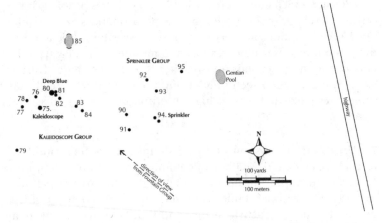

Map 6.7. *Kaleidoscope and Sprinkler Groups, Lower Geyser Basin*

is the large pool of Deep Blue Geyser, far across the sinter flats northwest of the Fountain Group. The group contains at least twelve active geysers plus numerous perpetual spouters. The area is known for exceptionally large geysers, some of which play from explosion craters that have formed during recorded history. That is part of the reason no trail leads into the area. The closest public approach and best view is from the "Fountain Overlook" within the Fountain Group, hundreds of yards away, and identification of the geysers from there is difficult even for experienced observers. To give a sense of scale, as viewed from the Fountain Overlook, if an eruption reaches as high as the top of the distant ridgeline, it is about 90 feet tall.

75. KALEIDOSCOPE GEYSER is the largest frequently active geyser of its group, but even it undergoes dormant periods. Historically, Kaleidoscope has been cyclic in its performances, with several hours of quiet passing between the eruptive series. Although barely visible at a distance, the later stages of the interval are marked by intermittent boiling in Kaleidoscope and small eruptions in a nearby feature called "Kaleidoscope's Indicator." The initial eruption is always Kaleidoscope's largest. Without apparent warning, explosive jets of water are rocketed to between 50 and 120 feet high for as long as 2 minutes. Subsequent eruptions in a series generally recur on intervals of only 1 to 3 minutes, last less than 1 minute,

Table 6.6. Geysers of the Kaleidoscope and Sprinkler Groups, Lower Geyser Basin

Name	Map No.	Interval	Duration	Height (ft)
Angle Geyser(s)	90	see text	sec–min	10–20
Blowout Spring	78	rare	2 min	5–40
Bridge Geyser	93	min–hrs	minutes	1–10
Deep Blue Geyser	80	6–90 min	seconds	3–40
Drain Geyser	76	minutes*	seconds	5–150
Earthquake Geyser	95	steady	steady	2–3
Ferric Geyser	92	steady	steady	1–8
Firehose, The	82	hrs–months	hrs–months	5–30
Honey's Vent	84	15–20 min	7–42 min	3–30
Honeycomb Geyser	83	hours–days	12–45 min	10–100
Impatient Miser Geyser	91	frequent	sec–min	5–25
Kaleidoscope Geyser	75	1–2 min*	20 sec–2 min	1–120
NTFL	79	3½–11 min	8–12 sec	30
Old Surprise Spring	85	[1880s?]	unrecorded	boil–150
Sprinkler Geyser	94	minutes*	1–2 min	8–15
"Three Vent Geyser"	77	3–30 min*	3 min	10–50
UNNG-KLD-12	81	minutes*	10–30 sec	50–200
UNNG-KLD-13	81	minutes*	seconds	30–75
West Sprinkler Geyser	94	5–12 min*	seconds	10

* When active.
[ ] Brackets enclose the year of most recent activity for rare or dormant geysers. See text.

and reach 40 to 70 feet high. A complete Kaleidoscope cycle may consist of anywhere from just one eruption (only the initial, as has been the general rule during the 2000s) to as many as thirteen eruptions. The final play of an extended series may last as long as 90 minutes but reaches only 1 to 20 feet high.

76. DRAIN GEYSER was named because most of the runoff from Kaleidoscope Geyser (75) runs into its crater. It is dormant on most occasions when Kaleidoscope is active. The overall activity is similar to Kaleidoscope's, being cyclic with pauses of several hours leading to series of eruptions only a few minutes apart. Like Kaleidoscope, more common during the 2000s have been single eruptions at erratic intervals of several minutes to 2 hours. Most of Drain's eruptions are 40 to 80 feet high, but it has been known to exceed 150 feet.

77. "THREE VENT GEYSER" plays from one of three openings within a single large crater a short distance beyond Kaleidoscope Geyser

Drain Geyser is one of several large pools in the Kaleidoscope Group. Although its eruptions are usually fairly small, it can reach over 100 feet high.

(75). The three vents are independent geysers, only one of which is at all frequent and large. At times, it has been the dominant geyser of the Kaleidoscope Group, and when it is active the only time it pauses is for a few minutes after a series by Kaleidoscope. "Three Vent's" intervals can be as short as 3 minutes, with durations fully as long. The height is usually 10 to 15 feet, but bursts as tall as 50 feet have been seen. Long dormant periods are known, as has been the case since 2003.

78. BLOWOUT SPRING is a rare performer. The crater is located at the edge of a large depression occupied by Kaleidoscope (75) and Drain (76) geysers, with which it can be confused since both are rendered dormant by Blowout's brief active phases. Eruptions consist of individual bursts of water several seconds apart distributed over a total duration of about 2 minutes. Some of the play reaches 40 feet high.

79. NTFL made its appearance as a prominent geyser during 2001. The name, now considered entrenched in use, stands for "New Thing Far Left." It is the southwestern-most geyser in the

Kaleidoscope Group, farthest to the left when viewed from the Fountain Group; people driving southbound on the highway across Fountain Flat see it on the far right of the group. The large jagged crater apparently blew out in a steam explosion sometime in early 2001. The geyser erupts from a water level 15 feet below ground level. Since the height reaches 30 feet aboveground, the full eruption is actually 45 feet tall. Intervals typically fell between 3½ and 11 minutes, while the durations often were shorter than 10 seconds, until the spring of 2007, when NTFL regressed to infrequent eruptions. Most of these were subterranean, so only intermittent puffs of steam were visible from a distance.

80. DEEP BLUE GEYSER, the largest spring in the Kaleidoscope Group, is very well named. The main crater measures 30 by 40 feet and is surrounded by an extensive area of shallow water. Deep Blue has what have been termed major and minor eruptions. It used to be felt that the majors culminated some kind of series and recurred about once per day. It is now known that the "majors" are really just random cases of extraordinarily large bursts. The intervals range from 6 to 90 minutes, but at times they are more regular within a span of 20 to 50 minutes. The eruptions involve individual bursts of water, sometimes just one but occasionally as a series that continues for several minutes. Most of the splashes are 3 to 15 feet high; extraordinary "blue bubble eruptions" can reach over 40 feet high and are equally wide. Deep Blue evidently can play much higher— washed areas and gravel berms far beyond the pool imply that rare eruptions never actually observed must exceed 100 feet high.

81. UNNG-KLD-K12 and **UNNG-KLD-K13** are huge geysers that erupt from vents separated by only 4 feet. They actually lie slightly within the east (or near) side of Deep Blue Geyser's pool (80), and the easiest way to distinguish the three is by the shapes of their eruptions: Deep Blue bursts from huge domes of water, K12 has a vertical column, and K13 has angled jets. With rare exceptions, K12 and K13 can erupt only when nearby Firehose (82) is not playing.

KLD-12 is the larger but less active of the two geysers. When active, it may have several eruptions over the course of a few minutes. Each eruption lasts 10 to 30 seconds and typically reaches 50 feet high. Exceptional bursts may be higher than 140 feet, and one in May 2004 clearly exceeded 200 feet tall. These jets tend to be massive, vertical columns of water.

KLD-13 often undergoes series of eruptions throughout an inactive period of Firehose. Much more common than those of KLD-12, these series can repeat as often as every 20 to 40 minutes, during which there may be numerous individual eruptions that last a few seconds each. The eruptions are fan-like, with numerous jets that reach as high as 30 to 75 feet at an angle toward the south.

82. THE FIREHOSE is controversial in its classification. It developed during 1988 along a fracture created by the 1959 earthquake. When Firehose is active, its play is steady jetting that reaches 30 to 45 feet high at a sharp angle toward the north. At these times, which may last for several months, Firehose behaves as Yellowstone's largest perpetual spouter. But sometimes it has geyser-like pauses. Quiet spans have been observed to range from just 9 minutes to nearly 2 months. This leads to the unanswerable question: Is Firehose an "intermittent perpetual spouter" or a "long-duration geyser"?

83. HONEYCOMB GEYSER, named in allusion to the decorative geyserite about its crater, is little-known. It erupted 10 to 30 feet high about twice per day for a short time following the 1959 earthquake but was mostly dormant until 1987. For the next two years, eruptions were erratic in time but major in scale. Lasting about 12 minutes (rarely as long as 45 minutes), the play was a violent surging and boiling mass of water that domed up to 50 to 70 feet high and was topped by some jets taller than 100 feet. Such eruptions have been rare since 1989. Now, several hours to many days pass between eruptions. These eruptions consist of "lazy" bursts reaching 10 to 30 feet high for durations of about 12 minutes that only infrequently culminate in a full major eruption.

84. HONEY'S VENT was created in 1960 when a steam explosion reopened an old crater. For its first decade of life it behaved as a geyser with both major and minor eruptions, but by about 1970 it had become a perpetual spouter. Most of the play was confined to the crater, but occasional surges produced jets as much as 12 feet high. The only time it would pause was for a few minutes following the rare major eruptions by nearby Honeycomb Geyser (83). More recently, Honey's Vent has again acted as a true geyser. Most intervals are between 15 and 20 minutes long. The durations range between 7 and 42 minutes, placing Honey's Vent into that category of uncommon geysers in which the durations are often longer than the intervals. The height reaches 3 to 15 feet above the pool, which normally lies several feet down inside the crater. Perhaps one or two times per day, the eruptions develop a larger scale that lasts as long as 20 minutes and repeatedly bursts water as high as 30 feet.

85. OLD SURPRISE SPRING was briefly one of Yellowstone's star attractions in the early 1880s, when it was visible from a spot near long-gone Marshall's Hotel (one of the Park's first lodgings). The oval crater presently contains three vents. Whether all or just one of those openings was active in Yellowstone's early days is unknown. Eruptions during 1881 reached at least 150 feet high, and in 1883 geologists with the U.S. Geological Survey described it as the third largest geyser in the Lower Basin. Unfortunately, the activity then decreased until only intermittent boiling was seen by the 1890s. The crater of Old Surprise did contain a small geyser for a short time after the 1959 earthquake, but its eruptions were completely confined to within the crater. Now, one of the three vents within the crater contains a boiling pool, and occasional puffs of steam seen from the distant road imply that bursting geyser action takes place there.

SPRINKLER GROUP

The Sprinkler Group (Map 6.7, Table 6.6, numbers 90 through 95) contains hundreds of hot spring vents and at least twenty-three geysers. The Sprinkler Group constitutes the springs that lie to the north-northwest of the Fountain Group, to the right of the Kaleidoscope Group as viewed from the boardwalk. Marking the eastern limit of the Sprinkler Group is Gentian Pool, 88 feet long and one of the largest hot springs in Yellowstone. As with the Kaleidoscope Group, no trail leads to these springs, and identification of the individual geysers from a distance is difficult even for the most experienced. Accordingly, only six of the larger geysers are described here.

90. ANGLE GEYSER, at the far north end of a long fracture zone, used to play from one of about a dozen vents within a single large complex of craters that is still evolving—steam explosions blew out new craters in 2003 and early 2004. Several of the vents have been active as angled geysers. The original Angle played frequently up to 20 feet high. Other geysers within this complex have informally been called "Vertical," "Acute," "Obtuse," "Horizontal," and so on. Perhaps none of these geysers is active now, though, as the eruptive action has shifted to the new craters. The 2004 blowout especially has had bursting eruptions as high as 30 feet and nearly that wide. Yet another, apparently new vent produced cyclic eruptions during 2007, when it had series of jetting eruptions as high as 40 feet.

91. IMPATIENT MISER GEYSER must have the strangest hot spring name. It came about because the eruptions produce virtually no external discharge. The geyser is located on the same fissure that includes the complex at Angle Geyser (90), and the frequent activity lasts only a few seconds. The splashing play is mostly less than 10 feet high but can be punctuated by jets that reach 25 feet.

92. FERRIC GEYSER, surrounded by sinter heavily stained with iron oxide minerals, is the only geyser of the Sprinkler Group to have formed a distinct, although small, geyserite cone. For a time following the 1959 earthquake, Ferric played every hour as high as 25 feet. The activity has long since regressed into weak perpetual spouting just 1 foot high. Infrequent, completely irregular surges may send some spray up to 8 feet. These bursts last only a few seconds. A spring with similar activity, called "Tangerine Spouter," erupts from a vent near Ferric.

93. BRIDGE GEYSER is near Ferric Geyser (92) in an area dotted by myriad hot spring vents. Most of these craters are collapse features. At Bridge, a remnant of the original geyserite roof still arches across the opening. The intervals are extremely erratic, ranging from minutes to hours, perhaps on a cyclic basis. Most of the play is too small to be seen at a distance, but some eruptions believed to rise from Bridge can reach at least 10 feet high.

94. SPRINKLER GEYSER lies within an iron oxide–stained crater of a rich red-brown color. Like many of the geysers in this area, it is cyclic in its action, with periods of relatively frequent eruptions separated by dormancies. The play consists of sharp jets of water 8 to 15 feet high.

More impressive, perhaps, is West Sprinkler Geyser, which rises from the same crater complex as Sprinkler. It erupts with a high degree of regularity and frequency. Recent intervals have been 5 to 12 minutes long, with water bursting up to 10 feet high for durations of 5 to 20 seconds.

95. EARTHQUAKE GEYSER merits a place in this book because of its size. Its entire history as a significant geyser spanned two weeks following the 1959 earthquake. From a point along an old fracture, Earthquake Geyser began to erupt as high as 125 feet. A tremendous volume of water was discharged. The eruptions ended when a steam explosion created a fumarole just a few feet away. Earthquake soon reverted to perpetual spouting about 3 feet high, which continues.

THUD GROUP

Once known as the "Hotel Group" because of the site of the old Fountain Hotel on the hill next to the springs, the Thud Group (Map 6.8, Table 6.7, numbers 100 through 105) reverted to its original name when the hotel was razed in 1923. Although nearly all the springs of this group are geysers, they are generally of minor size, character, and activity. They received much attention while the hotel was in operation but little since then, and geyser action may be more common than is recognized here.

Many visitors to and employees of the hotel used these springs as trash receptacles. Especially damaged was Thud Spring, the closest member of the group to the building. In 1948, Ranger George Marler made an attempt to clean it; the results were later published in Marler's classic *Inventory of Thermal Features of the Firehole River Geyser Basins*. The amount of material removed was astonishing. Here is the complete list of recovered trash: 3 1-gallon crocks, a frying pan, a duster, 7 soda pop bottles, 4 quart whiskey bottles, several beer bottles, a cog wheel, bones, 1 penny, 2 Colorado tax tokens, a 40-gallon drum, 2 wooden kegs, a bath towel, a bath mat, 1 rubber boot, a raincoat, some screen wire, 2 bricks, 1 horseshoe, 16 handkerchiefs, a copper plate, a pitchfork, 1 ladle, a large piece of canvas, a stew kettle, a gunny sack, 17 tin cans, 1 napkin, 1 pie tin, 1 window sash, 2 drawer handles, 1 cooking fork, 2 cake molds, 1 broom, 1 porcelain plate, 1 china plate, a surveyor's stake, 2 wagon braces, a blue dishpan, 2 knives, 1 fork, 1 spoon, a cigarette pack, a

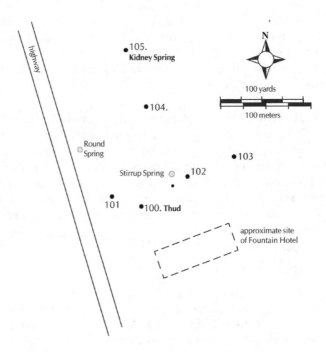

Map 6.8. *Thud Group, Lower Geyser Basin*

Table 6.7. Geysers of the Thud Group, Lower Geyser Basin

Name	Map No.	Interval	Duration	Height (ft)
Fungoid Spring (Thud Geyser)	101	[2005]	sec–5 min	2–12
Gourd Spring	103	[1988]	minutes	2–20
Kidney Spring	105	25 min	3–4 min	4
Oakleaf Spring	104	[2006]	unknown	2
Thud Spring	100	[2002]	2–4 min	2–15
UNNG-THD-1	102	5 min*	1½ min	1–7
UNNG-THD-2	102	[2001]	5 min	5

* When active.
[ ] Brackets enclose the year of most recent activity for rare or dormant geysers. See text.

1913 guidebook to Yellowstone, 2 marbles, 1 film box, 4 .22-caliber shells, 1 .45-caliber shell, 1 lightbulb, 1 apron, a large piece of pipe, a mixing bowl, a set of men's outer clothing, 1 butter tub, 1 kerosene lamp, 1 large copper lid, an oak evener, a single tree, several barrel staves, 2 ear tags for cattle belonging to a Rexburg, Idaho, rancher, several pieces of window glass, an oven rack, a cotton coat, miscellaneous pieces of iron, copper, and aluminum wire, paper, 1 Mason jar, 1 Vaseline bottle, *and* 1 seltzer bottle.

All of this was taken from a crater just 15 by 18 feet across and 12 feet deep. Please throw your trash into the real trash cans. After all the abuse experienced by Thud Spring, it's a wonder it was able to survive at all.

The Thud Group contains no trails. It is possible to view the springs from pullouts along the roadway or from the site of the old hotel. As is the case with the Kaleidoscope and Sprinkler groups, the intent is to keep this area inaccessible and therefore further untouched.

100. THUD SPRING, despite the treatment it received over the years, has rare active periods. However, because of confusing naming in this group (see Fungoid Spring, #101), the details are uncertain. When active, Thud Spring can play as often as every 3 to 4 hours, with most eruptions lasting 3 to 4 minutes and reaching 12 to 15 feet high. A weaker form of activity took place in December 2002, when 2-minute eruptions just 2 feet high recurred every 3 to 10 minutes.

101. FUNGOID SPRING has also been known as "Thud Geyser," and it is possible that at least some reports of eruptions by Fungoid, such as the play 2 feet high in July 1994, refer to Thud Spring (100) and vice versa. Fungoid received its name because of the small mushroom-like masses of geyserite about the rim of the crater. Its only certain years of major activity were 1929, 1948, and 1972, when regular intervals of about 1 hour resulted in 5-minute eruptions up to 12 feet high. It was also seen to splash 2 to 3 feet high during June and September 2005.

102. UNNG-THD-1 lies a few feet beyond non-eruptive Stirrup Spring, about halfway between Thud Spring (100) and Gourd Spring (103). During most years it behaves as a perpetual spouter 1 to 2 feet high. Infrequently, THD-1 acts as a geyser, with intervals typically about 5 minutes and durations 1½ minutes long. Water jets up to 7 feet high have been seen at these times. Because of its location on the far side of Stirrup's geyserite mound, THD-1 is not easily visible from the highway except when it is having its strongest jets. Another unnamed geyser, UNNG-THD-2, lies a few feet south of Stirrup Spring. It was known to be active only during 1988, 1993, and 2001, when eruptions lasting 5 minutes reached at least 5 feet high.

103. GOURD SPRING, nearly 400 feet northeast of Thud Spring (100), undergoes such infrequent geyser activity that few details of its action have been recorded. As with Thud Spring, Gourd's crater was filled with debris from the old hotel. Much of this was expelled during a single powerful eruption following the 1959 earthquake. During the 1970s it was seen to have infrequent eruptions that splashed 2 to 20 feet high over short durations, and washed areas implied additional action during 1988.

104. OAKLEAF SPRING, named because of the intricate shape of its crater rim, was reported to "boil vigorously" in 1883, but it was never known to actually erupt until June 5, 2006, when it was seen splashing 2 feet high on just that day.

105. KIDNEY SPRING is the northernmost member and the only regularly active geyser of the Thud Group. The crater is roughly kidney-shaped, 34 feet long and 6 feet wide. Kidney is a very regular geyser. Intervals average about 25 minutes, and any variation of more than 2 minutes from the average is uncommon. Each eruption lasts 3 to 4 minutes, during which the play is mostly less than 4 feet high. Kidney was inactive during the early years of the Park, but it has played without a known dormancy at least since the late 1940s.

QUAGMIRE GROUP

The Quagmire Group (Map 6.9, Table 6.8, numbers 110 and 111) occupies a small thermal pocket at the base of the Porcupine Hills near the northeast edge of Fountain Flat. Culex Basin and the Morning Mist Group are farther northeast beyond the same hills, which are composed of cemented thermal kame—that is, rocky debris carried by Ice Age glaciers that was dumped where the ice suddenly melted because of the hot springs. The Hayden Surveys of the 1870s often camped near the Quagmire Group, and several of the springs are named despite their small sizes, minimal activity, and remote locations. Among them are Lambrequin Spring and Tree Spring. The group has little to offer, but it does contain one true geyser.

110. SNORT GEYSER, named for the hissing, rumbling steam discharge, was apparently a true geyser until the time of the 1959 Hebgen Lake earthquake. Eruptions were a daily occurrence and sent steamy spray as high as 12 feet for durations as long as several hours. Since the earthquake, all observers have found Snort to be a perpetual spouter, sputtering about 2 feet high.

111. UNNG-QAG-1 plays from a small geyserite cone at the far south side of the group. Its manner of play seems to have been fairly consistent since at least the early 1980s, before which it had never been reported as a geyser. Intervals of 50 to 60 minutes separate durations that range from 13 to 17 minutes. The height is 2 feet.

MORNING MIST SPRINGS, BUTTE GROUP, MORNING MIST GROUP, AND CULEX BASIN

These groups of springs lie along and near the Mary Mountain trail (Map 6.9, Table 6.8, numbers 112 through 115), which departs from the highway at the north end of the Lower Geyser Basin. The trail is an old road open only to Park Service vehicles; bicycles are not allowed.

The first group, in which numerous springs are scattered throughout an open meadow area, is the Morning Mist Springs. It contains Porcupine Hill Geyser plus four or five small, infrequently active geysers.

Immediately north of Porcupine Hill are the few springs of the separately named Butte Group, where the Kitchen Springs included a small geyser (UNNG-KIT-1) active in the early 1970s.

Continuing along the trail through a grove of trees, the valley opens out again into the Morning Mist Group. Dominated by Morning Mist Geyser, the group encompasses a number of other small geysers. This area is unmistakable because to the south above the trail is a large rock quarry. It has been abandoned long enough to have trees growing among the loose boulders. The Morning Mist Group is east (left) of the quarry.

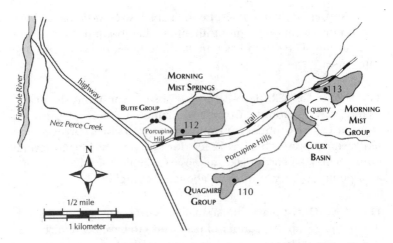

Map 6.9. *Quagmire, Morning Mist, and Culex Groups, Lower Geyser Basin*

Table 6.8. Geysers of the Quagmire Group, Morning Mist Springs, Butte Group, Morning Mist Group, and Culex Basin, Lower Geyser Basin

Name	Map No.	Interval	Duration	Height (ft)
Geyserlet	114	seconds	seconds	1
Morning Mist Geyser	113	days*	hours	1–6
Porcupine Hill Geyser	112	see text	seconds	25–30
Snort Geyser	110	steady	steady	2
UNNG-CLX-1	115	4–11 min	1–2 min	inches–2
UNNG-KIT-1		[1970s]	unknown	1–3
UNNG-QAG-1	111	50–60 min	13–17 min	2

* When active.
[ ] Brackets enclose the year of most recent activity for rare or dormant geysers. See text.

Culex Basin is out of sight, around and up the hill to the southwest (right) of the quarry. It consists mostly of quiet pools but also boasts at least four unnamed geysers and one perpetual spouter.

The hiking distance from the highway to the Morning Mist Group is a little more than 1 mile. Wildlife is abundant in the area and since this is grizzly bear country, hikers should travel in noisy groups.

112. PORCUPINE HILL GEYSER is the first spring encountered along the hike from the highway toward Morning Mist. Less than ¼ mile from the trailhead, the geyser is a superheated pool north of the trail at the summit of a wide, low geyserite mound. It is unlikely that Porcupine Hill Geyser has ever historically had a natural series of eruptions. Activity was first recorded in 1969, shortly after the U.S. Geological Survey completed a nearby research drill hole that probably induced the activity. The play was as much as 30 feet high but was very brief, usually lasting less than 30 seconds. The intervals were as long as a week. Since 1970, eruptions are known to have occurred in 1975, 1985, 1988, and 2001. They are suspected to have been artificially induced, a practice that is illegal and capable of destroying the spring. It is possible that Porcupine Hill Geyser undergoes occasional minor eruptions, but reports of major action must be discounted because the platform is always well covered by animal droppings and other debris that would be carried away by any significant overflow.

113. MORNING MIST GEYSER is the most significant geyser in the Morning Mist–Culex Basin area. It lies between the trail and the quarry, about 75 feet south of the road. It is the only feature in the area to have a large runoff channel. The eruption of Morning Mist is not particularly high, never reaching more than 6 feet with its biggest bursts, but it is significant in terms of discharge. Historically, a single eruption lasted about 12 hours. During that time more than 100,000 gallons of water poured down the channel, more than double the volume discharged by Old Faithful in a comparable period. Once the eruption ended, the water level in Morning Mist dropped about 12 feet within the crater. Refilling took place at about 1 foot per hour, and, once overflow was regained, from 12 to 36 hours of slight discharge took place before the next eruption. Thus, the intervals ranged from 24 to 48 hours and were fairly easy to judge on the basis of water level, overflow rate, and strings of small bubbles that rose through the pool in the last few hours before the play. Whether any of this pattern still holds true is unknown. Morning Mist's intervals since 1991 have been irregular and as long as 5 days, and the geyser has also had periods of complete dormancy. An eruption observed in September 2005 might have been the first actually witnessed in several years; its nature was of heavy, boiling surging that domed the water only 3 or 4 feet high. Eruptions were also reported in 2007.

114. GEYSERLET is the smallest geyser given a name by the early surveys of Yellowstone. It is within a small cluster of springs just around the hill to the left of the quarry. The play is frequent but only 1 foot high.

Across the stream and meadow east of Morning Mist (113) and Geyserlet are a number of additional springs and pools. One on the valley floor and two on the steep slope above have been seen to erupt as geysers 1 to 3 feet high. Additional springs about ¼ mile farther up the valley might also include a few small geysers.

115. UNNG-CLX-1 is the largest and only reliably active geyser in Culex Basin. It can be a bit difficult to find, since the crater is in an unlikely position very close to the eastern brink of the steep drop down to Culex Creek. The intervals are usually 4 to 11 minutes long. The 2-foot play, which lasts 1 to 2 minutes, often fails to fill the crater. Although there is a significant runoff channel, few eruptions produce any discharge. This clearly implies that greater eruptions (probably in terms of duration, not height) occasionally take place.

The other geysers in Culex Basin are more erratic and subject to dormant periods and are only inches to 1 foot high when active. On the west side of Culex Creek, on top of the flat geyserite platform, there used to be a perpetual spouter up to 4 feet high, but it has not been seen since the 1980s. *Culex*, by the way, is the scientific name for a genus of mosquitoes (*Culex tarsalis* is especially common in thermal marshes), and during warm, moist weather Culex Basin is very well named. More bothersome in the dry heat of summer, though, are deerflies, Genus *Chrysops*. Insect repellant provides little protection.

"MARSHALL'S HOTEL GROUP"

Several important hot spring groups are reached via Fountain Flats Drive. The "Marshall's Hotel Group" (Map 6.11, Table 6.9, numbers 120 and 121), so designated here, is the first of these groups. It consists of the scattered springs near the entrance of Fountain Flats Drive. Three of the springs are named, and two are geysers. Maiden's Grave Spring, a gently boiling pool surrounded by a log rail fence, lies next to the river to the right of the road; it has no record of eruptions. It was named in allusion to the grave of Mattie S. Culver, who died during the winter of 1889 at Marshall's Hotel. The actual grave is near the site of the hotel next to the Nez Perce

Picnic Area. Built in 1880 and in service until 1891, Marshall's was Yellowstone's second commercial lodging and first true hotel.

120. HYGEIA SPRING occupies the cone across the road near the entrance to the Nez Perce Picnic Area. It was named after the Greek goddess of health and cleanliness because its water was piped to the bathhouse and washroom at Marshall's Hotel. An 1882 visitor to the hotel referred to it as a "boiling spring" whose action "reached up level" with the surrounding ground. Whether this activity was intermittent or perpetual is not known, but evidently it was temporary. All other references have Hygeia as a quiet, relatively cool spring with only slight overflow, which is true of it today.

121. TWILIGHT GEYSER lies in the meadow several hundred feet across the river from Maiden's Grave Spring. Eruptions by this geyser were common in the 1970s, when they were reported almost daily by the bus drivers of "Twilight Tour" trips from Old Faithful Inn. Now the geyser is seldom seen, but it was active during June and July 2006. Most play is only 1 to 5 feet high, but exceptional bursts up to 20 feet are known. The eruptions have durations many minutes and possibly even a few hours long.

RIVER GROUP

Of all the major hot spring groups in the Lower Geyser Basin, only the River Group (Map 6.10, Table 6.9, numbers 122 through 138) lies along the banks of Firehole River. Thus, the setting of these springs is similar to that of the Upper Geyser Basin. The geysers, however, are relatively small, and there are wide spaces between some of the individual hot springs. Indeed, if this were a more heavily visited area, it would probably be considered as several hot spring groups rather than just one.

Much of the downstream part of the River Group is within an oval valley known as Pocket Basin. It was formed by a large hydrothermal explosion near the end of the last glacial episode of the Ice Age, roughly 10,000 years ago. The ridge that forms the rim around the basin is composed of the angular debris that was blasted out to form the valley. In terms of sheer power, the energy released during this explosion is said to have equaled that of the atomic bombs dropped at the end of World War II.

Over the ridge just east of Pocket Basin is an extensive area of mud pots. This is the largest assortment of mud pots in Yellowstone, and it alone is worth the hike to see. It is commonly called the Pocket Basin Mud Pots, but the actual official name is Mud Volcanoes. It naturally divides itself into two sections. The northern portion, which has been called "Lindern's Basin" after Baron F. H. von Lindern, a Dutch visitor to Marshall's Hotel in 1884, includes several impressively large, deep craters that are vio-

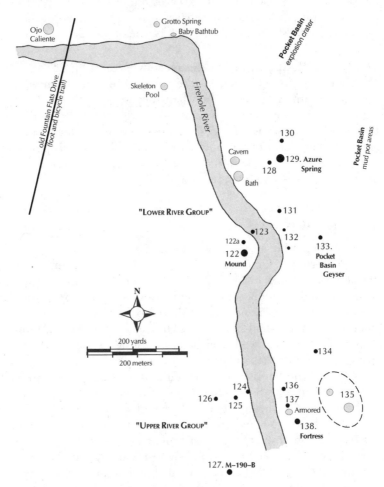

Map 6.10. *River Group, Lower Geyser Basin*

Table 6.9. Geysers of the "Marshall's Hotel" and River Groups, Lower Geyser Basin

Name	Map No.	Interval	Duration	Height (ft)
Azure Spring	129	rare	4–6 min	1–2
"Blurble Geyser"	136	erratic	seconds	1
"Brain Geyser"	136	minutes	1 min	1
Cone Spring	132	near steady	near steady	inches–3
Diadem Spring	131	[1976]	minutes	overflow
Fortress Geyser (Conch Spring)	138	near steady	near steady	5–10
Horn Spring	132	near steady	near steady	inches
Hygeia Spring	120	unrecorded	unrecorded	boil
"M-190-b"	127	[1993?]	minutes	1–50
Mound Geyser	122	11–63 min	5 min	6–12
Pocket Basin Geyser	133	14 min–hrs*	45 sec	15
Sand Geyser	123	frequent	sec–min	1–2
Twilight Geyser	121	rare	min–hrs	1–20
UNNG-RVG-1	124	minutes*	seconds	5
UNNG-RVG-2	125	[2004?]	2 min	2–7
UNNG-RVG-3	126	[1973]	10 min	15
UNNG-RVG-4	128	sec–min*	seconds	1–5
UNNG-RVG-5 ("Dark Pool")	137	8–60 min*	30 sec–2 min	8
UNNG-RVG-6	135	see text	see text	1–10
UNNG-RVG-7	130	[2004]	unrecorded	1
UNNG-RVG-8	134	rare	30 min	6
UNNG-RVG-9	122a	with Mound	minutes	1

* When active.
[ ] Brackets enclose the year of most recent activity for rare or dormant geysers. See text.

lently active. The one at the far north end of that group sometimes acts as a mud geyser—the "vertically gifted cyclic mud pot"—that intermittently throws thick mud as high as 30 feet. Near the south end of the mud pot area, close to where it opens out into the River Group, is Microcosm Basin, two clusters of small mud volcanoes with vigorously active craters and cones.

There are no established trails in the River Group, but it is traversed on both sides of Firehole River by paths worn into the ground by fishermen. From the end of Fountain Flats Drive, walk about ⅓ of a mile along the old road from the parking lot to Ojo Caliente Spring. From there, simply go to the appropriate side of the river and begin walking upstream. Remember the hazards of the geyser basins, and stay well back from all hot springs.

Ojo Caliente Spring is part of the River Group, but it is well separated from the rest of the group. It is superheated and smells strongly of hydrogen sulfide gas. Ojo Caliente had a brief episode of eruptions during 1968, at the same time a geothermal research well was being drilled nearby. The eruptions ended as soon as the drilling stopped and the well was capped. The group contains several other large springs, most notably Grotto Spring, Baby Bathtub Spring, Skeleton Pool, Cavern Spring, Bath Spring, and Armored Spring.

For more than 800 hundred feet south of Mound Geyser on the west side of the river there are only a few warm seeps until several small geysers are encountered. The same sort of gap between hot spring clusters is seen south of Pocket Basin Geyser on the east side of the river. These southern springs comprise the "Upper River Group." This designation is entirely unofficial but useful, given that the area is distinctly separate from the "Lower River Group" (see Map 6.10). However, in the descriptions here, the geysers of the entire River Group are discussed geographically from north to south—first all those on the west side of Firehole River, then those on the east.

Geysers on the west side of Firehole River

122. MOUND GEYSER is the most important named geyser in the River Group. Its crater measures about 25 by 10 feet and lies at the summit of a large sinter mound initially formed by a much older hot spring. Mound is quite regular in its activity. The intervals range from 11 to 35 minutes (up to 63 minutes in 2003), but they usually average about 22 minutes and often show only a few seconds in variation over considerable spans of time. The eruptions last about 5 minutes. The play takes the form of heavily boiling surges between 6 and 12 feet high. On rare occasions Mound undergoes powerful bursting, which was the case when a Park visitor was seriously burned there in 1993.

Mound itself discharges almost no water, but a collection of vents (UNNG-RVG-9, #122a) down the slope to the north gushes a heavy flow of water during Mound's play. Although obviously related to Mound, this geyser is known to have independent eruptions, too.

123. SAND GEYSER is little known because its vent is within the bed of Firehole River, where it is nearly always underwater. At those times, the eruptions simply roil the sand that fills the crater. However, in dry seasons when the river is exceptionally low, the vent can be exposed, and splashing eruptions then reach as high as 2 feet.

124. UNNG-RVG-1 is a large pool near the level of Firehole River in the west "Upper River Group," south of Mound Geyser (122). At times of high water the river covers this spring and renders it dormant. Water levels low enough to allow eruptions occur only during dry seasons. Then, the pool displays brief eruptions sometimes as high as 5 feet.

125. UNNG-RVG-2 had its first known activity during late 1985. After a year or two of play, it fell dormant. The eruptions were fairly regular at most times. The known range in intervals was from 20 to 50 minutes. Eruptions began when the water level within the square crater abruptly began to rise. First, a small subsidiary vent jetted 2 or 3 feet high, then the main vent would burst up as high as 7 feet. Several minutes of minor boiling and splashing followed the 2-minute play. Renewed activity in 1993 was similar but weaker. RVG-2 has not been reported since then, but action was inferred in June 2004 when the surrounding geyserite platform was found cleanly washed.

There is a small cone a few feet north of RVG-2. It was also active as a geyser in 1993, playing frequently about 6 inches high. About 15 feet down the slope from the cone was a small patch of gravel that independently sizzled every few minutes.

126. UNNG-RVG-3 is located in a grassy meadow area west of RVG-2 (125). Within this general area are two dozen or more hot springs, most of which show signs of having had geyser activity in the past. The only one of these with a known historical record of activity, however, is RVG-3, a small pool atop a low geyserite mound. During 1973, it erupted frequently. The play reached as much as 15 feet high with durations as long as 10 minutes.

127. "M-190-B" has come to be the accepted, although unofficial, name for a large pool at the far southern end of the River Group. The designation actually identified a water sample collected by the U.S. Geological Survey at a time when the pool was active as a geyser. During the late 1960s up to 1972, "M-190-b" was a vigorous geyser. Although the intervals were erratic, ranging from minutes to many hours, the play was spectacular, with some jets easily reaching 50 feet high. In 1993, although no eruptions were observed, the area surrounding the crater appeared washed, as if small eruptions had pushed waves of water across the rim of the pool.

Geysers on the east side of Firehole River

128. UNNG-RVG-4 is within the "Lower River Group," near Azure Spring (129). There is no record of this geyser having erupted prior to the mid-1970s. Its activity is highly variable, with periods of dormancy or quiet intermittent overflow more common than episodes of geyser action. During the active phases, RVG-4 has intervals of a few minutes to hours, with brief eruptions that reach 1 to 5 feet high.

129. AZURE SPRING is a large pool. Its irregularly shaped crater measures about 18 by 40 feet and is deep enough to produce a vivid blue color in the water. Azure is usually active as an intermittent spring. Periods of overflow are generally longer than those when the water level is low, but there is great variation as to how much time one of these cycles takes. On rare occasions, Azure has small bursting eruptions at the times of high water. These generally last 4 to 6 minutes, after which the water level drops a foot or more.

The jagged edges and broken layers of sinter within the crater are proof of an explosive origin. This may have happened sometime after the first mapping explorations of Yellowstone—the crater apparently did not exist in the 1880s, and Azure was not named until 1972. In an interesting case of serendipity, a different feature (probably in the "Upper River Group") was called "Azure Lake" in 1878.

130. UNNG-RVG-7 is located about 100 feet north of Azure Spring (129). An old feature whose geyserite mound is nearly covered with vegetation, it was never known to erupt before it was seen playing about 1 foot high in late May 2004. One month later it had cooled so much that dark orange cyanobacteria was growing in the crater.

131. DIADEM SPRING used to be one of the largest and prettiest pools in the River Group. Although never a geyser according to a strict definition, Diadem was a very impressive intermittent spring whose frequent overflow amounted to more than 2,500 gallons per minute. Just a few feet away, the stream dropped over a series of waterfalls, producing a remarkable steaming cascade into the river. Runoff channels suggest that Diadem has seen a lot of activity in its history. It fell dormant in 1976, when the water fell a foot below overflow and cooled so that dark brown cyanobacteria mats lined the crater. The pool partially recovered in 2006.

132. CONE SPRING sits in the top of a cone 4½ feet high perched on the steep bank of Firehole River. Similar Horn Spring is a short distance farther upstream. Both of these springs bubble and spout slightly. It is doubtful that they ever had large eruptions, as slender cones such as these are best formed by seeping flow running down their sides. Both, however, do erupt to a slight degree, with bubbling that reaches a few inches high. In 1972, Cone Spring was reported to throw steamy spray as high as 3 feet, but that would have been very exceptional behavior.

133. POCKET BASIN GEYSER lies near the natural drainage exit of the Pocket Basin Mud Pot area. It is not mentioned in any early reports, including some written as recently as 1973. It was, however, vigorously active in 1976, and the nature of its geyserite formations and deep runoff channel indicates that it is an old spring that reactivated after a long dormancy. Pocket Basin Geyser is cyclic in its activity. It has been known to go as long as several days without erupting, but at other times the intervals are as short as 14 minutes. When it is active, the water periodically rises and falls within the vent. Each rise brings the level a bit higher than before. Sometimes several preliminary overflow periods are needed to trigger the eruption, but at other times there are none. The play starts with bubbling that quickly grows into a series of vigorous bursts, with jets as much as 15 feet high. The entire play lasts about 45 seconds, whatever the interval.

134. UNNG-RVG-8 is the northernmost significant spring of the "East Upper River Group," about 700 feet south of Pocket Basin Geyser. A wide runoff channel leads away from the dark pool, but its eruptions are rare. One observed in progress on May 31, 2004, had a duration longer than 30 minutes, during which water was splashed as high as 6 feet.

Pocket Basin Geyser is one of the few geysers of size in the River Group. Although it sometimes goes hours to several days without erupting, it also can play as often as every 14 minutes.

135. UNNG-RVG-6 is this book's designation for the large area of hot springs on the flats above and east of Fortress Geyser (138) and Armored Spring. A number of these springs are vigorous perpetual spouters, and there are several large pools, one of which might be the "Azure Lake" of 1878. True geyser action here is extremely erratic, and it appears to be subject to long-term waxing and waning cycles that control the water levels throughout the area. As many as eight geysers have been active at once, but often none is seen anywhere in the complex. Play only 1 or 2 feet high is typical, but eruptions as high as 10 feet have been observed in one pool.

136. "BRAIN GEYSER" was named because of the convoluted shape of the dark gray geyserite that surrounds the crater. Only about 2½ feet across and plunging many feet straight downward, the pool is virtually black in appearance. Eruptions usually recur every few minutes. They last about 1 minute and consist of vigorous bubbling that can splash up to 1 foot high.

Equally small but less reliably active is "Blurble Geyser," just a few feet farther downstream.

137. UNNG-RVG-5 ("DARK POOL") is a blackish-green pool that has infrequent episodes of bursting eruptions that reach as high as 8 feet. It is often dormant. When active, the play usually takes place every 40 to 60 minutes, although intervals as short as 8 minutes were recorded during 2003 and 2006. The duration varies between 30 seconds and 2 minutes.

The first spring south of "Dark Pool" is **ARMORED SPRING**, named because of the massive geyserite formations around and within its crater. It has never been known to erupt as a true geyser but often exhibits intermittent variations in its rate of overflow, and sometimes it is hot enough to weakly boil around the edges of the crater.

138. FORTRESS GEYSER, also officially known as **CONCH SPRING,** looks as if it should be a powerful geyser. The superheated pool lies within a massive geyserite cone that rises more than 4 feet above its surroundings. Although periodic, the quiet intervals are only 2 to 4 seconds in length. The play is violent boiling that sends bursts about 5, rarely 10, feet above the rim of the cone. Wide and deep runoff channels lead away from Fortress, indicating that it did undergo major activity in the past. The only such play in recent history occurred the night of the 1959 earthquake. Although it wasn't seen, it was estimated from the area washed by the water to have reached 40 feet high.

FAIRY MEADOWS GROUP AND FAIRY SPRINGS

Along the broad valley of Fairy Creek, much of which is meadowland, are several small clusters of hot springs originally known collectively as the "Fairy Fall Group." There are few geysers of significance in these two groups (Map 6.11, Table 6.10, numbers 140 through 143). The easiest access to the area is by way of a trail that leaves the old Fountain Flats Drive about 1.3 miles south of Ojo Caliente Spring. This trailhead also provides access to the Imperial and Sentinel Meadow groups.

140. COLUMN SPOUTER is a superheated pool just a few feet east of the trail, about ½ mile from the old road. Intervals of complete quiet are uncommon, but at those times the pool level drops 1½ to 2 feet. It then refills at a rate as fast as several inches per minute, with progressively stronger boiling. The time of the first overflow following such a pause is when Column Spouter can have true bursting eruptions, which can reach 6 to 8 feet high. The play rapidly declines to vigorous boiling that persists for many hours.

141. UNNG-FCG-2. The Fairy Meadows Group proper lies west of Column Spouter (140), west of Fairy Creek. A trail to Sentinel Meadow passes just north of this area, but the springs have never received extended study. Among them are some impressively large and deep pools, several perpetual spouters, and at least eight geysers. The largest is the impressively deep "Rhinoceros Spring," which in 1992 erupted as high as 10 feet. The eruptions are generally small, however, and often the largest active geysers play only a foot or two high. "Tremor Spring," which audibly thumps the surrounding ground but never erupts, and the perfectly symmetrical "Trumpet Pool" are located in the central part of this area.

142. UNNG-FCG-3. Up the valley from the Fairy Meadows Group, isolated by several hundred yards of nonthermal meadow, is Fairy Springs, also known as the Fairy Creek Group. It lies west of a stand of trees and extends onto the hillside to the west. FCG-3 is the closest spring to the trees. Eruptions observed during the early 1970s were as high as 30 feet with durations of 5 minutes. However, this spring dramatically changed its appearance around 1980, when a steam explosion enlarged the crater. What once was a vent only 4 feet across is now a pool with an irregular diameter of 10 feet surrounded by tilted, broken blocks of geyserite. No eruptions have been observed since the explosion was discovered.

143. LOCOMOTIVE SPRING lies a short distance up the hillside on the far western side of Fairy Springs. It is nearly a perpetual spouter, with only the briefest of pauses interrupting its 6-foot jet. Water discharge is copious. Just down the slope from Locomotive is another spring that, based on extensive washed areas, had some powerful eruptions during the early 1980s, possibly at about the same time a steam explosion disrupted FCG-3 (142).

IMPERIAL GROUP

The most popular access to the Imperial Group (see Index Map 6.1, Table 6.10, numbers 150 and 151) is from the south end of the Midway Geyser Basin. The trail crosses Firehole River at Steel Bridge, follows the old road along the west side of Midway, and then passes through the woods to Fairy Falls. Fairy Falls is one of the nicest waterfalls in Yellowstone, with a sheer drop of 200 feet. It is unfortunate that this area was one of the most severely burned during the forest fires of 1988. From the waterfall the trail continues westward, joins the path that crosses Fairy Meadows, and passes near Imperial Geyser. The round-trip distance from Steel Bridge to Imperial Geyser is 7.2 miles. Access is also possible via the trail that starts at the end of Fountain Flats Drive and passes by the Fairy Springs; that round-trip is about 9 miles.

150. IMPERIAL GEYSER did not erupt before 1927, but it probably developed from a much smaller, quiet pool recorded in 1911. The first detailed observations were conducted in July 1928 when Drs. Allen and Day of the Carnegie Institute of Washington began observations on the development and behavior of Imperial for their classic work, *Hot Springs of the Yellowstone National Park.* They were instantly impressed by the geyser, and its fame spread rapidly

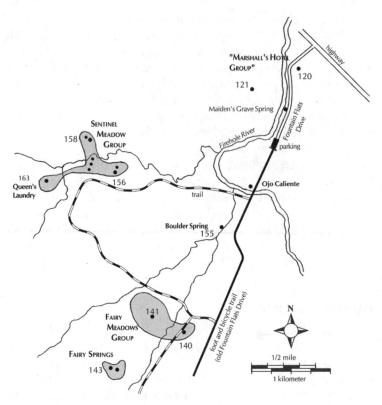

Map 6.11. *"Marshall's Hotel," Fairy Meadows Group, Fairy Springs, and Sentinel Meadow Group, Lower Geyser Basin*

Table 6.10. Geysers of the Fairy Meadows Group, Fairy Springs, and Imperial Group, Lower Geyser Basin

Name	Map No.	Interval	Duration	Height (ft)
Column Spouter	140	hrs–days	very long	boil–8
Imperial Geyser	150	[1929]	2 hrs	25–80
Locomotive Spring	143	near steady	near steady	6
"New Imperial Geyser"	150	20–60 sec	40 sec	20–60
"Rhinoceros Spring"	141	[1992]	unrecorded	10
Spray Geyser	151	near steady	near steady	2–30
UNNG-FCG-2	141	min–hrs*	minutes	1–2
UNNG-FCG-3	142	[1980?]	5 min	30

* When active.
[ ] Brackets enclose the year of most recent activity for rare or dormant geysers. See text.

outside the Park. It received its name as the result of a contest at a newspaper editors' conference.

Erupting from an enlarged crater nearly 100 feet across, Imperial would shoot bursts 30 feet wide to as high as 80 feet. A single eruption would last 1 to 5 hours, throughout which about 3,000 gallons were discharged each minute (over 360,000 gallons during the entire eruption). "Quiet" intervals of 12 to 20 hours were periodically interrupted by brief splashing to 25 feet, and even then the discharge amounted to 700 gallons per minute. In every way, Imperial was one of Yellowstone's most significant geysers. But after barely two years of activity, Imperial fell dormant in October 1929. Because of many apparent steam leaks in the floor of the crater, Allen and Day surmised that Imperial had ruined its plumbing system, never to erupt again. And perhaps it hasn't. Today's eruptions rise from a smaller vent at the edge of Imperial's large crater, which is filled with a beautiful blue-green pool.

Although Imperial's high water discharge continued and there were periodic episodes of small splashing, major activity did not take place again until sometime in August 1966, when the vent at the edge of Imperial's crater began playing up to 40 feet high. The new vent is distinctly different from the Imperial of the 1920s and has been interpreted as a delayed effect of the 1959 earthquake. Eruptions by "New Imperial Geyser" (as designated here) were nonstop and of different form from those of the 1920s, being rocketing jets rather than massive bursting. Through the succeeding years, "New Imperial" gained in strength. In 1973, measurements proved some eruptions to be more than 70 feet high. Although sometimes nearly steady, most activity was intermittent, with intervals of 20 to 60 seconds and durations of 40 seconds. There was a short span of near-dormancy during the mid-1980s, but "New Imperial" continues to erupt, with intervals and durations both only seconds long. The height varies between 20 and 60 feet. All of the water discharge is into the old Imperial, which still produces a large runoff stream.

151. SPRAY GEYSER has been known since the earliest days of the Park and is one reason a pre-1927 existence for Imperial Geyser (150) is unlikely. When the first huge steam clouds of Imperial were seen at a distance, it was thought that the source was Spray. It sits back in the woods, up a small tributary a few yards from the runoff from Imperial, perhaps ¼ mile downstream from that geyser.

Spray is one of the few geysers whose intervals are consistently shorter than the durations. Although the length of these times can vary greatly from one year to the next, the relationship between the two is always the same. Until the mid-1980s, Spray's eruptions lasted 3 to 5 minutes and were separated by intervals of less than 1 minute. The eruption includes two main jets of water. When Spray is at its best, one jet is nearly vertical and 25 to 30 feet high; the smaller, angled column reaches about 12 feet. Several other openings play water between 2 and 6 feet. Unfortunately, for the past twenty years Spray has regressed to near perpetual spouting, and the greatest height has been only about 6 feet.

SENTINEL MEADOW GROUP

The hot springs of the Sentinel Meadow Group (Map 6.11, Table 6.11, numbers 155 through 164) lie scattered about a broad valley. Three of these have formed large geyserite mounds. From northwest to southeast, they are Flat Cone, Steep Cone (or Sentinel Cone), and Mound Spring. Each is capped by a deep, boiling pool, and the springs atop Flat Cone and Steep Cone have acted as geysers in recent years. A fourth major spring is The Queen's Laundry, which acted as a geyser during the 1870s and early 1880s.

The Sentinel Meadow Group contains at least nine geysers. Little is known about them, since this is a fairly remote area seldom

Table 6.11. Geysers of the Sentinel Meadow Group, Lower Geyser Basin

Name	Map No.	Interval	Duration	Height (ft)
Boulder Spring	155	steady	steady	1–6
Bulgers, The	161	frequent*	sec–min	1–3
Flat Cone Spring	158	hours*	minutes	boil–15
Iron Pot	162	6–9 hrs	17–33 min	3–6
Queen's Laundry, The	163	steady	steady	1–2
Rosette Geyser	159	[1970s]	sec–min	2–35
Steep Cone Geyser	160	infrequent	seconds	boil–4
UNNG-SMG-1	156	[1993]	20 sec	2–20
UNNG-SMG-2 ("Convoluted")	157	sec–hrs*	sec–min	2
UNNG-SMG-3	164	[1993]	minutes	1

* When active.
[ ] Brackets enclose the year of most recent activity for rare or dormant geysers. See text.

visited by researchers. The Sentinel Meadow trailhead is directly across Firehole River from Ojo Caliente Spring. From there it is 2.1 miles to Queen's Laundry, so the round-trip from the parking area at the end of Fountain Flats Drive is about 5 miles.

155. BOULDER SPRING is immediately at the base of a low, boulder-strewn hill about 300 yards southwest of Ojo Caliente Spring. Not a part of the Sentinel Meadow Group, it is included here since access to it is via the Sentinel Meadow trail (see Map 6.11). Initially named "Perpetual Geyser," Boulder Spring is a perpetual spouter that plays from two vents within one pool. The height of the eruption constantly varies in a pulsating motion, and it also slowly waxes and wanes in overall force. The height ranges from less than 1 foot to more than 6 feet.

In 1991 another vent, among the boulders just outside the pool of Boulder Spring proper, was observed to have periodic eruptions. The play was frequent and up to 4 feet high but persisted for only a few weeks.

156. UNNG-SMG-1 is located on the western flank of the wide, low mound of Mound Spring, the first of the large superheated pools encountered when entering the area along the trail from Ojo Caliente Spring. The geyser plays from a beaded vent stained orange by iron oxide minerals. Active periods are extremely rare and are known to have been witnessed just two times, in 1972 and 1993. The play was a series of distinct squirts of water, at first reaching up to 20 feet but quickly dying down to just 2 or 3 feet at the end of the eruption that lasted no more than 20 seconds. In 2005 the vent was nearly filled with debris, and the small runoff channels had almost disappeared.

157. UNNG-SMG-2 ("CONVOLUTED GEYSER") lies down the slope from Mound Spring toward Sentinel Creek. It was named after the exquisitely ornate geyserite surrounding the crater, which looks much like the surface of a brain. The vent is about 6 inches in diameter. "Convoluted" is highly variable in its activity, sometimes erupting every few seconds for a few seconds but on other occasions going hours between plays that last a few minutes. The height is always about 2 feet. Dormant periods have been recorded.

158. FLAT CONE SPRING is at the summit of Flat Cone, the only bare geyserite flat north of Sentinel Creek. A description written in 1878 seems to imply that there were no runoff channels on the cone at that time. Eruptions were finally observed in the late 1980s. Intervals were generally a few hours long. The beginning of the play was marked by sharp pounding of the ground that could be both felt and heard at a distance of several hundred feet. The eruption consisted of boiling surges, with some bursts as high as 15 feet. Although still active, the frequency and force of the eruptions have decreased considerably.

159. ROSETTE GEYSER, located just northwest of Flat Cone, was named because of the numerous decorative rosettes of geyserite (similar to the silver rosettes on horse saddles) that used to decorate the platform surrounding it. Research has shown that during the first fourteen years of Park history, Rosette was a well-known geyser. Eruptions were frequent and up to 35 feet high. A long dormancy began in 1886. Rosette was not described again until 1929, when play lasting a few seconds reached 10 to 15 feet as often as every 1½ minutes. Again, there is no further record of eruptions until the 1970s, when the intervals were 2 or more hours long and the play just 2 to 4 feet high. When that active phase ended is unknown, but Rosette has now been dormant long enough for the beautiful rosettes to have decayed and weathered away.

160. STEEP CONE GEYSER is the spring atop the prominent geyserite structure of Steep Cone, which rises immediately next to Sentinel Creek. The water is superheated, so there is constant sizzling around the rim of the pool's crater. Although bursting eruptions are unknown, Steep Cone Geyser infrequently undergoes violent boiling that domes the water as high as 3 or 4 feet.

161. THE BULGERS is the collection of small springs between Steep Cone (160) and Iron Pot (162). They were named in 1878 when it was their beautifully beaded geyserite, rather than eruptions, that attracted attention. Although the name implies eruptive action, no play was described at that time, and they have been dormant so much of the time that the ornate geyserite has weathered away. There have, however, been occasions when The Bulgers erupted so frequently that they were nearly perpetual spouters, splashing up

Iron Pot, one of the main attractions in the Sentinel Meadow area, is an odd feature in that its eruptions never produce any overflow. Here, it was being watched by members of The Geyser Observation and Study Association.

to 3 feet. In 2005 they were intermittent but reached only about 1 foot high.

162. IRON POT is a fascinating feature. It lies within the cluster of trees that stands alone on the valley floor, south of Steep Cone. The crater is oval, about 15 feet in diameter, and lined with smooth, tan geyserite. The crater tapers downward to a depth of 15 feet, where there is a narrow opening approximately 6 feet long. Iron Pot never overflows, and the lack of runoff channels indicates that it never did. During the interval, which is 6 to 9 hours long, the water level slowly rises within the crater. When it reaches about 6 feet below the rim, the eruption begins suddenly. A violently roiling boil domes the water 3 to 6 feet high and usually lasts about 25 minutes. The play ends abruptly, and the crater drains within a few minutes.

163. (THE) QUEEN'S LAUNDRY was named by Park Superintendent Philetus T. Norris in 1880, when bathers draped their colorful clothing on the nearby trees. At that time, the runoff fed a natural "bathing pool" that measured 30 by 20 feet and 5 feet deep and was far enough from the spring for the water to have cooled to a comfortable temperature. That pool has disappeared. Also gone are most of the wide overflow terraces covered with brilliant orange-red cyanobacteria that led to an alternate name, Red Terrace Spring. Nearby is a wooden structure, the remains of a bathhouse started by Norris in 1881 but never completed. As the oldest existing structure in Yellowstone and, in fact, the oldest building ever intended for public use in any national park, the bathhouse was declared a separate National Historic Site within the national park on July 25, 2001.

The Queen's Laundry is a large oval pool, extending 30 by 50 feet, that constantly boils and surges above several separate vents. Some of this action reaches up to 1 to 2 feet. Descriptions from 1880 and 1881 imply that it was a significant geyser. Although the only eruption specifically described was just 5 to 6 feet high, different authors referred to the spring as "a great boiling fountain," as "an immense fountain," and as "the largest and most complete washing machine in the world." Perhaps, though, those eruptions were from a different spring somewhere near but outside the south edge of The Queen's Laundry. Separately described as a vigorous geyser in the early 1880s, its vent was noted as "collapsed and decaying" in 1886, and it cannot be positively identified today. Clearly, it was not the same as cool, quiet Dumbbell Spring, about 100 feet farther south.

164. UNNG-SMG-3 is a small spring southwest of The Queen's Laundry (163) and beyond Dumbbell Spring. A buildup of sinter along its runoff channel has formed an interesting flume-like form. The spring itself usually only overflows a heavy but steady stream, but one time in 1993 it was seen splashing up to 1 foot high.

CHAPTER SEVEN

Norris Geyser Basin

The Norris Geyser Basin is very different from the other geyser basins in Yellowstone in several ways. Only the geyser eruptions look the same. The casual visitor to Norris immediately notices that the scene is stark. The Porcelain Basin is a barren depression almost totally devoid of plant life, drab gray without the pastel shades of other areas. Runoff channels are sporadically lined with mats of true algae, such as purple-brown *Zygogonium* and green *Cyanidium*, *Galdiera*, and *Chlorella*, but not with the yellow, orange, and brown cyanobacteria typical of the other geyser basins. The few other visible colors are dominated by the rich orange-brown of iron oxide minerals, occasional spots of brilliant red and orange arsenic sulfide compounds, and rare olive-green patches of iron arsenate hydrate (the mineral scorodite).

Norris's unique appearance results from the presence of acid water, which is not common in the other geyser basins. Large amounts of sulfur are brought to the surface. In the springs it forms sulfuric acid when it is oxidized by the metabolism of another primitive form of life, a member of Archaea called *Sulfolobus*. The siliceous sinter deposited in the acid water is spiny and does not form the thick masses so common in the alkaline areas such as the Upper Geyser Basin. Norris does have some springs of alkaline water, and they tend to deposit geyserite at rates greater than elsewhere.

Norris is also the hottest geyser basin in Yellowstone and one of the hottest in the world. The water temperatures are generally higher and the geyser activity is more vigorous, while quiet pools are less common, than in other areas. One research drill hole here reached a temperature of 459°F (237°C) at a depth of only 1,087 feet (330 meters) below the surface. Why this is so is clear. Norris lies at the intersection of three major geologic structures: the rim of the Yellowstone Caldera, the largest single volcanic structure in the world; the Norris-Mammoth Corridor, a fault zone that may be a northern extension of the Teton Fault at Grand Teton National Park; and an eastward extension of the Hebgen Lake Fault Zone, the source of the 1959 earthquake.

One curious aspect of the activity at Norris occurs almost every year, most often during the late summer or fall. Formally called the "widespread contemporaneous changes" but popularly known simply as the "disturbance," its exact cause is uncertain but apparently includes several conditions. There is evidence that the hot springs at Norris are served by two separate geothermal reservoirs at depth. According to geochemical studies by the U.S. Geological Survey, one is deep and hot at about 640°F (340°C), and the other is shallower and somewhat cooler, with a temperature near 520°F (270°C). Because of the lack of extensive geyserite deposits, which might otherwise seal the upper part of Norris's geothermal system from inflowing surface water, there is an inflow of cold groundwater as well as some recycling of hot spring water back into the shallow parts of the geothermal system. As summer progresses, the supply of surface water decreases and allows a higher temperature to develop in the upper reaches of the regime. The result is that quite suddenly, often within a few hours, a great many of the Norris springs and geysers become muddy. Pools that are normally quiet spring to life as geysers, and existing geysers' activity becomes much more frequent. The disturbance usually occurs just once per year, most often during August or September, but series of more localized disturbances have been seen at all other times of the year. Disturbances were first recorded in 1878 and have been noted in at least seventy-nine of the years since then. They unquestionably also occurred in many of those early years for which there are no written records, so a year in which there is no disturbance (such as 2005) is unusual. In any situation where an individual spring is especially affected by the disturbances, that fact is noted in the following descriptions.

As with geysers anywhere, those at Norris are affected by earthquakes. There is little information about specific changes in 1959, but the shock of June 30, 1975 (magnitude 6.2), caused both immediate and long-term alterations in many of the geysers. The earthquake of March 26, 1994 (magnitude 4.9), may have caused changes, too, but the 1994 adjustments were difficult to infer since significant disturbance effects were already under way when that quake took place. Since then, activity at Norris has declined. It may not be that much of a joke when people say: "What Norris really needs is another good earthquake."

The only public facility at Norris is a small museum. It contains displays about the different kinds of hot springs and the life of the geyser basins. A book sales area is nearby, and restrooms and vending machines are located at the parking lot.

PORCELAIN BASIN

The Porcelain Basin (Map 7.1, Table 7.1, numbers 1 through 39) comprises the northern half of the Norris Geyser Basin. It was named for one of the few masses of alkaline sinter found at Norris. Porcelain Basin is a relatively small area, but the geyser and other hot spring activity is highly concentrated. A single glance from the museum patio used to easily take in a dozen or more erupting geysers. Unfortunately, trees allowed to grow overly tall have almost eliminated the view, so to clearly see Porcelain Basin one must walk some distance down the trail.

Porcelain Basin is an extremely changeable area. The larger, long-lived geysers—most of those that have names—are never quite the same from one year to the next. Most of the geysers are small and very short-lived, and many erupting vents act as perpetual spouters rather than as true geysers during their brief existences. Because of the internal deposition of minerals (mostly silica along with some clay), they soon seal themselves in. The hot water they gave off eventually finds exit through new geysers and spouters.

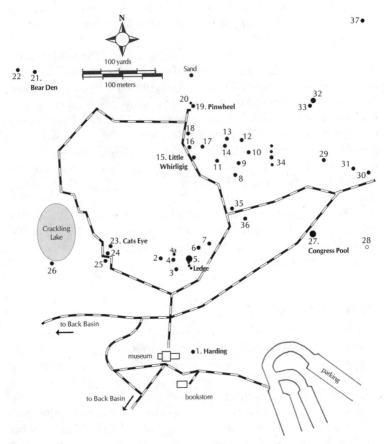

Map 7.1. *Porcelain Basin of Norris Geyser Basin*

Such features, possibly numbering fifty or more at any given time, obviously cannot be included in the descriptions here. Look for them in the central part of Porcelain Basin, especially on the wide flats east of Pinwheel Geyser and below Porcelain Terrace at the far eastern end of the area.

For reasons not known, the southwestern and central portions of Porcelain Basin substantially declined in vigor during the 1980s. There was a simultaneous increase in the action of the eastern part of the basin, but then it declined, too. Perhaps this was because of a large-scale exchange of function. There was some rejuvenation of the major geysers during the mid-1990s, but that action lasted only a few months. Porcelain Basin has perhaps seen less activity during the 2000s than in any previous decade on record.

The Porcelain Basin trail is a combination of dirt, asphalt, and boardwalk paths. The entire system covers about 1 mile.

Table 7.1. Geysers of Porcelain Basin, Norris Geyser Basin

Name	Map No.	Interval	Duration	Height (ft)
Africa Geyser	11	[1981]	90 min–steady	20–45
Arsenic Geyser[†]	9	min–hrs*	5 min–hrs	3–35
Basin Geyser	7	[1970s]	near steady	6
Bear Den Geyser	21	extinct?	sec–min	10–70
Blue Geyser[†]	32	disturbance	15 min–hrs	15
Carnegie Drill Hole	28	artificial	steady	1–2
Cats Eye Spring[†]	23	disturbance	near steady	2
Collapsed Cave Geyser	39	see text	5–10 secs	sub–20
Congress Pool[†]	27	rare	minutes	4–20
Constant Geyser	17	min–hrs*	seconds	30
"Crackling Spring"	26	near steady*	near steady	1–4
Dark Cavern Geyser, normal	2	infrequent	seconds	5–20
Dark Cavern Geyser, steam phase	2	[1997]	17–45 min	40–50
Ebony Geyser	22	[1974]	sec–min	6–75
Fan Geyser[†]	13	irregular	10 min	10–25
Feisty Geyser	29	see text	minutes	25
Fireball Geyser[†]	14	5 min–days	5–20 min	12–25
Geezer Geyser[†]	8	rare	8–15 min	8–15
Glacial Melt Geyser[†]	24	disturbance	near steady	5
Graceful Geyser	38	irregular*	min–hrs	20
Guardian Geyser	4	variable	sec–min	8–30
Harding Geyser	1	[1982]	5 min	50
Hurricane Vent	36	[1991]	unrecorded	3–30
Incline Geyser	30	[1995]	minutes	30–110
Iris Spring	33	near steady*	near steady	5–15
Jetsam Pool	6	[1994]	sec–min	65
Lava Pool Complex[†]	10	rare (extinct?)	hours	10
Ledge Geyser	5	[2007]	2 hrs	30–125
Little Whirligig Geyser	15	[1991]	min–hrs	15–20
Pequito Geyser	20	[1993]	minutes	2–5

continued on next page

Table 7.1—*continued*

Name	Map No.	Interval	Duration	Height (ft)
Pinto Geyser†	12	disturbance	2–30 min	20–40
Pinwheel Geyser†	19	[1974]	5 min	6–20
Primrose Spring	34	steady*	steady	3
Ragged Spouter	37	buried?	minutes	10–50
Splutter Pot†	18	min–steady	min–steady	1–6
Sunday Geyser†	35	[1982]	5 min	30–50
Teal Blue Bubbler	25	rare	hours	1–6
UNNG-NPR-5 ("Lambchop")	31	steady	steady	10–12
UNNG-NPR-6	4a	rare	19 min	6
Valentine Geyser	3	rare	5–22 min	20–75
Whirligig Geyser	16	hrs–days	3–4 min	6–15

* When active.

† Indicates geysers that show more frequent and/or more powerful activity at the time of a disturbance. In the interval column, "disturbance" indicates geysers that are active essentially only at the time of a disturbance.

[ ] Brackets enclose the year of most recent activity for rare or dormant geysers. See text.

1. HARDING GEYSER probably had some eruptions as early as 1916, but it wasn't named until 1923, the year President Warren G. Harding visited Yellowstone. The vent is almost invisible from the trail, hidden in a gully a few feet east of the museum building. Harding Geyser rarely erupts. Its best seasons on record were 1974 and 1982, with several eruptions each. The play lasts about 5 minutes, and the 50-foot water jet is concluded by a short steam phase.

2. DARK CAVERN GEYSER used to be one of the most faithful and frequent performers at Norris. It issues from a cave-like opening in a pile of dark gray geyserite-coated boulders. The water jet would be higher and more pronounced if it didn't strike an overhanging rock. Deflected as it is, the maximum height is still as much as 20 feet. Dark Cavern is known to have both minor and major eruptions. When active, eruptions recur every 17 to 25 minutes.

During the winter of early 1994, Dark Cavern began having eruptions of unprecedented force. Durations were highly variable and as long as 45 minutes. Angled water jets more than 45 feet high would periodically yield to loud steam phases whose roar could be heard nearly a mile away. The intervals between these eruptions ranged from 12 hours to 5 days. The frequency of this activity began to decrease in late 1994, and it came to a complete end near the end of 1997, about the time when Ledge Geyser (5) also fell dormant. Even weak, minor eruptions are now rare.

3. VALENTINE GEYSER was named because of an eruption on St. Valentine's Day in 1907, but either it or an adjacent vent had been active as "Alcove Spring" in 1890–1891. Valentine's geyserite cone, about 6 feet high, is the largest at Norris, but it is difficult to see because of its location in a deep alcove below the trail. Its history since 1907 has been highly variable. In good years the intervals can be as short as a few hours. At other times days to weeks may pass between eruptions, and Valentine is dormant more than it is active. Valentine's eruptions are strangely quiet. Little more than the hissing sound of falling water can be heard. Near the beginning of the play the steady jet may reach 75 feet high, although 20 to 50 feet is more typical. The water rapidly gives way to steam (sometimes within a few seconds), and the remainder of the play is an impressive cloud of steam and fine spray shooting to 40 feet. Most durations are in the range of 5 to 22 minutes. The most recent vigorous cycle of activity took place during 1989, when a series of minor eruptions on intervals of 8 to 20 hours led to major eruptions 3 to 5 days apart. A few additional eruptions took place in the days following a minor earthquake swarm in June 1990, but the geyser was apparently unaffected by the March 1994 tremors. Valentine currently has minor eruptions on long, erratic intervals.

4. GUARDIAN GEYSER is one of the few Porcelain Basin geysers to have increased its activity during the 2000s. It lies at the narrow exit of the alcove containing Valentine Geyser (3) and spouts from a small pile of rocks stained orange-brown by iron oxide. Guardian has shown several different modes of activity. It sometimes acts as a precursor to Valentine, preceding that eruption by a few moments. Such action is 8 to 25 feet high and lasts only a few seconds. When Valentine is dormant, Guardian shows different manners of play. Independent eruptions have been observed to last as long as 4 minutes, with steady jets that reach up to 30 feet, but in 2004 the activity had intervals of 1 to 4 minutes, durations only a few seconds long, and play that burst water only as high as 8 to 10 feet. Guardian has also been known to undergo noisy steam-phase eruptions with little or no water ejection, and since 2005 it has been the noisiest steam vent at Norris. In the long run, however, Guardian has been dormant more than active.

Another geyser, UNNG-NPR-6 (4a), located just outside the alcove opening a few feet from Guardian, made its appearance during 1984. The activity coincided with steam-phase eruptions by Guardian. The play recurred every 9 to 26 minutes, lasted as long as 19 minutes, and very much resembled the action of a squirt gun. The height was about 6 feet. NPR-6 has rarely been seen since 1984.

5. LEDGE GEYSER is, aside from Steamboat (48), the largest geyser at Norris, but its activity is highly irregular and it is dormant during most years. Its best year on record was 1974, when during part of the summer season it erupted every 14 hours with predictable regularity. It continued frequent action until the earthquake of June 30, 1975, then gradually declined into dormancy.

When active, there is an extraordinary amount of pre-play from Ledge. It erupts from five vents aligned so the activity somewhat resembles the human hand. Three of the central vents jet water a few feet high; they are known as the "finger vents," and their activity is almost constant during the buildup to an eruption. At the point nearest the trail is a deep cavity with a small vent at the bottom. This pressure pool (the "little finger") slowly fills with water during the pre-play. When Ledge was frequent and regular, as in 1974, experienced people could use the water level in this pool to predict the time of an eruption within minutes. The main vent is difficult to see from the trail. It lies to the far right, in the area where the bench containing the finger vents drops off to lower ground. This opening, the "thumb," penetrates the hillside at an angle. During the pre-play, water occasionally splashes out of this vent. When Ledge is ready to erupt, one of these splashes becomes a steady surge and then suddenly bursts into a tremendous eruption, reaching full force in a matter of seconds. The water shot from the main vent reaches a vertical height of 80 to 125 feet, but, because of its 40-degree angle, it can land more than 200 feet outward from the vent. Meanwhile, the fingers play slender columns of water to 30 feet and more, and the pressure pool bursts more than 60 feet high. The booming and roaring spectacle cannot be matched anywhere. Ledge's major activity lasts about 20 minutes, throughout all of which the maximum force is maintained by the main vent. Thereafter, the eruption slowly subsides. After about 2 hours it is finished, although Ledge never falls completely silent.

Ledge apparently entered dormancy in early 1979. Whatever affected it might well be the same as what started a long-term dor-

Ledge Geyser is the largest in Porcelain Basin, but its eruptions have been rare since the 1970s. It is at the edge of Porcelain Basin, a barren area studded with numerous other geysers of erratic activity.

mancy in Valentine Geyser (3) and a general decline throughout the western portion of Porcelain Basin that began in 1978. In 1989 and the early 1990s there were some signs of recovery. Jetting by the finger vents was sometimes quite strong, and there was occasional overflow from the main vent. However, because a number of noisy steam vents had developed on the slopes near Ledge, some observers felt they were robbing Ledge of eruptive energy. Finally, though, Ledge did erupt during the last week of December 1993. All of the old vents were active as before, and one of the new steam vents joined in by arching 100 feet high up and over Ledge as far as the asphalt walkway. Eruptions at intervals of 9 to 14 days continued until April 1998. With the exception of single, disturbance-related eruptions on June 10, 2006, and September 6, 2007, Ledge has been very quiet since 1998.

6. JETSAM POOL is a few feet northeast of the main vent of Ledge Geyser (5). It had small, bursting eruptions in 1970–1971, probably in conjunction with Ledge, so it was not a historical surprise when it joined the early 1994 action by Ledge, with erratic bursts up to 65 feet high. It had only a few of those eruptions before lapsing back to the murky, quiet pool of before. More recently the crater has been nearly empty, with only a bit of tepid water in its bottom.

7. BASIN GEYSER used to erupt from the center of its pool. The play was frequent but small. Near the time of the 1975 earthquake, some new vents developed at the left end of the crater, and most of the activity shifted to them, jetting up to 6 feet high. In the late 1970s the crater drained, and all eruptive activity stopped. Basin again contains a bubbling pool, but a lack of eruptions has allowed wildflowers to grow deep within the crater.

8. GEEZER GEYSER erupts from an unimpressive little hole among some jagged rocks. It is active essentially only at the time of a disturbance, but even then the action is rare. The play can last as long as 15 minutes and reach up to 15 feet high. The only known non-disturbance play was during one day in July 2004, when Geezer had three observed eruptions. Each reached about 8 feet high and lasted 8 minutes.

9. ARSENIC GEYSER poses some mystery. Through the years the name has apparently been variously applied to the geysers now known as Pinto (12), Fan (13), one of the vents of the Lava Pool Complex (10), and perhaps others. Accordingly, many historical references to "Arsenic" actually apply to those other geysers. Further confusing the situation, it is possible that the modern Arsenic is what was referred to as "Moxie Geyser" from the mid-1970s into the 1980s. Arsenic plays from a small, conical, sinter-lined vent slightly south of the others. Sometimes it is nearly dormant, the eruptions little more than intermittent bubbling. At other times true bursting plays up to 20 feet high recur as often as several times per day and last as long as 20 minutes. During the disturbance events, Arsenic is much more active, and durations as long as 3 hours with heights of 35 feet have been seen.

"Moxie Geyser" was first described in 1975, and in 1976 it underwent a few eruptions as high as 20 feet. Thereafter, the play was infrequent and usually only 1 to 2 feet high. It is likely that "Moxie" is the same as Arsenic Geyser, and the name has not been used since 1984.

10. LAVA POOL COMPLEX is a group of springs at the base of the low hillside beyond Arsenic Geyser (9). In the 1970s there were at least seven vents, and there was nearly always some spouting action among them. During the disturbances, all of the vents could erupt simultaneously for durations several hours long and to heights as great as 10 feet. Now, however, it is possible that the vents have been clogged with mineral deposits, as no water was visible within the complex even during the disturbances of 2006 and 2007.

11. AFRICA GEYSER developed during 1971 in a previously inactive crater shaped like the continent of Africa. At first the eruptions were 45 feet high and very regular, but as time progressed the intervals gradually shortened and the force weakened. By 1973, Africa had become a perpetual spouter perhaps 20 feet high. Change continued. The water gave out in the late 1970s, and a steady steam phase ensued. At times, the roar could be heard from the far end of the Back Basin nearly a mile away. But even that was temporary. Africa quit playing entirely when the sinter sheet ruptured nearby in 1981. The vent is now just another hole in the ground, partially filled with new mineral deposits.

12. PINTO GEYSER had early records of activity under the names "Twentieth Century Geyser" and "Arsenic Geyser" (see #9), but it spends most of its time as a quiet pool. During the disturbances, Pinto can become a geyser of considerable power. Muddy water is thrown as high as 40 feet in eruptions that last as long as 30 minutes. The irregular intervals range from 5 minutes to several hours during the few days of activity.

13. FAN GEYSER is located a few feet north of Fireball Geyser (14). The pearly, orange vent is almost impossible to see from the boardwalk. Fan is often dormant for long periods but sometimes becomes a very regular geyser. Then, it may have 10-minute eruptions as high as 10 feet several times per day. During the disturbance episodes, Fan develops a nearly perpetual bubbling and splashing punctuated by frequent, brief eruptions 15 to 25 feet high.

14. FIREBALL GEYSER shoots out of several small vents among a low pile of red, iron-stained rocks about 100 feet beyond Little Whirligig Geyser (15). Each of the openings jets water at a different angle and height. The largest is vertical and is 12 feet high. Eruptions by Fireball are normally frequent, with durations of around 5 minutes.

During the early 1990s the intervals were as much as several hours long, but the play lasted as long as 20 minutes and reached up to 25 feet. As with many other geysers in this part of Porcelain Basin, Fireball becomes much more active at the time of the disturbances. During 1984, when a series of minor disturbances was recorded, it was in eruption about half the time. Unfortunately, it participated in the slowdown of the early 2000s, and it often goes several days between eruptions.

There has been discussion as to which geyser, #13 or #14, is Fan and which is Fireball. The play of #14 (Fireball) is distinctly fan-shaped, while that of #13 (Fan) certainly is not. There is little doubt that the names were inadvertently switched some time in the past, so we now have a non–fan-shaped geyser named Fan and a fan-shaped geyser named Fireball.

15. LITTLE WHIRLIGIG GEYSER erupted from a crater colored orange-yellow by iron oxide minerals. Although known to have been active in 1887 and 1922, Little Whirligig rarely erupted prior to the early 1930s. During that decade, nearby Whirligig Geyser (16) declined in activity, and Little Whirligig became nearly a perpetual spouter. The eruptions were a squirting sort of action, jetting water at an angle toward the boardwalk and as much as 20 feet high. It wasn't until 1973 that it began taking occasional rests, pausing for about 20 minutes just two or three times per day. These pauses gradually grew more extreme. By 1975, play by Little Whirligig was infrequent and brief, and it has been completely dormant since 1991. Occasional strings of small bubbles that rise through the sediment in the crater mark the location of the vent.

16. WHIRLIGIG GEYSER was named because of the way the water swirls around in the crater during an eruption. After many years of dormancy, Whirligig rejuvenated in the summer of 1974, corresponding with a decline in vigor in nearby Little Whirligig Geyser (15). Eruptions were frequent and regular into the 1980s, but then Whirligig became erratic, with intervals as long as 12 hours. It fell completely dormant in June 2004 but has since reactivated to a limited degree. The play begins suddenly with a rapid filling of the crater. The largest portion of the activity comes from the central vent, where the water is tossed 6 to 15 feet high by a series of rapid bursts. This is joined by play from the "rooster-tail vent" on the east side of the crater. It jets water in a series of puffs, looking much like

Whirligig Geyser bursts from a deep crater and simultaneously jets out of a slit-like vent to one side.

the bird's tail plumes and sounding like an old steam engine. This chugging sound can easily be heard from the museum. Most durations are 3 to 4 minutes long.

17. CONSTANT GEYSER has never seen anything approaching perpetual activity, but at one time it had such a high degree of regularity and frequency that it was the one reliable "constant" at Norris. That has certainly changed, as has its relationship to the two Whirligigs. Historically, Constant was most active when Whirligig Geyser (16) was also active and nearly dormant if Little Whirligig Geyser (15) was playing. Yet in 2004, when both Whirligig geysers were dormant, Constant continued to erupt with little apparent change. The eruptions usually occur in series, typically with three to five eruptions separated by a few minutes each. Intervals between active phases range from only 20 minutes to a few hours. One sign to look for is a pulsation of the surface of the shallow pool, located a few feet beyond the Whirligigs. Watch closely! With only a few seconds of warning, the 30-foot play lasts just 5 to 10 seconds.

18. SPLUTTER POT (once known as the "Washing Machine") plays from a small crater at the far side of the runoff from Pinwheel Geyser (19). It has had many long dormancies. When active, the intervals are normally about 4 to 6 minutes long. The play is a chugging splashing 1 to 6 feet high that lasts 1 to 2 minutes. In some seasons and during disturbances, Splutter Pot acts as a small perpetual spouter.

Along the runoff between Splutter Pot and Pinwheel Geyser are two other vents that occasionally play as small, irregular geysers.

19. PINWHEEL GEYSER was once one of the stars at Norris. Its eruptions were frequent and over 20 feet high, and, in allusion to Grand Geyser in the Upper Basin, it was known as "Baby Grand." During the late 1960s the ground on the far, upstream side of the crater settled a few inches. Cooled runoff from other springs, which had previously flowed away from Pinwheel, was able to run directly into the pool. This water lowered the temperature of Pinwheel enough to stop almost all eruptive activity. During the disturbance of August 1974, this flow decreased markedly, and Pinwheel had a few minor eruptions 6 to 8 feet high. There has been no action since 1974, and it is possible that enough clay and silt have now washed into the vent to effectively seal Pinwheel off. If so, it may never again erupt as it did in the past.

20. PEQUITO GEYSER (Spanish for "little one") erupted from a tiny vent near the left edge of Pinwheel Geyser's (19) sinter shoulders. It was seen during only a few seasons, most recently 1993, when the irregular eruptions squirted water up to 5 feet high for a few minutes. The vent is now filled with mud.

At the base of the hillside about 50 yards north of Pinwheel and Pequito is Sand Spring. It sometimes erupts as a perpetual spouter, with splashes up to 3 feet high.

21. BEAR DEN GEYSER has been dormant since 1984. Its site is at the base of a cliff, where it issues from a narrow defile among some rhyolite boulders. Even when it was at its best, Bear Den never attracted much notice because it was partially hidden by trees and had no pool to emit steam. Therefore, the amazing eruptions came as a complete surprise to the few people who managed to see one from close at hand. Water was squirted by a series of distinct pulses to heights as great as 70 feet and angled so the water landed well downhill from the vent. A single eruption consisted of as many as 35 individual bursts over the course of several minutes. Intervals varied between 3 and 7 hours. Unfortunately, the vent has become thoroughly clogged with rocky debris fallen from above, and Bear Den is unlikely to ever again erupt as it used to.

22. EBONY GEYSER erupted from a yawning crater lined with dark gray geyserite. Its existence was not recorded until it was shown on a 1904 map, and details of its activity are unknown until the late 1940s. Ebony was then one of the largest and most faithful geysers at Norris. Eruptions recurred every few hours and reached as high as 75 feet. Unfortunately, debris thrown by Park visitors into the ready target below the old trail apparently spelled its demise. Even though the vent appears to be clear of rubble, it probably is thoroughly choked at depth. Major eruptions have not been seen since the 1950s. The most recent activity of any note occurred during 1974, when a few surges 10 feet high were seen. In 2005, Ebony was a small pool of lukewarm water well hidden from the trail, which has been rerouted several hundred feet away.

23. CATS EYE SPRING is a few feet northeast of Glacial Melt (24). Like Glacial Melt, it is most often active at the time of a disturbance, when play can reach about 2 feet high. Otherwise, its crater is often empty.

24. GLACIAL MELT GEYSER was named because of an opalescent appearance to the water, caused by a high content of colloidal silica particles suspended in the water. The result is a milky-blue color similar to that of glacial meltwater. This geyser is most active during the disturbances, when individual splashes several seconds apart reach 5 feet high. Other eruptive episodes are known, but Glacial Melt is usually a quiet pool when it contains any water at all.

25. TEAL BLUE BUBBLER was a sometimes-spouter, sometimes-geyser that played from a small pool perched on the hillside. The boardwalk stairway passes immediately next to it. Although not active during most seasons, eruptions as high as 6 feet were occasionally recorded, and, once triggered, the play could last as long as several hours. However, when a number of small spouters called the Milky

Complex formed downslope a few feet away, Teal Blue drained. It may never be active again.

26. "CRACKLING SPRING." The large body of water at the base of the hillside west of Teal Blue Bubbler (25) is called Crackling Lake. "Crackling Spring" is on the lake's south shore, against the hillside where it is slightly elevated within a geyserite formation. It is variably active as a small perpetual spouter or frequent geyser. Most of the play is 1 to 4 feet high. The name is an allusion to the sound of steam bubbles that implode within the pool, causing a popping sound. This spring should not be confused with noneruptive Crackling Lake Spring, a small pool near the north side of Crackling Lake.

27. CONGRESS POOL began life in 1879 as a powerful steam vent (the "Steam Whistle"), then was a mud pot before it became a geyser in 1891. Within weeks it had become a large and regular performer. Because of a visit to Yellowstone that year by members of the 50th International Geological Congress (which led to the geyser's name), the geyser gained considerable attention and fame. Much to everybody's disappointment, though, it soon stopped erupting and became a quiet pool. From that time into the 1970s it sometimes acted as a geyser, but overall there was little activity from Congress other than infrequent "almost eruptive" behavior that caused intermittent roiling of the pool's surface. Finally, in 1974 it again became a true geyser. This activity began earlier than that year's disturbance but otherwise coincided with the disturbance. Congress would sometimes burst muddy water as high as 20 feet, and the intervals were as short as 4 minutes. As this activity continued, the water level slowly dropped until it was down fully 5 feet. After about two weeks, Congress began to decline in force, the crater refilled, and the pool was soon back to normal. Since then, Congress Pool has occasionally drained entirely and acted as a weak steam vent, yet at other times it can be an overflowing pool. It continues to show disturbance-related behavior, but never to the degree seen in 1974.

28. CARNEGIE DRILL HOLE. Across the flat barren area east of Congress Pool (27) is what looks like a conical rock pile. Water sometimes spurts from the top of the "cone" and from a small pool at the base of the mound. What you cannot see from the trail is that the entire area—rocks, pool, and all—pulsates up and down as much as several inches every second or so, pulsations that indicate that this could someday be the site of a substantial steam explosion.

Actually, the rock pile is cemented together with concrete, marking the site of a drill hole. The work was done in 1929 by Dr. C. N. Fenner of the Carnegie Institute of Washington. The project was part of one of the earliest efforts to do deep drilling in a geothermal basin to gather data about rock alteration and subsurface temperatures. The results were surprising. First, all notable rock alteration was confined to the upper few feet of the hole, showing that Norris's strongly acid condition is a surface phenomenon. Second, the hole had to be abandoned at a depth of only 265 feet—the steam pressure was so great it threatened to blow up the drill rig. At that shallow depth the temperature was 401°F (205°C). After abandonment the hole was filled with cement, but the hot water soon found a way around the plug. Now active as a perpetual spouter, the eruption from the top of the rock pile sometimes reaches as high as 2 feet.

Numerous perpetual spouters lie along the base of the hillside behind Congress Pool and the Carnegie Drill Hole. Some have names, such as Vermilion Springs, White Bubbler, and Locomotive Spring, and a few of these vents may act as intermittent geysers at times.

29. FEISTY GEYSER was once one of the finest geysers in Porcelain Basin, but if it still exists, it cannot be distinguished from the many other geysers and spouters that have developed in its vicinity. Feisty's site is on the sinter shield below the Porcelain Springs, which flow alkaline water and are depositing geyserite at the fastest rate ever measured in Yellowstone (several inches per year compared with the more typical 1 or 2 inches per century). As a result, existing springs change rapidly. Feisty was always somewhat irregular, but it usually played a few times per hour. Lasting several minutes each time, the play was as much as 25 feet high. No such geyser exists in the area now, but at least a dozen vents are known to play up to about 10 feet. One of them may be Feisty.

30. INCLINE GEYSER made its first appearance near the base of Porcelain Terrace during February 1990, and for about five years it alternated dormant episodes with short periods as a vigorous geyser. Its jagged crater, proof of an explosive origin, is immediately at the bottom of the hillside, below the old road near the end of the trail. The eruptions were some of the largest ever seen in Porcelain Basin. Angled against the hillside so as to cause tremendous erosion, the play uniformly reached 30 to 70 feet high, and some of the first recorded eruptions hit at least 110 feet and spanned the old roadway. As a member of the Porcelain Springs, where the rapid deposition of fresh geyserite tends to quickly seal new springs, Incline was expected to have a short lifetime, and it did. The last known eruption was in early 1995. However, the crater is still there and contains superheated water, so future activity is not out of the question.

31. UNNG-NPR-5 ("LAMBCHOP GEYSER") came to life in 1990, about the same time as Incline Geyser (30) just a few feet away. The informal name was from the shape of the crater as it appeared in 1992, but continuing evolution quickly destroyed that outline. Because of its location among the Porcelain Springs, "Lambchop" was expected to have a short life, but as of 2007 it still played as a powerful perpetual spouter, noisily jetting superheated water 10 to 12 feet high.

32. BLUE GEYSER lies far out on the barren flats in the north-central part of Porcelain Basin. The only large pool in that area, it is most active during disturbances, sending up periodic large domes of water as high as 15 feet. The name is something of a misnomer, as Blue is gray in color much of the time.

33. IRIS SPRING occupies a small crater immediately in front of Blue Geyser (32). When active, it acts as a perpetual spouter. Most of the splashes are quite small, but jets reaching 15 feet are occasionally seen. Dormant periods are common.

34. THE PRIMROSE SPRINGS lie along a shallow "valley" extending from below the trail northward toward the main flats of Porcelain Basin. Within this group are several perpetual spouters. The largest, Primrose Spring itself, has been known to act as a geyser, splashing up to 3 feet high. It and the other craters have usually been empty since the late 1990s.

The area between the Primrose Springs and Blue Geyser used to be occupied by numerous large pools given names such as Apple Green Geyser and "Norris Geyser." These pools occasionally had massive bursting eruptions as wide as their 30-foot heights during the disturbances of the 1970s. They are now gaping holes partially filled in with geyserite deposited by the runoff from the Porcelain Springs.

35. SUNDAY GEYSER got its name because it had its first spectacular eruptions on Sunday, July 12, 1964. It is dormant nearly all the time. The last episode of large eruptions was in 1981–1982, when it was a frequent performer. Eruptions 30 to 50 feet high recurred every 15 to 20 minutes. The vent is a rather small hole at the south end of a shallow, pale-blue pool just below the boardwalk. This pool is a separate spring that rarely has small eruptions during disturbances.

36. HURRICANE VENT is a deep, sinter-lined hole on the uphill side of the boardwalk near Sunday Geyser (35). The reason for the name was the cyclone-like whirling of the water in the crater during early eruptions. It formed during the winter of 1885–1886, probably as a result of an earthquake. Hurricane Vent became quite a tourist attraction during 1886, with frequent eruptions up to 30 feet high. It was even stronger the next summer, when a viewpoint constructed the previous year was frequently inundated by bursts of muddy water. But that was it. Hurricane Vent has been almost completely inactive since 1887. A few small, erratic splashes were seen during 1991.

An unnamed hole that sometimes intermittently issues steam under pressure is located high on the east wall of Hurricane's crater. Viewers often mistake it for Hurricane Vent itself.

37. RAGGED SPOUTER was a 1968 development in the far northeastern part of Porcelain Basin. The vent was simply a wide spot along an old fissure. In 1981 it was described as having "Echinus-like" bursts approaching 50 feet high (see Echinus Geyser [50]), but clearly the geyser was dooming its own future. As the crater enlarged itself, runoff flowing back from the adjacent hillside carried debris into the vent. The activity rapidly died down, and the crater is no longer clearly identifiable.

38. GRACEFUL GEYSER became active in late 1981 or early 1982. At first, the eruptions played from a small cone. However, being old and badly weathered, the cone was quickly torn apart. The original slender, graceful jets were replaced by a bursting action out of a jagged crater. Graceful is one of the few Porcelain Basin geysers to have continued vigorous activity since the early 1990s. Some eruptions reach at least 20 feet high and continue for an hour or longer. However, since it is nearly 1,000 feet from the nearest boardwalk, Graceful is difficult to appreciate.

39. COLLAPSED CAVE GEYSER is another little-known geyser in the far northeastern corner of Porcelain Basin. It lies within a large, cavernous opening. Eruptions in the early 1980s, although lasting no more than 10 seconds, reached as high as 20 feet and recurred on intervals as short as 5 minutes. Part of the cave's roof collapsed inward in the late 1980s. Although the geyser is still active, its eruptions are blocked by the debris and are impossible to see from a distance.

Note: Many other geysers are known to have existed within Porcelain Basin. Most have been active only once or twice during disturbance events, but some have been more persistent. Those that have had active phases that lasted as long as a year or two have often been given informal names. Some of these are "Junebug Geyser," "Ramjet Springs," "Blowout Geyser," and "Christmas Geyser." Whether any of these geysers will be active in the future is, of course, unknown.

BACK (OR TANTALUS) BASIN

Hot spring activity in the Back Basin (Map 7.2, Table 7.2, numbers 45 through 84), the larger, southern portion of the Norris Geyser Basin, is less concentrated than in the Porcelain Basin. Pine trees separate small groups of hot springs from one another, often even isolating individual springs. These springs are affected by disturbances as are those of Porcelain Basin, but overall they tend to be more stable, with fewer features appearing and disappearing over the years. The exception to this is the area between Green Dragon Spring (see map) and Porkchop Geyser (70). Disturbances always have especially dramatic effects in that zone, and the development of new hot ground during 2003 required that a portion of the trail be rerouted with a new boardwalk that opened in 2004.

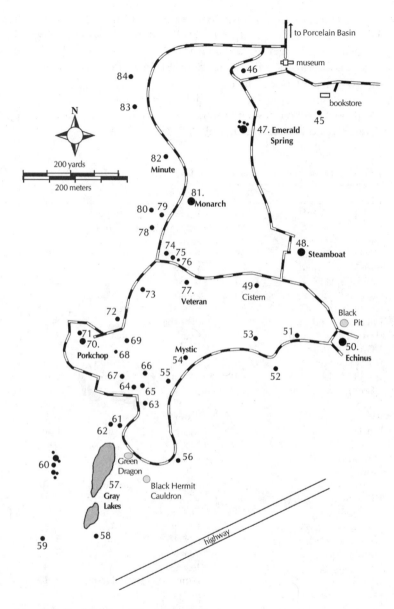

Map 7.2. *Back Basin of Norris Geyser Basin*

An alternate name for the Back Basin is "Tantalus Basin," which is how the area was identified by Yellowstone's earliest geological surveyors and is also the name of the hot water stream that drains both it and Porcelain Basin. Tantalus was one of the sons of Hades, the god of the Underworld.

The loop trail around the Back Basin is about 1½ miles long. For those who do not wish to walk the entire distance, the "Cistern-Veteran Cutoff" shortens the hike to about ¾ mile.

Table 7.2. Geysers of Back Basin, Norris Geyser Basin

Name	Map No.	Interval	Duration	Height (ft)
Arch Steam Vent	52	[1974]	unknown	40
Bastille Geyser	68	see text	1½ min	1–3
Bathtub Spring	46	steady	steady	3
Big Alcove Geyser	58	seconds	near steady	10–25
Blue Mud Spring	61	steady	steady	sub–10
Cistern Spring	49	w/ Steamboat	minutes	20
Corporal Geyser†	75	20 min*	1–3 min	1
Crater Spring†	51	[1984]	minutes	10–15
Dabble Geyser†	66	disturbance	4 min	8
"Daughter of Green Dragon Spring"†	65	steady	steady	1–3
Dog's Leg Spring	76	rare*	sec–hrs	1–3
Double Bulger Geyser Complex†	69	see text	steady	5–10
Echinus Geyser†	50	[2007]	3–5 min	20–75
Emerald Spring†	47	disturbance	steady	3–75
Fearless Geyser	79	steady	steady	boil
Forgotten Fumarole	83	[2008]	seconds	6–30
"Grandson of Green Dragon Spring"	64a	steady*	steady	1–3
Gray Lakes, The†	57	erratic	hrs–days	1–6
Hydrophane Springs†	60	see text	min–steady	1–10
Medusa Spring†	59	variable	sec–hrs	12
Minute Geyser	82	steady	steady	3–8
Monarch Geyser	81	[1994]	minutes	15–200
Mud Spring†	55	40–60 min*	12–20 min	1–30
Mystic Spring	54	1–1¼ hrs*	10 min	1–4
Orby Geyser†	67	minutes*	1–2 min	3–10
Palpitator Spring	78	hours	hours	1–3
Pearl Geyser†	72	frequent*	seconds	1–8
Porkchop Geyser†	70	infrequent	minutes	6
Puff-n-Stuff Geyser	56	near steady	near steady	spray–10

continued on next page

Table 7.2. Geysers of Back Basin, Norris Geyser Basin

Name	Map No.	Interval	Duration	Height (ft)
Rediscovered Geyser	84	minutes*	minutes	1–4
Rubble Geyser	74	rare	6 min	10
"Second Erupter"	71	hours	hours	2–3
Son of Green Dragon Spring	64	steady	steady	1–6
Steamboat Geyser, major	48	[2005]	hours	250–386
Steamboat Geyser, minor	48	sec–min	seconds	20–100
Steamvalve Spring†	45	[2006]	15 min	12
Tantalus Geyser†	53	disturbance	minutes	3–35
UNNG-NBK-3†	80	steady	steady	1–3
UNNG-NBK-4 ("Spearpoint")	62	minutes	seconds	3
Veteran Geyser	77	frequent	sec–2½ min	3–50
Vixen Geyser (major)	73	[2003]	min–1 hr	35–40
Vixen Geyser (minor)	73	infrequent	seconds	1–10
Yellow Funnel Spring†	63	steady	steady	1–6

* When active.
† Indicates geysers that show more frequent and/or more powerful activity at the time of a disturbance. In the interval column, "disturbance" indicates geysers that are active essentially only at the time of a disturbance.
[ ] Brackets enclose the year of most recent activity for rare or dormant geysers. See text.

45. STEAMVALVE SPRING was known during the early days of the Park, but a dormancy of many decades left it all but forgotten except as a "dead" crater in an old parking area. During the late 1970s Steamvalve reappeared as a geyser. For several years its activity was fairly predictable. About 1½ hours before an eruption, bubbles began to rise through the water as the crater slowly filled. The play began well before overflow was reached. Some of the bursts were as much as 12 feet high, and the biggest usually occurred near the end of the 15-minute eruption. Steamvalve returned to dormancy in the mid-1980s. Weak activity was recorded in late 2006, but the geyser was dormant in 2007.

46. BATHTUB SPRING is usually an acid spring that splashes muddy water from a crater lined with massive geyserite formations. The sinter formed at times when Bathtub held alkaline rather than acid water. It has been known to switch its chemistry several times. Bathtub is probably identical to a geyser of the 1800s known as the "Schlammkessel." In more recent years, Bathtub has acted as a muddy perpetual spouter, most play reaching 3 feet high.

47. EMERALD SPRING spends most of its time as a beautiful yellow-green pool. The color is from the combination of the blue of the water and the yellow of sulfur lining the crater walls. The color is intense and quite different from that produced by yellow cyanobacteria.

Emerald Spring calmly bubbles above its vents. The bubbling is not boiling but is a result of mixtures of steam, carbon dioxide, and other gasses. At the time of a disturbance, Emerald commonly becomes turbid gray-brown, undergoes true boiling, and sometimes acts as a geyser. Most such eruptions are 3 to 6 feet high, but in 1931 it had an extraordinary episode of activity, when it was in eruption at least 87 percent of the time with bursts reaching 60 to 75 feet high.

Around the northern sides of the shallow alcove that includes Emerald Spring is a series of depressions. These mark the vents of several ephemeral springs and spouters. First active in the 1930s, they were not reported again until some brief action in 1971. Then, in September 1974, they became active again. It was remarkable how within 30 minutes the area went from no springs at all to more than a dozen vigorous spouters. Their development was the first sign of a localized disturbance. Although some of these springs played as high as 10 feet and their total water discharge was large, most completely disappeared within 6 weeks. Two or three continued their action into 1975. None of these springs has been active since then, and small trees now grow in some of the craters that formed in 1974.

48. STEAMBOAT GEYSER is the tallest and physically largest geyser in the world—when it is active. The only geyser known to have played to greater heights was New Zealand's Waimangu, which has been dead since 1904. Most of the time, Steamboat's only eruptions are a continuous series of "minor" splashes every few seconds to minutes. Actually, Steamboat is farther from the boardwalk than it looks, and some minor eruptions reach as high as 75 feet. For Steamboat, such bursts are almost insignificant.

Steamboat has an interesting history. According to some descriptions, there was no spring of any kind here before 1878. It is more likely that something did exist but that it was small and/or inactive. In any case, 1878 is the year of Steamboat's first known series of major eruptions. The new geyser was called both Steamboat and New Crater. (Both names are officially approved.)

The destruction the eruptions brought to the surrounding forest was unmistakable: trees were killed, huge rocks thrown about, and plants covered with sand and mud by eruptions well over 100 feet high. Through the next three decades most of Steamboat's eruptions were minor in size, but eruptions at least 100 feet high were recorded in 1888, 1890, 1891, 1892, 1894, and 1902. Although there may have been action between 1903 and 1910, the next year in which there was a series of major eruptions was 1911. The height was said to be over 250 feet. Then Steamboat fell quiet and had only minor activity for almost exactly 50 years. Starting in 1961 and continuing into 1969, Steamboat had its best active period on record. All told, there were 90 major eruptions during those 9 years, 29 of them in 1964 alone.

It is difficult to adequately describe a major eruption of Steamboat. Even photographs do not do it justice, for they always seem to impart a sense of beauty rather than violence. With the proper study, Steamboat's minor eruptions are seen to pass through a relatively clear-cut progression of types of play, defined mostly on the basis of which one of the two vents initiates the eruption, whether there is simultaneous action, the duration, and so on. To those not familiar with this sequence, though, a major eruption seems to begin with a normal-looking minor, except that the action persists longer than a few seconds and continues to build in height. More often than not, these "superbursts" (or "silver bullets") fail to trigger a full eruption. But there is simply no describing the scene when the action proves to be the start of a major. The height continues to grow, almost beyond belief. A number of eruptions have been measured at about 380 feet high, more than three times the average height of Old Faithful, and none reaches less than 250 feet. The water phase of the eruption lasts from 3 to 20 minutes, during which several hundred thousand gallons of water are discharged. Then comes the steam phase. The roar is tremendous and may last several hours. There is a rumor that one eruption, on a cold, crisp, and windless winter day, was heard in Madison Junction, 14 miles away! That is unlikely, but at times it is literally impossible for people on the boardwalk to yell at one another and be heard.

After its major eruption on March 20, 1969, Steamboat failed to play again for another 9 years. There is no chance that any eruptions were missed in the interim, since any eruption by Steamboat leaves abundant signs. So, it was hoped that the eruption of March 28, 1978, presaged a new series of activity. And so it did, although

Steamboat Geyser is the largest geyser in the world, with eruptions that have been measured as tall as 380 feet. Unfortunately, its eruptions are sometimes years apart. Credit, NPS photo by Robert Lang and William Dick.

eruptions were sparse until 1982. There were 23 eruptions that year; 12 more occurred during 1983, but there were only 5 in 1984. Since then, Steamboat has played just 12 additional times: 3 eruptions in 1989, 1 in 1990, 1 in 1991, 1 more in 2000, 2 in 2002, 3 in 2003, zero in 2004, and 1 in 2005 (May 23).

On three known occasions—January 1994, April 1995, and February 2007—the Steamboat Geyser–Cistern Spring system acted as if there had been an eruption even though there was none. As if "a drain plug had been pulled" so the water drained away, Steamboat entered a powerful steam phase and Cistern Spring gradually drained to a low level, yet there was no unusual melting of ice and snow near Steamboat or along the runoff channels. These events apparently resulted because of midwinter disturbances.

Steamboat's activity has revealed some things about the water circulation within the Norris geothermal system. A geyser as large as Steamboat must have access to a huge supply of very hot water. It seems likely that Steamboat's deeper plumbing taps rather directly into a second, hotter and deeper geothermal aquifer than do many of the other hot springs at Norris. At the times of the disturbances, however, there is a mixing of the two waters. Recently, the onset of a disturbance has first appeared in Steamboat, sometimes as long as two days before it was evident elsewhere. Often this happens when Steamboat's minor play appears to indicate the approach of another major eruption. The disturbance ends any such possibility until well after its effects have passed. Known as the "zap," the effect is also seen in Echinus Geyser (50), Emerald Spring (47), and Steamvalve Spring (45), suggesting that these geysers may be associated along some sort of subsurface fracture system. Indeed, it has even been suggested that Steamboat's action somehow causes the disturbances that sometimes affect all of Norris.

So, for both geyser gazing and geothermal research, Steamboat is a very important geyser. It is not likely that you will see a major eruption. Then again, somebody will.

49. CISTERN SPRING is the only spring known to be directly connected with, and therefore affected by, an eruption of Steamboat Geyser (48). Every major eruption causes Cistern to drain by as much as 12 feet, and Steamboat's "silver bullets" have also been observed to have slight effects. Cistern itself has been known to erupt on just two occasions. Both followed Steamboat majors, in 1978 and 1982. The height was about 20 feet.

50. ECHINUS GEYSER used to be a favorite of nearly everybody. Very regular and predictable, it was also large and beautiful. Nowhere could you get so close to a major geyser as at Echinus. Unfortunately, Echinus began to change in 2001.

The name "Echinus" comes from the Greek word for spiny. The same root gave the sea urchins and starfish their collective name of *echinoderms*. The name was applied here because an early visitor thought some of the stones about the crater resembled sea urchins. Indeed they do. They are rhyolite pebbles that have been coated with the spiny sinter typical of acid water conditions. This is one of the better examples of such geyserite. The prominent reddish and yellow-brown colors are caused by iron oxide minerals deposited along with the sinter. This is also one of the rare cases where arsenic compounds are deposited by the hot water—the sinter at Echinus contains about 5 percent arsenic pentoxide.

When active, Echinus's activity begins with its large crater slowly filling with water. As the filling progresses and the eruption time nears, the filling becomes faster. Sometimes the play starts with the level still a few feet below the rim, but usually Echinus fills completely and overflows before the eruption. Either way, the bubbling above the vent becomes a boiling and the boiling a heavy surging, and the eruption is triggered within a few seconds. Echinus is a typical fountain-type geyser, throwing its water in a series of closely spaced bursts. Each burst is different from every other—some small, others nearly 75 feet tall, some straight up, and others so sharply angled that they harmlessly soak people on the benches with warm water.

Historically, Echinus's average interval has always varied. Prior to 1948, Echinus was often dormant, and the best of its infrequent eruptions seldom reached 35 feet high. It then became progressively more active, and throughout the 1970s and 1980s the intervals generally ranged between 40 and 80 minutes. During the 1990s, Echinus began to suffer hours-long intervals. In 1998 it began to have short episodes of dormancy at the same time significant new activity began at The Gap (discussed later) more than half a mile to the west. When it recovered, intervals often fell in the range of 4 to 8 hours, but instrumental monitoring proved some to be as long as 3½ days. Minor eruptions, which lasted only about 1 minute and failed to fill the crater, occasionally took place during these long intervals. Echinus finally entered complete dormancy in December 2005. It briefly recovered in July 2006, with intervals of 2

Echinus Geyser has sometimes been both the largest and the most predictable geyser at the Norris Geyser Basin, but in recent years its performances have been erratic, with several months-long dormant periods.

to 3 hours and durations of 3 to 5 minutes, but was again dormant within a few weeks. Echinus's only action since then has been a few eruptions associated with the same February 2007 disturbance that caused the odd, steam-phase–only eruption by Steamboat Geyser plus a few short episodes of weak activity in late 2007.

Echinus has always been affected by the disturbances, when it may undergo what have been termed "super eruptions." The water turns from nearly clear to muddy during play that lasts as long as 70 minutes and is of considerable power. Heights in excess of 100 feet have been reported. Like the "zap" at Steamboat (48), the super is often one of the first indications that a disturbance is starting.

51. CRATER SPRING has also been called "Collapse Crater Spring" because sometime in the distant past part of the crater's undercut rim collapsed inward. The rubble still lies there. During most of Crater Spring's known history, a steady eruption jetted above the boulders. During an exceptionally early disturbance in May 1983, Crater began playing as a true geyser. Eruptions lasted as long as

10 minutes and recurred as often as every 30 minutes. The main jet was angled to the north and reached over 20 feet high, arching over the trail so that a new route had to be constructed on the other (south) side of the pool, where it remains. By midsummer 1983 Crater had regressed to a quiet pool, but unlike before it was full and overflowing and of a rich blue color. Gradually, the water level dropped, and Crater has returned to its previous state as a small perpetual spouter. In 2007 the crater was lined with a deposit of bright yellow sulfur.

52. ARCH STEAM VENT is capable of major eruptions. This is known from splashed areas, runoff channels, and killed vegetation, but apparently no reporting observer has ever seen one of the plays except at a distance. The greatest number known for any single year is four, in 1974, and none has likely taken place since then. The evidence is that an eruption jets at about a 45-degree angle away from the slope, reaching perhaps 40 feet high for durations of only a few seconds. Deep runoff channels below the vent imply that Arch had considerable eruptive activity in the prehistoric past.

53. TANTALUS GEYSER had significant activity only during 1969, and, in fact, it is probably no different from the many other ephemeral, disturbance-related geysers and spouters so common to Norris except for its size. During its few days of activity in 1969, Tantalus played as high as 35 feet, with massive, widely spraying bursts of muddy brown water. Usually a quiet pool, Tantalus still tends to become muddier and to splash weakly during most disturbances.

54. MYSTIC SPRING was known to have true geyser eruptions during some disturbance events, but in August 2004 the gently boiling pool of yellow-green water was found playing up to 4 feet high. There was significant overflow, and the adjacent slope was progressively eroded to reveal old geyserite deposits. Following an eruption, the water level in the crater rapidly dropped about 2 feet, then immediately began to slowly rise. Eruptions began every 1 to 1¼ hours when the water was still several inches below the rim of the pool. The duration was 10 minutes. This activity continued through 2006 but had decreased to perpetual splashing just 1 foot high before the end of summer 2007.

55. MUD SPRING is, contrary to its name, clear and quiet most of the time. When active, however, it is very muddy and once upon a time was known as "Chocolate Fountain." Most active periods are associated with disturbances, when bursts up to 30 feet high have been recorded. The more typical sort of action is steady surging to just 1 or 2 feet. During 1984, Mud Spring performed as a small but regular geyser for most of the summer. Intervals were 40 to 60 minutes, durations 12 to 20 minutes, and the height about 1½ feet.

56. PUFF-n-STUFF GEYSER is right next to the trail. Named in allusion to the talking dragon of the old *H. R. Pufnstuf* television show, the dissected cone constantly rumbles and gurgles violently at depth, sending a fine spray of water a few feet high. Puff-n-Stuff looks and acts as if it could do more at any time, and a runoff area between the cone and the boardwalk implies that it does, but only one significant eruption has ever been recorded. That occurred during a disturbance in January 1994, when Puff-n-Stuff sent strong jets as high as 10 feet.

57. THE GRAY LAKES are the large, muddy pools beyond the perpetual spouters of Green Dragon Spring. The lakes contain numerous hot spring vents where eruptions are occasionally seen, as during 2005 and 2007 when a vent in the north lake splashed as high as 6 feet.

The next three geysers are all but impossible to see from the trail. Since the forest fires of 1988 cleared much of the growth in the southern part of Norris, Big Alcove Spring (58) and Medusa Spring (59) are visible from the highway at a point about ½ mile south of Norris Junction. Portions of the Hydrophane Spring Complex (60) can be seen from that same point as well as from the trail near Blue Mud Spring (61).

58. BIG ALCOVE SPRING is, as seen from the highway, near the front-right side of the large pools known as the Gray Lakes. It plays from a crack in the volcanic bedrock within a deep, alcove-like crater. The pulsating jet is nearly steady and has been known to reach over 25 feet high, although most bursts are much smaller. Big Alcove is never completely quiet, but the jetting pauses for a few seconds often enough to qualify it as a geyser.

59. MEDUSA SPRING is most likely to be active during a disturbance, when it can erupt as high as 12 feet. Otherwise, it is a nondescript pool about 10 feet in diameter. Viewed from the road, it lies south of the Gray Lakes next to the forest on the far side of the thermal tract.

60. HYDROPHANE SPRINGS is a complex of pools, spouters, and geysers largely hidden from view from any trail or road access. When one is standing at Blue Mud Spring (61), they are the features across the stream and beyond a low ridge. The activity within the complex is extremely variable. Often, but not exclusively, active at the time of a disturbance, a number of the vents may act as perpetual spouters or geysers, and eruptions of substantial size are sometimes seen. A pool here had bursts 30 to 50 feet high in late 1972, and a small vent jetted constantly as high as 8 or 10 feet during 2004. Most often, though, all that can be seen of the Hydrophane Springs is a bit of blue reflected in the steam rising from the deeper pools.

61. BLUE MUD SPRING is a perpetual spouter immediately below the boardwalk. Nearly all the action is subterranean, but sometimes the play will briefly spray muddy mist as high as 10 feet.

62. UNNG-NBK-4 ("SPEARPOINT GEYSER") is on the flat between Blue Mud Spring (61) and the stream. Its crater is shaped like a stone arrowhead, but that seemingly good name is used in the Upper Geyser Basin. "Spearpoint" erupts every few minutes. Most play lasts between 20 and 30 seconds and is 2 or 3 feet high.

NBK-4 and the other features here are members of the Muddy Sneaker Complex. They began to appear in previously inactive ground during 1971, when a small mud pot opened right in the middle of the old trail. Since then, several springs have come and gone frequently, and a few additional temporary geysers have been observed. "Spearpoint" is the only one of these that has formed a solid geyserite lining in its crater.

The Tangled Root Complex occupies the bottom of a shallow draw a few feet north of Blue Mud Spring. It also developed during the early 1970s, and, again, some brief geyser activity has been observed. Sometimes these small pools are of a beautiful opalescent blue color because of high content of colloidal silica in the water. Most commonly, the craters are empty.

63. YELLOW FUNNEL SPRING deserved its name in the past, when it was a quiet pool whose symmetrical crater was lined with yellow sulfur. However, it always became muddy during the disturbances, and in 1989 and 1992 it had eruptions several feet high. Smaller eruptions, 1 to 2 feet high, occurred during subsequent disturbances. In the early 2000s the water level dropped, and constantly bursting play over 6 feet high has eroded the crater into a jagged hole.

64. SON OF GREEN DRAGON SPRING bears little resemblance to the much larger, cavernous Green Dragon Spring a few hundred feet away (see Map 7.2). Son of Green Dragon is usually active as a small perpetual spouter, so weak that it is difficult to see from the boardwalk. There are exceptions, though. In 1993 it acted as a true geyser, playing clear water as high as 8 feet with both intervals and durations of about 5 minutes. During disturbances it throws muddy water 2 to 6 feet high, and for much of 2007 it was a noisy fumarole.

A few feet west of Son of Green Dragon is "Grandson of Green Dragon Spring" (64a). It usually acts as a muddy perpetual spouter 1 to 3 feet high.

65. "DAUGHTER OF GREEN DRAGON SPRING" is one of the features that broke out as completely new springs during 2003. Since its creation, it has acted as a small perpetual spouter less than 3 feet high. Adjacent to and eventually consuming some of the old dirt trail, it is one of the developments that required that the trail be closed in 2003 and replaced by a rerouted boardwalk. Unfortunately, "Daughter of Green Dragon" is almost impossible to see from the new walk.

66. DABBLE GEYSER is active almost exclusively at the time of the disturbances, although it has been seen a few other times as well. Its action is highly variable. At its uncommon best, intervals as short as 30 minutes result in vigorous 8-foot splashing for as long as 4 minutes. When inactive, it is a quiet and rather cool pool that is invisible from the trail.

Someplace within a few feet east or northeast of Dabble—exactly where is uncertain—used to be Pebble Geyser. In its time it was one of the larger geysers at Norris, with play as high as 50 feet between 1878 and 1887. It has not been seen since then, and which among the many eroded vents in this area it erupted from is unknown.

67. ORBY GEYSER plays from a shallow, round ("orbicular") crater about 100 feet north of the boardwalk, which provides a much better view than the old trail did. During most of its known history it has been an insignificant spring, sometimes even filled with cold water. When active, though, as it has been much of the time since the late 1970s, it has frequent and vigorous eruptions. Most intervals are shorter than 5 minutes in length. Water is splashed 3 to 10 feet high over durations of 1 to 2 minutes. At the times of the disturbances, Orby often approaches perpetual activity, being in eruption more than 90 percent of the time, with intervals of seconds and durations as long as 10 minutes. There are numerous small spouters and geysers in the vicinity of Orby, but Orby is the only geyser there with a distinct, symmetrical crater.

68. BASTILLE GEYSER made its appearance during a disturbance on July 14, 1992, Bastille Day in France. The crater developed immediately next to the boardwalk, covering it with sandy debris and leading to a temporary closure of the area. Eruptions were frequent and vigorous, bursting 1 to 3 feet high from several small vents. However, geyserite was deposited rapidly in and about the crater, and the vents were completely sealed by 1995. A brief history such as this is typical in this portion of the Back Basin, where several similar developments have taken place since 1992. Bastille's history is one of the reasons the old trail through this area was closed in 2003—whenever a new vent opens and evolves in this fashion, it is possible that a substantial steam explosion might ultimately result.

69. DOUBLE BULGER is only half alive now—if that. There were once two small, symmetrical geyserite vents adjacent to each other. The larger of the two acted as a murky perpetual spouter barely large enough to throw water from the crater. The second, smaller opening probably used to perform in a similar manner, but years ago it sealed itself with geyserite and then became filled with rocks and gravel. Located in the area where many changes have taken place since 2003, it is likely that Double Bulger no longer exists in recognizable form.

70. PORKCHOP GEYSER used to play from a crater that was indeed shaped like a pork chop. The vent was a 2-inch hole at the narrow end of the crater. It was a vigorous geyser, sometimes highly regular in both its intervals and durations. The typical play sounded like an old steam engine, with several chugs per second sending jets of water 15 to 20 feet high. Through the years, Porkchop showed a tendency toward progressively steadier, stronger, and steamier eruptions. By the late 1980s it had become a perpetual spouter that sent spray over 30 feet high. The culmination happened on September 5, 1989. During a disturbance, Porkchop literally exploded. Its beautiful crater was replaced by a ragged pile of broken geyserite boulders tumbled around a pool that measures 6 by 12 feet. The blast threw rocky debris as far as 220 feet away. Porkchop continues to have infrequent boiling eruptions that are seldom higher than 6 feet.

71. "SECOND ERUPTER" has an odd name of uncertain origin. Its play bursts out of a cavernous vent to reach only 2 or 3 feet high but outward as far as 6 feet. The quiet intervals tend to be much shorter than the durations, and both may be hours long.

Several hundred feet to the west, two large steam clouds rise from the Ragged Hills. The one to the left is from Psychedelic Steam Vent. The other is produced by Recess Spring, a pool that in recent years has erupted as high as 15 feet. Neither of these features is approached by any trail.

72. PEARL GEYSER is one of the few hot springs at Norris to resemble those of the other geyser basins. The wide crater of smooth geyserite is centered by a symmetrical vent filled with a pool of clear water. Pearl is highly variable, though. In many years the water level is well down inside the vent, where nearly constant bubbling gives rise to frequent but erratic splashes 1 to 2 feet high. Only on rare occasions will Pearl completely fill and overflow, but then it tends to act as a regular geyser. During such episodes, cyclic activity produces intervals ranging from seconds to minutes in length, and some bursts reach 8 feet high. Pearl is also known to completely drain and behave as a weak steam vent.

73. VIXEN GEYSER plays from a reddish vent close to the trail and attracts a considerable crowd when it is active. Vixen has two types of play. Minor eruptions are generally the rule. They last only a few seconds but recur every few minutes and reach up to 10 feet high. Major eruptions tend to occur during brief episodes months to years apart, most recently during 2003. They have durations of only a few minutes (a few that lasted as long as 1 hour were seen many years ago). This jetting reaches 35 to 40 feet high. It is only during major eruptions that Vixen discharges enough water to produce a runoff stream away from the vent. The channel leading to Tantalus Creek is shallow and nondescript, an indication that major eruptions have always been rare. Unfortunately, during most recent seasons even minor play has been uncommon.

One of the displays at the United States Centennial Exposition in Philadelphia in 1876 included a beautiful geyserite cone. It is possible that this cone belongs to Vixen, summarily chopped off as a special exhibit from America's then-new national park. The unlabeled cone now resides in the archives of the Smithsonian Institution.

74. RUBBLE GEYSER first erupted in 1972. Before then there was no evidence of any spring having existed at the site, but its coming was foretold by the small lodgepole pines in the area. In 1970 they began dying as the ground warmed up and exceeded their temperature tolerance. The ragged vent formed when Rubble broke out in a small steam explosion. At first, the geyser played frequently and regularly. The eruptions began after 1 to 2 minutes of heavy overflow. Lasting about 6 minutes, some of the bursts reached 10 feet high. Rubble rapidly declined in frequency and force, and eruptions are now rare.

75. CORPORAL GEYSER hardly merits either attention or the name "geyser." It is a fairly cool pool containing much silt and debris and is decidedly not a pretty feature. The eruptions consist of nothing more than intermittent overflow accompanied by bubbling and sometimes a few splashes perhaps 1 foot high. During active years the intervals tend to be regular at about 20 minutes. The duration is 1 to 3 minutes.

76. DOG'S LEG SPRING is located at the break in slope about 6 feet east of Corporal Geyser (75). When active as a geyser, it far overshadows Corporal. It has been known to play up to 3 feet high for durations of several hours. More often, its water level drops when Corporal erupts, and it refills quickly when Corporal stops.

77. VETERAN GEYSER is a spellbinder. People have been known to sit here for long periods, waiting for an eruption that always seems about to happen. Often they are not disappointed, but Veteran is a long-term cyclic geyser that can go hours to several days between major eruptions.

Veteran's main vent is the large pit on the far side of the deep crater. Connecting the main vent and the crater is a small hole. A third vent is outside the crater, within a jagged cavern next to the trail. Subsurface churning and splashing within the main vent is almost constant and causes water to intermittently gush through the hole into the crater. When it is time for an eruption, the turbulence becomes violent. Water is shot through the hole into the crater, rapidly raising the water level while bursts gush out of the main vent. Once the pool is high enough to thoroughly cover the hole, the action is forced largely to the main vent. Most often the action stops at about this point. The less common major eruptions continue, jetting water with increasing force at an angle away from the trail. The best bursts reach as much as 25 feet high and 50 feet outward. It is only during a full, major eruption that the third vent joins in, shooting from the cavern toward the trail at a very low angle. The eruption ends abruptly after durations that range from only 10 seconds to 2½ minutes. During Veteran's long cycles, there is a gradual increase in the frequency and force of the minor eruptions, which in turn leads to a greater frequency of major eruptions. A cycle often ends with a closely spaced series of frequent and powerful majors. This was especially common in 2003, when some intervals were shorter than 20 minutes and durations exceeded 7 minutes.

A few feet east (left) of Veteran's pool is another crater with a similar appearance. This is known as "Veteran's Auxiliary Vent." As Veteran approaches the start of a new cycle, this spring undergoes a simultaneous rise in water level. Seldom is the water actually visible within the small vent, but on rare occasions the Auxiliary has its own, mostly subsurface eruptions.

78. PALPITATOR SPRING is a geyser. It apparently had some rather strong eruptions during the 1880s, but in modern times it was not known as a geyser until 1974. Since then its action has stabilized, and it is now a fairly regular performer. As a quiet pool prior to 1974, and now during the quiet intervals, the pool constantly bounces (palpitates) over the vent, sending small waves over the rim of the crater. The eruptions consist of splashes up to 3 feet high. The crater slowly drains during play that lasts as long as 3 hours. The intervals between eruptions are hours long.

79. FEARLESS GEYSER has a confused history. It may have had some eruptions as high as 30 feet during the 1880s, but it is also possible that those early reports referred to Monarch Geyser (81) or to some ephemeral hot spring in this area. A few decades ago the intervals were listed as "several per day" with a height of 3 feet. The present Fearless barely qualifies as a perpetual spouter that domes and surges boiling water about 1 foot high.

80. UNNG-NBK-3 is a set of perpetual spouters that made their initial appearance during the 1970s. Spouting muddy brown water, some of the play can reach 6 feet high. The activity has decreased during the 2000s and is most vigorous during disturbances.

81. MONARCH GEYSER, as the name implies, was once a major geyser. Its play reached as high as 200 feet. Lasting around 10 minutes, the eruption would throw out so much water that the old road through Norris had to be closed each time the geyser played. The last few eruptions were muddy, an indication that Monarch might have damaged its plumbing system. It has also been conjectured that the "death" of Monarch in 1913 was in some way related to the dormancy in Steamboat Geyser (48) that began in 1911. The two geysers are quite near one another, so there could be a relationship, but Monarch underwent minor activity in 1920, 1923, and 1927 without any reported changes in Steamboat.

In recent years, the low-temperature acid pool was so unlike a geyser in its appearance that most people felt Monarch was truly dead. That word should never be used for geysers. In the fall of 1993, Monarch began having occasional "hot periods," in which heavy overflow accompanied episodes of superheated boiling. In early March 1994 there was an unseen eruption, and continuing hot periods finally resulted in fairly frequent play by the end of April. The activity was cyclic. A few days separated active episodes, when several eruptions occurred at intervals as short as 1 hour. The action consisted primarily of superheated boiling, but infrequent bursts could throw water as high as 15 to 20 feet. Although that was a far cry from the action of a century ago, it shows that there is no such thing as a "dead" geyser. Unfortunately, though, Monarch has not had one of those eruptions since 1996. Weak intermittent bubbling and variable overflow continued through 2007.

82. MINUTE GEYSER received its name because of the very regular but brief nature of its eruptions many years ago. Some of the eruptions were reported to be as high as 60 feet. The geyser later changed so thoroughly that the name appeared in literature as "mi-NUTE," meaning something very small. What happened is that a vandal threw a large boulder into the vent—the rock could not have gotten there any other way. Unable to remove it, Minute shifted its activity to another opening a few feet away. The play became almost constant and only 4 feet high. Rare episodes of stronger eruptions were recorded in the 1980s, when intervals of about 20 minutes separated 5-minute play that reached 10 feet high.

83. FORGOTTEN FUMAROLE was named because until 1993 it was located away from any trail. It erupts through openings in an old crater of decaying geyserite and very likely was a significant geyser in the distant past. Historically a gentle steam vent, its first known eruptions were seen during May 2004, when spray was jetted up to 6 feet high. Later that summer, bursts of black, mud-laden water reached as high as 30 feet. That action persisted for only two days before Forgotten Fumarole lapsed back into quiet steaming. However, considerable activity is believed to have taken place during the following winter because the water was largely clear during the summer of 2005. That action repeated every few days with series of two to five eruptions that reached 20 feet high. The next action took place in early May 2008.

84. REDISCOVERED GEYSER was well away from the trail until 1993, when a new path was opened along the old roadway between Minute Geyser (80) and the museum area. Even now, Rediscovered is largely hidden from view by a dense growth of young trees below the trail, but the commotion it raises during an eruption is easily heard. Rediscovered is mostly active as a perpetual spouter only about 1 foot high. During some years it becomes intermittent and then, with both intervals and durations several minutes long, it erupts up to 4 feet high.

Note: As is the case with the Porcelain Basin, many other geysers have been observed in the Back Basin. Most of these have been of small size with temporary existences related to the seasonal disturbances.

One Hundred Spring Plain is the northwestern part of Norris Geyser Basin. Not accessible by any trail, the great number of hot spring vents and extensive mud flats make this a dangerous area. It is the site of several geysers. Most are quite small, but one, informally called "Breach Geyser," had eruptions 20 to 25 feet high during 1974.

The Gap, a southwestern extension of One Hundred Spring Plain, underwent significant heating starting in 1998. Dozens of springs came to life as perpetual spouters and mud pots, and a few have acted as significant, although short-lived, geysers. "Elk Geyser" played as high as 50 feet from what is normally a quiet pool. "Gap Geyser," near the southwestern-most limit of the area, has been active on several occasions, with powerful bursts that reach well over 30 feet high from one vent and simultaneously up to 20 feet from two other openings within a single large crater. The nature of the activity here has led geologists to speculate that this could be the site of Yellowstone's next large-scale hydrothermal explosion.

Immediately adjacent to a highway pullout in Elk Park, about 1 mile south of Norris Junction, are the small, sinter-lined "Elk Park Springs." Part of the Norris geothermal system, two of these springs have records of infrequent, small eruptions. In the forest near the still-visible route of the old highway leading north from Elk Park is large and hot Big Blue Spring. This spring is not described in early reports about the park, and it probably formed around 1920, when it was named.

CHAPTER EIGHT

West Thumb Geyser Basin

Compared with the other areas of clear, alkaline water, geyser activity at the West Thumb Geyser Basin (Map 8.1, Table 8.1) is limited and generally weak. Only two dozen geysers have been observed within the developed part of the basin, which is properly known as the Lower Group. The geysers currently of greater importance lie to the north, near one another in the narrow tract of the Lake Shore Group between the lake and the highway. Still farther north is the Potts Hot Spring Basin, originally called the Upper Group, in which there are numerous small geysers but only infrequently any eruptions of significant size.

West Thumb may be better known for its pools. Some are among the largest in Yellowstone, and two of them, Abyss Pool and Black Pool, underwent powerful eruptions in 1991–1992. The Thumb Paintpots were first described by Daniel T. Potts way back in 1826. Well-known as the "Mud Puffs" in the early days of the Park because of their variety of pastel colors, these mud pots are

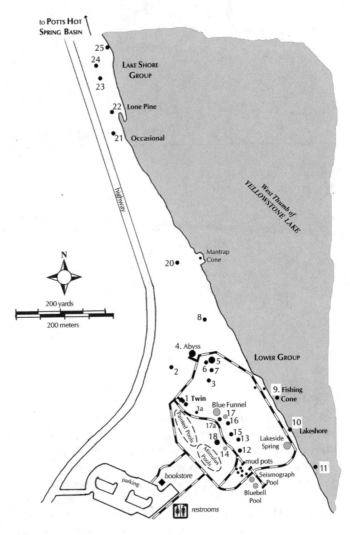

Map 8.1. *West Thumb Geyser Basin*

now a few low, gray mud cones far less active than they used to be.

For unknown reasons, much of the Lower Group suffered a sudden and drastic decline in water levels during the late 1970s, and geyser activity decreased in hand. Except for occasional brief

Table 8.1. Geysers of the West Thumb Geyser Basin

Name	Map No.	Interval	Duration	Height (ft)
Lower Group				
Abyss Pool	4	[1992]	1–3 min	50–100
Black Pool	5	[1991]	minutes	1–3
Collapsing Pool†	14	[2001]	minutes	15
Ephydra Spring†	17	with WTL-2	10 min	boil
Fishing Cone	9	[1928?]	minutes	4–40
Hillside Geyser	3	[2005]	5–15 min	30–100
King Geyser	8	[1997]	5–10 min	6–8
Lakeshore Geyser	10	[2004]	sec–10 min	2–25
Ledge Spring†	13	minutes*	3 min	2–6
"New Twin Geyser"	1a	[1997]	minutes	10
North Star Geyser	7	[2000]	seconds	2–4
Percolating Spring†	15	[2005]	sec–min	1–2
Perforated Pool†	16	[2002]	minutes	1–2
Roadside Steamer	2	frequent	seconds	3–40
"Skinny Geyser"	6	days*	15–20 min	15
Surging Spring†	12	infrequent	sec–min	3
Thumb Geyser	18	erratic*	4–5 min	6–10
Twin Geysers	1	[1999]	3–4 min	12–120
UNNG-WTL-1 ("Beach")	11	rare	minutes	4–6
UNNG-WTL-2 ("Footprint")†	17a	[2002]	10 min	10–15
Lake Shore Group				
Blow Hole	23	rare	minutes	4–20
Goggle Spring	20	infrequent	sec–min	4–6
Guidebook Spring	24	2 hrs*	minutes	10
Lone Pine Geyser	22	23–60 hrs	30 min	25–75
Occasional Geyser	21	20–35 min	30 sec–2 min	10
Overhanging Geyser	25	2–4 hrs	3–25 min	3–10
Potts Hot Spring Basin				
"Explosion Pool"	27	[2003]	unknown	unknown
"Mercurial Geyser"	28	minutes*	seconds	8–10
(Potts) Geyser #38	29	hours*	1½–2 min	20–30
"Resurgent Geyser"	26	3–5 hrs*	8 min	4–12

* When active.
† Geysers most likely to be active only during "energy surge" events.
[ ] Brackets enclose the year of most recent activity for rare or dormant geysers. See text.

recoveries known as "energy surges," there is rather little geyser action there.

Numerous hot springs and extensive geyserite deposits lie drowned a few tens of feet out from the lakeshore. These sinter shields and cones could not have formed underwater, as they are now. In fact, they have only been recently submerged. Because of a shallow intrusion of volcanic magma in the north-central part of the Park, east of the Grand Canyon of the Yellowstone, the entire basin of Yellowstone Lake is being tilted toward the south. The north shore is getting higher, whereas forests and the hot springs at West Thumb are progressively inundated along the southern parts of the lake. These drowned springs are still active, and when the lake is calm, bubbles can be seen rising from their vents. Several geysers near the shoreline, including Lakeshore Geyser and Beach Geyser, erupt only when their craters are exposed at times of low lake levels.

The Lower Group is serviced by a boardwalk system. The Lake Shore Group and Potts Basin are not accessible except to view from the side of the road or by boat on the lake. A Yellowstone Association bookstore and restrooms are the only facilities at the West Thumb Geyser Basin. The full range of visitor services, including a hotel, store, gas station, campground, and visitor center, is available at Grant Village, 2½ miles to the south.

LOWER GROUP

The Lower Group (Table 8.1, numbers 1 through 18) is the "main" portion of the West Thumb Geyser Basin and the only part traversed by a boardwalk system. Geyser activity has always been an on-again, off-again affair here, but in recent years it has decreased so that on most occasions only a few small geysers are active. In similar fashion, the once beautiful Painted Pools and Mimulus Pools now are largely low ponds of tepid water. At long and unpredictable intervals, there is a temporary increase in water levels and temperatures—an energy surge—and then there may be several active geysers. In late 1991 and early 1992, first Black Pool and then Abyss Pool underwent episodes of explosive eruptions among the largest ever seen at West Thumb.

1. TWIN GEYSERS (a single geyser with two distinct vents) is the largest and most spectacular geyser at West Thumb, when it is active. During most of its history, however, it has been dormant. When in such a state, the water often boils in the northern of the two craters; the other basin is active only just before and during a major erup-

Twin Geysers, although seldom active, is the most important geyser in West Thumb's Lower Group, the only portion of the geyser basin threaded by board-walk trails. Credit: NPS photo by Lawrence Boe.

tion. The first recorded activity in Twin was probably during 1910, when a single eruption described as 200 feet high was recorded. It also played in 1912, 1916, 1932–1934, and from 1948 into 1952. It was then dormant until 1971, when it began its best active episode on record.

The initial 1971 activity was 40 to 60 feet high, but by 1973 many of the eruptions were estimated at 120 feet high. The geyser became known as "Maggie and Jiggs" after the old *Life with Father* comic strip characters. First one crater would erupt, and as its water jet reached 50 feet the other would begin spouting. The two columns converged far above the ground, and, just as Maggie often ended Jiggs's great plans, the second column would take over the eruption, reaching the maximum height as the other died. There were at least ninety-nine eruptions during 1973. At times the eruptions recurred every 4 to 8 hours, and for a few weeks Twin was regular enough to be predicted. The play lasted 3 to 4 minutes.

Twin's activity ended abruptly in August 1973, and it has been largely dormant since then. There were a few eruptions in the first two days following the earthquake of June 30, 1975. There were five eruptions as high as 75 feet on August 23, 1998, and two minor eruptions just 12 feet high were witnessed on February 25, 1999. Unfortunately, Twin has now cooled so much that dark orange-brown cyanobacteria is growing in the southern crater.

In September 1997, a geyser referred to as "New Twin Geyser" (1a) appeared next to the boardwalk about 135 feet southeast of Twin Geysers. Both the intervals and durations were a few minutes long and some eruptions reached as high as 10 feet, but the action lasted only a few days. The geyser's site is now only a muddy depression.

2. ROADSIDE STEAMER is a pool at the base of the slope immediately below the old highway route. It began having eruptions during 1948, some of which reached 30 to 40 feet high. Never regular and dormant much more often than active, Roadside Steamer continues to have occasional active phases. Few recent eruptions have exceeded 3 feet high, and they are not visible from the boardwalk.

3. HILLSIDE GEYSER has an uncertain history. Some 1948 records apparently refer to it, but no further action was seen until August 1995. Never either frequent or regular, Hillside's intervals ranged from a few hours to many days in length. The eruptions lasted as long as 15 minutes and reached as high as 50 feet; an eruption in June 2001 is said to have exceeded 100 feet tall. Hillside returned to dormancy in November 2005, and it now bubbles with slight overflow.

4. ABYSS POOL is one of the larger, deep blue pools in the Park. The fact that it was active as a major geyser in 1904–1905 was one of those points of almost forgotten history until recent research turned up some old U.S. Army reports. Active in July 1904 to 20 feet, Abyss (then called "Elk Spring") began an episode of large, explosive eruptions in early 1905. Bursts reached 80 to 100 feet high over typical durations of 20 to 40 seconds. It was common for chunks of geyserite weighing up to 30 pounds to be thrown out of the crater. Between April 1 and May 31, 1905, fifty-seven eruptions were recorded. There the records stop, because the army sergeant who made them was arrested to face court-marshal. The geyser's action undoubtedly continued for some time, but there are no further records.

Abyss Pool did not play again until a single eruption occurred in 1987. Perhaps that was a prelude to a substantial active phase that began in September 1991. Over a one-month period it had a few eruptions up to 100 feet high. These initial plays lasted 45 seconds. After a brief dormancy, Abyss renewed activity in December and then continued to play for six months. At its best during the winter of early 1992, the intervals ranged between 45 minutes and 5 hours. With durations of 1 to 3 minutes, the rocketing bursts reached 50 to 70 feet high. Unfortunately, by the summer season of 1992 the activity had declined in frequency and force. The last known eruption took place on June 7, 1992. For several weeks thereafter, Abyss continued to boil around the edges of the crater, but it soon returned to the beautiful but quiet pool of before.

5. BLACK POOL usually had a water temperature low enough for dark orange-brown cyanobacteria to grow throughout its crater. The color was never really black, but it was an exceptionally dark green. During 1991, an exchange of function transferred energy from the central part of the Lower Group toward Black and Abyss (4) pools. A small spring at the edge of Black Pool gradually heated up and began having small eruptions. With this, Black also warmed up until the cyanobacteria had been killed and Black assumed a rich blue color. Then, on August 15, 1991, the small spring exploded and became a part of Black Pool itself. The explosion was not witnessed by any reporting observer, but it must have been impressive, since blocks of geyserite several feet in dimension were blasted loose. Over the next few hours, Black Pool had frequent boiling eruptions that domed the water 1 to 3 feet high and produced heavy runoff. The activity was confined to that single day, but only about three weeks later eruptions began at nearby Abyss

Pool. Two more brief episodes of small eruptions by Black Pool took place in early 1992 while Abyss Pool was active. Black Pool remains very hot, and it is now one of Yellowstone's most beautiful blue pools.

6. "SKINNY GEYSER" (also known as "Skinny Man Geyser"—the expression that originated the name apparently went "Wow, that's skinny, man!") erupts from a small vent about 20 feet west of Black Pool (5). Never seen before the mid-1990s, it is often dormant and the play is rather uncommon even when the geyser is active. The intervals are typically several days long, when thin, steamy jets of water reach up to 15 feet high for durations of 15 to 20 minutes. In the same area as "Skinny" are at least two other vents that occasionally erupt as much smaller geysers.

7. NORTH STAR GEYSER is a bit of a mystery whose location is somewhat uncertain. It received its name for an unrecorded reason in the early 1950s when eruptions several feet high were sporadically recorded. In 2000 a geyser described only as "near Black Pool" played infrequently and briefly 2 to 4 feet high. There is a jagged vent about 50 feet upslope from Black Pool, and that is probably North Star.

8. KING GEYSER played as a major geyser in 1905, at the same time Abyss Pool (4) was active. Intervals were days long and the eruptions lasted only a few minutes, but some bursts were 60 feet high. At that time it was called "Lake Geyser." The modern name is probably a result of the visit to Yellowstone by the future King Gustaf of Sweden in 1926. That implies that there were some eruptions during that year, but none were specifically recorded until 1933. King played erratically through the rest of the 1930s, generally lasting 5 to 10 minutes and reaching 6 to 8 feet high. Only a small number of similar eruptions have been recorded since 1940, the most recent being several seen during May–June 1997.

9. FISHING CONE is one of the most famous hot springs in Yellowstone. As the story goes, it was possible to stand on the cone while fishing in the lake, then cook the catch in the boiling hot spring without having to remove it from the hook. At times this actually could have been done, but Fishing Cone is usually too cool to cook in efficiently. Also, its crater is sometimes under the water

It has been many years since Fishing Cone erupted as a geyser. Now it is so cool that it could hardly be used to cook fish, and the entire cone is often under the water of the lake.

of Yellowstone Lake. Fishing Cone has had two recorded episodes of vigorous geyser activity. In 1919 the play reached as high as 40 feet, and in the late 1920s it was no more than 4 feet tall. More recently, Fishing Cone has sometimes splashed weakly as a near perpetual spouter. During the exchange of function that caused eruptions in Abyss (4) and Black (5) pools in 1991–1992, Fishing Cone completely drained, and a small hole in its side acted as a weak steam vent. This condition lasted into July 1994, when Fishing Cone began to slowly refill.

North of Fishing Cone are two smaller cones, Big Cone and Little Cone, respectively. Although usually cool, quiet, and with only seeping discharge, Big Cone has been known to undergo rare eruptions 1 foot high or less. Eruptions are not known in Little Cone.

10. LAKESHORE GEYSER has a cone much like that at Fishing Cone (9). It sits at the edge of the lake, too, but unlike Fishing Cone its vent is often covered by water. It can erupt only when the lake is low enough to completely expose the crater, and even then Lakeshore is usually dormant. Historically, when in an active phase it erupted every 30 to 60 minutes. The play lasted about 10 minutes and reached as high as 25 feet. During the winter seasons of 2000 and 2003, however, brief hourly eruptions reached just 2 to 6 feet high, and nearly perpetual splashing to 2 feet was seen during October 2004.

11. UNNG-WTL-1 ("BEACH GEYSER") lies near the lakeshore south of Lakeshore Geyser (10). Not known as a geyser until 1974, it is active only when the water level of Yellowstone Lake drops to exceptionally low levels, late in the season or during drought years. Eruptions are uncommon even then. The bursting play reaches 4 to 6 feet high for durations of several minutes.

12. SURGING SPRING is the southernmost of a series of springs that have developed along an alignment that might represent a subsurface fracture that extends as far as Twin Geysers (1). On rather infrequent occasions, the central portion of the Lower Group undergoes an energy surge during which several springs are able to erupt. These include Surging Spring, Ledge Spring (13), Percolating Spring (15), Perforated Pool (16), "Footprint Geyser" (17a), and Ephydra Spring (17) and, somewhat off the main alignment, Collapsing Pool (14) and Thumb Geyser (18). The most recent energy surges occurred separately in 2001 and 2002. At other times, including during exchanges of function such as that toward Abyss (4) and Black (5) pools during 1991–1992, these springs cool and drop to low water levels.

When active, Surging Spring fits its name. The eruptions consist of strong pulsations of the pool, with heavy surges doming the water and producing voluminous overflow. Some bursting may occur, splashing water as high as 3 feet. Most commonly, Surging Spring is a quiet pool below overflow and serves as a drain hole for runoff from Collapsing Pool.

13. LEDGE SPRING drains across a very wide runoff area. It was first witnessed as a geyser during 1928 and has been seen during only a few years since, including 2004 and 2005. During its episodes of activity, eruptions can occur as often as every few minutes. The action produces a huge discharge, while bursting as high as 6 feet high can continue for as long as 3 minutes.

14. COLLAPSING POOL was named when a portion of the pool's sinter rim collapsed inward at the time of the 1959 earthquake. It has apparently had only one active phase as a geyser. In June 2001, a brief exchange of function south from Percolating Spring (15) allowed Collapsing to erupt as high as 15 feet. The action lasted a few days before the energy shifted back to the north.

15. PERCOLATING SPRING was probably named because of the steady streams of bubbles that rise from several small vents at the bottom of the crater. At the times when an energy surge leads to eruptions in the nearby hot springs, Percolating sometimes joins by splashing 1 to 2 feet high. The only known year of consistent, season-long activity was 1987. Eruptions were also seen during an energy surge in June 2001, and erratic but frequent bursts up to 1 foot high took place during July 2005.

16. PERFORATED POOL was active as a geyser during an energy surge in June–July 2002. The eruptions were of long durations but were not vigorous, consisting of occasional splashes 1 to 2 feet high separated in time by several seconds each.

17. EPHYDRA SPRING was named in reference to the *Ephydra bruesi* brine flies that inhabit the mats of cyanobacteria throughout Yellowstone. While UNNG-WTL-2 (17a) was actively erupting in 2002, Ephydra Spring underwent surges of superheated boiling that domed the water as high as 1 foot. Next to Ephydra is quiet Blue Funnel Spring. During the 1991–1992 activity by Abyss Pool (4) more than 400 feet distant, Ephydra, Blue Funnel, and Perforated Pool (16) cooled significantly and dropped to low water levels. They recovered quickly when Abyss stopped having eruptions.

17a. UNNG-WTL-2 ("FOOTPRINT GEYSER") made its initial appearance during an energy surge in June 2002, at the same time nearby Perforated Pool (16) was active. The crater developed beneath the boardwalk, which was temporarily closed and then moved because of eruptions that reached 10 to 15 feet high and lasted as long as 10 minutes. Only a few eruptions were recorded.

18. THUMB GEYSER can have very impressive eruptions. During 1925 it played every 6 to 12 hours, bursting up to 20 feet high for as long as 15 minutes. More recently, it has been active only during energy surges, and even then the intervals have been long and erratic. The play lasts 4 to 5 minutes and reaches 6 to 10 feet high.

LAKE SHORE GROUP

The Lake Shore Group (Table 8.1, numbers 20 through 25) includes West Thumb's most active geysers. The group extends nearly half a mile northward from the Lower Group along the shoreline of Yellowstone Lake. The hot springs are confined to a narrow tract of ground mostly less than 100 feet wide sandwiched between the lake and the highway. Unfortunately, there is no trail here, and direct access is not permitted. Since the forest fires of 1988 cleared out the growth here, some of the geysers can be seen from the edge of the road. Boaters on the lake enjoy clearer and often closer views.

20. GOGGLE SPRING lies directly inland from the prominent geyserite peninsula called Mantrap Cone. The reason for the name is unknown, but it appears in several National Park Service reports about the area. Most of the eruptions are brief and just 4 to 6 feet high, but extensive washed areas in the surroundings imply that eruptions of greater size and power occur on infrequent occasions.

21. OCCASIONAL GEYSER is, in spite of its name, a frequent and regularly active geyser. It plays from a complex of vents, one of which is an extraordinarily smooth and perfectly round crater lined with tan, beaded geyserite. It is a beautiful sight. Eruptions generally recur every 20 to 35 minutes. The water rises steadily in the craters and begins to bubble during the few minutes before an eruption, which begins abruptly. Eruptions last 30 seconds to 2 minutes, and some of the vents outside the round crater may reach a height of 10 feet. Occasional's crater sits atop a high sinter platform, and the runoff flows only a few feet before dropping over a fall into the lake.

In 1897, observers described a "Lake Geyser" in this area. This was definitely not the "Lake Geyser" in the Lower Group of West Thumb, now known as King Geyser (8). Although the eruptions were 100 to 200 feet high, the modern identity of that geyser is simply unknown. Later reports listed an "Occasional Geyser" as playing up to 60 feet, but it is believed that was actually nearby Lone Pine Geyser (22).

22. LONE PINE GEYSER is currently the largest active geyser at West Thumb. It erupts from a small round crater that is less than obvious. Lone Pine received its name in 1974, the year of its first modern activity, because of the single pine tree that stood on a lakeside peninsula of sinter prior to a forest fire in 1988. The initial activity consisted of eruptions recurring every 20 minutes to heights as great as 25 feet. Through time, the intervals grew dramatically longer, but the height became greater. By 2007 the intervals varied between 23 and 60 hours, with an average over 30 hours. There is little warning of the initial, major eruption—slight bubbling in the small pool is of no help, since it can be seen several hours before the eruption. The eruption's trigger is a sudden, heavy overflow. Lone Pine rapidly develops strong jetting that can exceed 75 feet high. This eruption lasts as long as 30 minutes, about half of which is steam phase. The major eruption is followed by a series of minor eruptions. These recur at intervals of a few minutes, last only a few seconds, but still reach up to 40 feet high. There may be as many as five such follow-up eruptions in one series. Lone Pine then falls silent and begins to refill, requiring several hours to resume light overflow.

23. BLOW HOLE is not visible from the highway and is only barely so from the lake. It is usually a quietly bubbling spring, but it has played as a geyser during rare and short active phases. The most recent recorded activity was in 2001. The eruptions send steamy spray 4 to 6 feet high. Blow Hole was named in 1954 when more powerful eruptions reached 20 feet high and ended with noisy steam phases.

24. GUIDEBOOK SPRING was named without explanation by geologist Walter Weed during the 1880s. Nothing was known of its activity from then until 1998, when eruptions recurred every 2 hours and reached as high as 10 feet for several minutes. The crater is an exceptionally deep, symmetrical funnel in which no water is visible until immediately before the play begins. Like Blow Hole (23), Guidebook is almost impossible to see from any legal viewpoint, but it is regularly checked by researchers.

The crater of Overhanging Geyser is actually perched out over the water of Yellowstone Lake, in a location where it can be viewed only by boaters.

25. OVERHANGING GEYSER is so named because erosion caused by waves on Yellowstone Lake has undercut the crater so that it is actually perched out *over* the water. It is a miracle that the erosion did not tap into any part of the plumbing system. Although Over-hanging has shown widely variable intervals over the years, it most often plays with considerable regularity about every 2 to 4 hours. The eruptions last 3 to 25 minutes and burst the water 3 to 6 (rarely 10) feet high. The runoff drops in free fall over the edge of the over-hang, a unique sight when the geyser is in eruption. Overhanging can be viewed only by boat.

POTTS HOT SPRING BASIN

Initially known as West Thumb's "Upper Group," the Potts Hot Spring Basin (Table 8.1, numbers 26 through 29) was renamed in 1956 in honor of Daniel T. Potts, an early trapper who described his 1826 visit to the West Thumb area in a letter to his brother the next year. That letter contains the earliest known written description of Yellowstone's geysers and mud pots. The highway used to pass directly among these springs, and for a short time decades ago there was a boardwalk. However, even then these springs received little attention. The area was never mapped in detail until after the 1959 earthquake, when fourteen geysers were noted. Curiously, a 1961 report stated that there were no geysers there, but that could be correct—Potts Basin, like the rest of the West Thumb Geyser Basin, is affected by occasional "energy surges" that can dramatically alter the geyser activity. Small geysers were reported in 1983, but it was left to detailed studies conducted by volunteers for Yellowstone's Resource Management Division starting in the late 1990s to reveal that Potts includes more than forty geysers. Their activity is highly complex because of the energy surges plus wide-spread and frequent exchanges of function. Nearly every visit to the area reveals a "new" geyser or two. Thus, only a few of the more important geysers are described here.

The road through the area was removed in 1970 and rerouted across the hillside to the west, and Potts Basin is closed to public entry. It can partly be seen from a viewpoint next to a large high-way pullout at the top of the hill or from the road along the open lakeshore to the north.

The Empty Hole Group encompasses the southern portion of the Potts Hot Spring Basin and is all but impossible to see from the highway. The road passed directly among these springs until late 1970. It's good that it was moved, because later a rather large steam explosion excavated a good-sized crater here. Exactly when that happened is unknown, but part of the old roadway was involved. The Empty Hole Group received its name because of a large crater among numerous "empty holes" that perforate the sinter platform between the old road and the lake. Four of these are small geysers of frequent but erratic performances. Another geyser across the old road is the most important member of the group and is the largest regularly active geyser in the Potts Basin.

26. "RESURGENT GEYSER," just west of the old road, is the largest geyser in the Empty Hole Group. Fairly irregular for most of its history, ranging from minutes to days between eruptions, "Resurgent" became a reliable performer in the mid-1990s, with intervals of 3 to 5 hours. The play is a steady jet between 4 and 12 feet high that lasts as long as 8 minutes. Although about a mile distant, it can be seen from the highway along the lakeshore north of Potts Basin. Unfortunately, "Resurgent" fell dormant during 2007.

27. "EXPLOSION POOL" was recorded as having three eruptions during August 1962 and again in 1965 from a vent at one end of a cool pond that had supported yellow pond lilies. Despite that explosive action, it evidently continued to be a cool, quiet pool until sometime after 1972, when it was thoroughly disrupted by a steam explosion that years earlier would have included some of the roadway, which had been closed and removed in 1970. More explosions dramatically enlarged the crater in 1983 and 1984, another "disruptive event" took place during the winter of 2002–2003, and some sort of eruptive surging occurred in July 2003. "Explosion Pool" is now about 50 feet in diameter. Since 2003 the water level has dropped several feet and cooled to less than 140°F (60°C), but several perpetual spouters erupt nearby.

The "Mercurial Group" and "Beach Group" of the Potts Basin lie at a lower elevation and comprise the northernmost part of the West Thumb Geyser Basin. This is the part of Potts Hot Spring Basin that is visible from a large highway pullout on the hill above, as well as from the highway north of the area. These groups encompass several dozen hot springs. There are at least thirty-five geysers, but most of them erupt only a foot or two high during short active phases. Many of the features are aligned along a series of fracture zones. The "Mercurial Group" received its name because of the highly variable nature of its activity, which is a result of exchange of function among these fracture-related springs.

28. "MERCURIAL GEYSER" is the best-known and most visible large geyser in the "Mercurial Group," but its intervals are always erratic, and long dormant periods are known. During the 1980s it had massive bursts up to 10 feet high, but since 2000 it has not been seen higher than about 2 feet.

29. (POTTS) GEYSER #38 is described here as an example of the vigorous, but usually temporary, geyser activity typical of the Potts Basin. It is located in the "Beach Group," adjacent to the old highway route along the lakeshore. Starting in July 2007, this pool began a series of powerful eruptions that washed wide areas around the crater with enough force to roll large rocks. The play of massive, doming bursts reached as high as 30 feet for durations of 1½ to 2 minutes. The intervals were apparently several hours long, but eruptions were continuing as the 2007 season ended. Several nearby springs, including "Centerline Spring" in the middle of the old roadbed, reactivated as small geysers after having been dormant since 1988.

CHAPTER NINE

Gibbon Geyser Basin

The Gibbon Geyser Basin (Map 9.1 and Map 9.2, Table 9.1, numbers 1 through 22) includes all the hot spring clusters that lie around the perimeter of Gibbon Meadows, about 5 miles south of Norris Junction and south from there into Gibbon Canyon. The main highway runs across Gibbon Meadows near the middle of the Gibbon Basin. Trails proceed to some of the hot spring areas. Geysers are comparatively few and small, but they exist in seven of the eight groups of springs. The only group without geysers is the Chocolate Pots, along the river next to the road in the small canyon at the north end of the basin, where relatively cool springs are forming unique sinter deposits.

ARTISTS' PAINTPOTS

The Artists' Paintpots is the best-known group in the Gibbon Geyser Basin. The trail to the Paintpots begins at the end of a side road

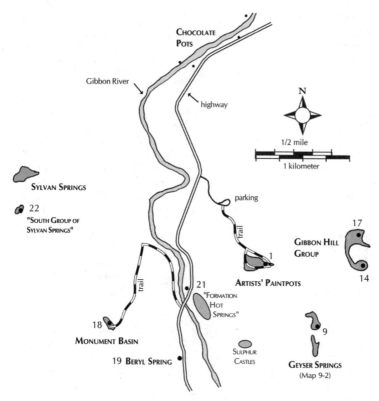

Map 9.1. *Index Map of the Gibbon Geyser Basin*

east from the middle of Gibbon Meadows. The path is well maintained and incorporates some boardwalks where it passes among the hot springs. The round-trip to the area is less than 1 mile. The springs are at the bottom of Paintpot Hill, a volcanic dome that may be as little as 70,000 years old.

The group was named for the numerous bright colors that characterize the springs. Most dominant are oranges and reds created by iron oxide minerals. Clay minerals form pastel pinks and blue-grays, while hot water cyanobacteria add other oranges plus browns and greens. Add to these the usual blues of the pools, abundant wildflowers, and the forest, and the overall effect is striking. Like artists' palettes, it is one of the most colorful places in Yellowstone.

Table 9.1. Geysers of the Gibbon Geyser Basin

Name	Map No.	Interval	Duration	Height (ft)
Artists' Paintpots				
Blood Geyser	1	near steady	near steady	2–6
UNNG-GIB-2	2	minutes	seconds	4
Geyser Springs				
Anthill Geyser	5	[1990s]	2–9 min	4
Avalanche Geyser	9	8½–9½ min	2½–3 min	25–30
Bat Pool	12	sec–3 min	seconds	1–3
Big Bowl Geyser	10	seconds	seconds	boil–30
Bull's Eye Spring	4	seconds*	seconds	4–6
Subterranean Blue Mud, minor	8	2–5 min	seconds	subterranean
Subterranean Blue Mud, major	8	10–25 min	1 min	sub–3
Tiny Geyser	13	2–3 min*	seconds	inches
UNNG-GIB-4	7	unknown	seconds	4
UNNG-GIB-8	6	rare	1–2 min	10
UNNG-GIB-9	11	minutes*	30–60 sec	2
UNNG-GIB-14	3	[2004]	minutes	3
Gibbon Hill Group				
Gibbon Hill Geyser	14	[1989]	20–50 min	6–25
Phoenix Geyser	15	2–6½ hrs	5–80 min	2–8
UNNG-GIB-10	16	infrequent	unknown	1–2
UNNG-GIB-11 ("Aquamarine Spring")	17	steady	steady	1
Monument Geyser Basin				
Monument Geyser	18	steady	steady	spray
Beryl Spring and "Formation Hot Springs"				
Beryl Spring	19	steady	steady	3–4
UNNG-GIB-12	20	[2002]	minutes	2–6
UNNG-GIB-15	21	rare	minutes	4
Sylvan Springs				
UNNG-GIB-13	22	10–12 min	8–9 min	5

* When active.
[ ] Brackets enclose the year of most recent activity for rare or dormant geysers. See text.

The Artists' Paintpots themselves are a pair of mud pots located up the hillside, somewhat separated from the other hot springs. Although restricted to just two small basins, they tend to be very active and at one point in 2006 forced a temporary closure of the trail. At times, the gray mud is tossed as high as 20 feet. Despite their proximity to one another, the two sets of mud pots are always of different thickness, and the one to the west sometimes dries completely during the summer season. Also on the hillside are several perpetual spouters, "frying pans," and steam vents—springs typical of acid thermal areas.

The important springs of clear water, including two geysers, are located at the base of the hill. Several other springs in the area appear to be geysers but actually are not. Their spouting is perpetual and largely a result of the evolution of carbon dioxide in water cooler than boiling. The best-known of these is Flash Spring, immediately next to the easternmost point of the trail where it begins the steep climb up the hillside. The two geysers are short distances to its southwest (right).

1. BLOOD GEYSER plays from a shallow basin at the east end of the group. Possibly the spring named "Red Geyser" in 1878, it was described as a perpetual spouter in 1882, and, indeed, it only briefly and infrequently pauses its play. Most quiet intervals last less than 1 minute, and they apparently can be as long as several hours apart. The bursting reaches up to 6 feet high.

The water discharge of this one spring amounts to 150 gallons per minute, about half that of the entire group. The small alcove surrounding the spring is highly colored by iron oxide, and, in fact, the water contains a large amount of iron in solution. A sample, allowed to cool and sit quietly, will develop a precipitate of reddish iron oxide within a few minutes. This is what led to the name Blood Geyser. The only other Yellowstone springwater known to do this comes from the Chocolate Pots, elsewhere in the Gibbon Basin.

2. UNNG-GIB-2 is difficult to see from the trail. Its vent is nothing more than a crack in a large rhyolite boulder, near the base of the hillside a short distance to the right (west) of Blood Geyser (1). Every few minutes a brief squirt of water rises from the crack, reaching as high as 4 feet. A small pool at the base of the rock constantly pulsates and splashes a few inches high.

"THE SULPHUR CASTLES"

"The Sulphur Castles" area is high on the southwest side of Paintpot Hill, basically over the mountain from the Artists' Paintpots. Access can be gained by a rigorous hike up the canyon immediately west of the paintpots; electrical power lines follow this same canyon. The area was named after old deposits of pure sulfur that have eroded into fanciful shapes that resemble miniature castle turrets and ramparts. Around the hill from there, several small springs play as perpetual spouters, and one of them is believed to act as a true geyser at times.

GEYSER SPRINGS (GEYSER CREEK GROUP)

The Artists' Paintpots trail ends at this group. Signs admonish the visitor to stay on the trail, so continuing onward can be tricky. What appears to be a good trail beyond the signs rapidly deteriorates to nothing. It used to be maintained through the forest and around the east side of Paintpot Hill another half mile to the Geyser Springs, but since the forest fires of 1988 that route has become impassible because of downed timber and a dense growth of young trees. Access is now via a much longer and strenuous bushwhacking trek that gets more difficult every year.

The Geyser Springs (Map 9.2) is the site of the most vigorous geyser activity in the Gibbon Basin. Eleven geysers are described here, and at least as many other small, irregular geysers have also been observed. Dozens of additional springs, ranging from clear pools to mud pots and sulfurous cauldrons, make this a dangerous area. Perhaps no place in Yellowstone contains more fragile crust, and the temperatures are high. Any exploration of the area should be with groups of people and confined to the hillsides as much as possible.

The Geyser Creek area naturally divides itself into three sections. First is an open valley, the "Lower Basin," in which most of the larger springs are found. Farther upstream, separated from the lower area by a forested ridge, is the smaller "Upper Basin" that includes several pools and at least three geysers. The uppermost set of hot springs is on the precipitous slopes of a narrow canyon south of the "Upper Basin," where there are no geysers but several mud pots and some of Yellowstone's largest perpetual spouters, one of which has been called "The Monster."

This is one of those Yellowstone areas where you should definitely notify somebody of your planned trip, and then check back in afterward. Carefully watch where you step, and make noise—the canyon beyond the Upper Basin was the residence of a grizzly bear during 1992. As always, your safety is your own responsibility.

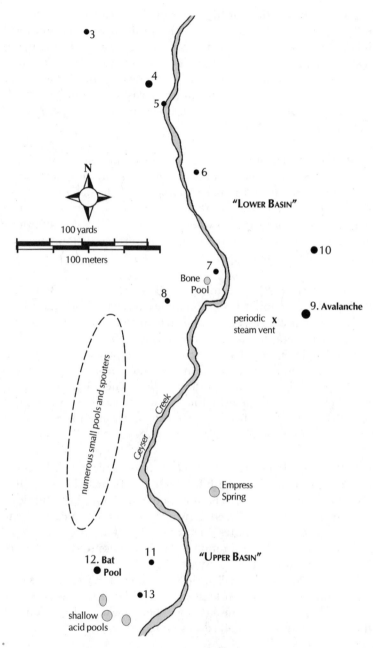

Map 9.2. *Geyser Springs, Gibbon Geyser Basin*

3. UNNG-GIB-14 is the northernmost of all the Geyser Springs. The old trail entered the area right next to this spring, but now most hikers first find themselves near Bull's Eye Spring (4) about 150 feet southeast of GIB-14. Long suspected of being a geyser because of beaded geyserite within the oval crater, GIB-14 was never observed in eruption until 2003, when splashing about 3 feet high was entirely confined to the crater. It was still weakly active in 2004 but was dormant in 2007.

4. BULL'S EYE SPRING is a large pool near the lower end of the "Lower Basin." Located within a massive geyserite rim, this pool is most commonly active as a perpetual spouter. The water is acid and rich in free sulfur, but occasionally it clears. At such times Bull's Eye may have periodic eruptions when both the intervals and durations are a few seconds long. With substantial overflow, the bursting can reach 4 to 6 feet high.

5. ANTHILL GEYSER is next to the creek a few feet from Bull's Eye Spring (4). It erupted from a complex of vents dominated by a small symmetrical geyserite cone that looked something like an anthill. The duration was commonly longer than the interval, but both ranged between 2 and 9 minutes. The play was a noisy mixture of steam and spray jetted from the cone, sometimes reaching 4 feet high. Additional vents sputtered a few inches high, and a small pool below the cone splashed to about 1 foot. Always variable in its activity, Anthill has been dormant since the mid-1990s, and the weathered cone has all but disappeared.

6. UNNG-GIB-8 is a large, muddy pool almost at creek level a short distance upstream from Anthill Geyser (5). Berms of loose gravel around the crater imply relatively frequent action, but eruptions have seldom been witnessed. During one such episode, the play repeated every few minutes. Black, muddy water was burst as high as 10 feet for as long as 2 minutes. Most commonly, the pool bubbles and pulsates slightly.

There are several small springs on the wet geyserite flats east and southeast of GIB-8. At least three have been seen as geysers 1 to 3 feet high, but it is unusual to see them playing. The ground in that area is extremely fragile and should not be entered.

7. UNNG-GIB-4 is near Bone Pool, the only clear-blue pool in the lower portion of Geyser Springs. First observed in 1985, GIB-4 erupts from a jagged crater that appears to have been formed by a recent steam explosion. The play splashes about 4 feet high for a few seconds. Nothing is known of the intervals, and the spring is often dormant.

Bone Pool has no record of eruptions. During the 1990s its water level dropped to several inches below overflow, but it remains hot and a lovely pale-blue color.

8. SUBTERRANEAN BLUE MUD GEYSER is unique among Yellowstone geysers. It is located up a narrow draw west of Bone Pool, a short distance above some small mud pots. Its vent is pipe-like, about 1½ feet across, and plunges vertically downward to a water level probably 10 feet below the ground surface. There are both minor and major eruptions. All begin with subterranean surging, which pounds the ground and creates a deep echoing sound. Minor eruptions last a few seconds and are entirely confined to the subsurface. They repeat every 2 to 5 minutes. Two to four minor eruptions usually take place between the majors. They begin the same way as the minors but continue to gain strength until a fine spray of water charged with blue-gray clay is jetted out of the opening for as long as 1 minute.

9. "AVALANCHE GEYSER." A second name, Oblique Geyser, is officially approved for this spring. It is seldom used because it probably originally applied in 1878 to a geyser in Gibbon Canyon near Beryl Spring (19) and was inadvertently moved to this geyser in the 1880s. The debate about the correct location has been contentious. This geyser has also been referred to under a series of other, informal names such as Rockpile, Talus, Marvelous, Geyser Creek, and Spray Geyser. All of these names are applicable, providing good descriptions of the setting and activity. "Avalanche" spouts from no fewer than seventeen separate vents, each producing a steady jet with its own character. The vents open among a pile of boulders coated with spiny, pale-brown geyserite.

"Avalanche" is a frequent and highly regular performer, a geyser that would do justice to any geyser basin. Active at all known times, it has shown a gradual increase in the intervals through the years. During 1928, for example, the eruptions recurred every 6 minutes. By 1974 the intervals had increased to 7 to 8 minutes, and

"Avalanche Geyser" is the most vigorous geyser in the Gibbon Geyser Basin. It spouts water from at least seventeen vents, each of which jets in a different direction.

since 1985 most observers have reported average intervals of 8 to 9½ minutes. The duration is 2½ to 3 minutes.

The eruption begins with sudden puffs of steam from the two main vents. With a series of rapidly stronger gushes, the steam is quickly followed by water, and before many seconds have passed, "Avalanche" is in full eruption. Both main vents shoot water at slight angles to 25 or more feet high. A third important vent erupts out away from the others at a low angle, with the water falling 30 feet from the opening. Most of the other vents play only 2 to 5 feet or noisily emit only steam under pressure. The display is very impressive, roaring and spraying water in all directions. The end of the eruption is sudden. The water abruptly gives out, and "Avalanche" falls quiet after two or three dying gasps of steam. Occasional gurgling interrupts the quiet interval.

Near a huge boulder on the hillside west of "Avalanche" is a periodic steam vent. At times it seems to play in sympathy with "Avalanche," but this may be sheer chance. The steam vent also has independent activity and is sometimes dormant.

10. BIG BOWL GEYSER is about 150 feet north of "Avalanche Geyser" (9), near the base of the slope at the eastern side of the valley. The crater and its surroundings are exquisitely decorated with pearly geyserite beadwork that gave it an alternate name, "Necklace Geyser." Activity in the deep basin is nearly constant. Highly super- heated, the water boils continuously, punctuated by truly periodic eruptions. The intervals are usually less than 15 seconds long, and the durations are equally short. Most of the play is confined to the crater, but occasional heavy surges send sharp jets of water as high as 30 feet.

Access to the "Upper Basin" is gained by way of two routes. One is to climb up to a bench above Subterranean Blue Mud Geyser (8). Along this direction are several small pools and perpetual spouters. Alternately, over the slope above Avalanche Geyser (9) is an open terrace. At its southern end is "Empress Spring," a dark pool that sizzles because of the evolution of gas bubbles other than steam. These routes converge at the "Upper Basin."

11. UNNG-GIB-9 is a pair of small connected pools down the slop- ing geyserite platform toward the creek from Bat Pool (12). It is usually observed only as a weakly intermittent spring, but eruptive periods in which splashing 2 feet high repeats every few minutes are known. In addition, the runoff channels are sometimes found to be extensively washed, indicating that larger eruptions or gush- ing discharges sometimes take place.

12. BAT POOL is the largest of several deep, clear-blue pools among the "Upper Basin" springs. Its water is superheated. The boiling is steady but normally rather gentle. Every few seconds to minutes it increases in vigor, and bursting can then throw water up to 3 feet high.

One of the boulders near Bat Pool was fractured in its fall from the hillside above. A colony of bats usually nests inside the crack. Take a close look. They'll scurry about, twitter complaints, and glare at you with beady eyes. These bats are insectivores, harmless to people. Obviously, though, the crack is not a place for fingers.

13. TINY GEYSER is located among a cluster of springs (some of which are surrounded by dangerously thin crust) a few feet southeast of

Bat Pool (12). It can be difficult to find except when actually erupting, when it can be heard from many feet away. Tiny is extremely small, perhaps the smallest true geyser in Yellowstone. The vent is within a shallow depression in the platform and is surrounded by beaded, yellowish geyserite. Water rises within the 1-inch hole, sputters a bit of fine spray a few inches high, and then drops following about 5 seconds of play. Eruptions usually recur every 2 to 3 minutes. Tiny has had some known dormancies, but the solid geyserite and repeated rejuvenations indicate that it is a long-term feature despite its minuscule size.

GIBBON HILL GROUP

Gibbon Hill lies east of Artists' Paintpots and northeast of Geyser Springs, about 1 and ¾ mile, respectively, from those groups. Like Paintpot Hill, Gibbon Hill is a youthful volcanic dome, perhaps as little as 70,000 years old. The hot springs of the Gibbon Hill Group lie along the western base of the hill and include at least four geysers. No trail leads to the area. Access is by bushwhacking through dense forest and wet meadows, at first via the same rigorous route that leads toward Geyser Springs.

14. GIBBON HILL GEYSER is difficult to find for anybody who has not been to it before, unless nearby Phoenix Geyser (15) happens to be in eruption and its steam cloud acts as a guide. Gibbon Hill used to be a significant geyser. The jetting, 25-foot bursts were continuous throughout durations of 20 to 50 minutes. Eruptions ended without warning and the crater drained, often with a loudly sucking whirlpool over the vent. Slow refilling usually began before the crater was completely empty. At those times the intervals were regular, with an average of about 5 hours. Occasionally, especially during the early 1980s, Gibbon Hill acted as a perpetual spouter, with weaker play not more than 6 feet high.

Large portions of the Gibbon Geyser Basin, including most of the Gibbon Hill–Paintpot Hill area, burned in the forest fires of 1988. The barren hillsides were subject to severe flooding and erosion until new plants gained a foothold. Thunderstorms in August 1989 produced a number of muddy debris flows. One of these completely inundated Gibbon Hill Geyser. Given the geyser's former vigor, it seems remarkable that it appears to be permanently sealed in and that it took six years for some part of the water and energy to finally find a new exit at Phoenix Geyser (15). The crater of Gibbon Hill remains all but obliterated, filled with debris that yields a slight trace of steam on cold days.

15. PHOENIX GEYSER apparently was active as a small geyser during the 1920s, but at an unknown time it fell into a long dormancy. The crater was difficult to find because it was filled with debris and covered by fallen logs and brush. In August 1989 a flash flood washed debris from the 1988 forest fires into the crater of nearby Gibbon Hill Geyser (14), filling it with mud and causing it to cease all activity. Six years later, in 1995, Phoenix made its modern appearance. It was named in allusion to the mythical bird that arose anew from the ashes.

At first, Phoenix erupted about every 2 hours with durations of 5 to 10 minutes. The intervals have gradually increased since then, but so have the durations. Play now recurs every 3 to 6½ hours and lasts as long as 80 minutes. Much of Phoenix is covered by a roof of geyserite, but some spray reaches as high as 8 feet through a hole in the roof. The crater drains completely after an eruption and then begins to slowly fill. It overflows for as long as 2 hours before the next eruption takes place.

16. UNNG-GIB-10 is the largest member of a cluster of hot springs about ¼ mile north of Gibbon Hill Geyser (14). This area has been referred to as the "Gibbon Hill Annex." It is a more significant thermal area than is the immediate area at Gibbon Hill Geyser. The important springs are north of a warm stream lined with dozens of frying pan springs and acid spouters, one of which in 2003 reached over 10 feet high. The stream cannot be crossed, so access to GIB-10 and GIB-11 (17) must be roundabout.

The pool of GIB-10 is about 20 feet in diameter and very deep. It is atop a broad sinter mound and has massive but badly weathered geyserite shoulders. It is clear that it used to be considerably more active than it is now. The spring bubbles at several points, and on what are probably irregular intervals this can develop into splashing 1 to 2 feet high.

17. UNNG-GIB-11 ("AQUAMARINE SPRING") is a gorgeous spring on top of a conical mound near the edge of the thermal area, up the slope northeast of GIB-10 (16). The intricate scalloping of geyserite around the crater is silvery gray. The basin is filled with shimmering, pale-blue crystal-clear water, and some of the surrounding catch basins are tinted pastel shades of blue, green, and yellow. The spring, though, is a perpetual spouter of rather low temperature. The 1-foot splashes are primarily a result of gasses other than steam, probably mostly carbon dioxide.

MONUMENT GEYSER BASIN

High on the hill above the south side of Gibbon Meadows, 600 feet above the valley floor, sits the Monument Geyser Basin. It is a long, narrow area of limited activity. Most of the springs are acid mud pots, sulfurous pools, and steam vents. The thermal activity extends southeast from Monument down the precipitous slopes of Chromatic Canyon, where it actually includes Beryl Spring next to the highway in the bottom of Gibbon Canyon.

Monument was named for the weird sinter cones scattered along the southwestern margin of the basin. Some have names, such as Sunning Seal, Sperm Whale, and Dog's Head. Built up either by geysers with spraying eruptions or perhaps by deposits formed along subsurface plumbing channels and since exhumed by erosion, the cones are totally unlike those of other areas. Only one of these—Monument Geyser—is still active. No other spring in the group contains alkaline water. Were it not for these odd formations, the Monument Basin would have attracted little attention.

A 1-mile trail leads to the basin. It starts next to the highway bridge that crosses the Gibbon River between Beryl Spring and the Gibbon Meadows. The entire 600-foot climb is accomplished in the second half-mile. It is a rigorous hike, but Monument is a unique area, and the view from the mountaintop is terrific.

18. MONUMENT GEYSER, also called Thermos Bottle Geyser, spouts from one of the tallest cones in the Monument Geyser Basin. Unlike all the others, it has not quite sealed itself in by internal deposits of sinter. That time is not far off, though. During the 1930s and before, Monument played most of the time, with fine spray jetted as high as 15 feet. Now it steams gently, hissing under slight pressure, but almost no liquid water is ever ejected.

The Thermos Bottle, also called Monument Geyser, is the only one of the old geyserite cones in the Monument Geyser Basin that still shows a bit of thermal activity.

BERYL SPRING AND "FORMATION HOT SPRINGS"

The hot springs in the bottom of Gibbon Canyon about 1 mile south of the Gibbon Meadows are dominated by Beryl Spring, which is next to the highway and served by a remarkably large parking area. Several springs near Beryl were apparently buried by the road construction, and one of these might be the original "Oblique Geyser" of the 1870s (see Avalanche Geyser, #9). Upstream near the highway bridge across the Gibbon River, the "Formation Hot Springs" include numerous acid spouters, small mud pots, steam vents, and at least two geysers.

19. BERYL SPRING is superheated, constantly boiling and surging so as to perpetually throw water as high as 3 or 4 feet. Beryl has appeared in some lists as a geyser, but there is no evidence that it has ever had intermittent activity as a true geyser except perhaps on a few occasions shortly after the 1959 earthquake. The name is an allusion to the pale-blue color of the pool that resembles aquamarine, a gem form of the mineral beryl.

20. UNNG-GIB-12 is located among the "Formation Hot Springs" on the hillside just north of the Gibbon River highway bridge. All of these springs are small, but most have high temperatures and occasionally act as perpetual spouters. One of the closest to the highway reached 6 feet high during 1985. In 1991 and again in 2002, it was smaller in size but truly periodic in its performances. Eruptions were frequent and 2 feet high.

21. UNNG-GIB-15 erupts from a small crater just north of the highway bridge, between the road and the Gibbon River. Little is known about it, but eruptions are certainly rare. One seen in October 2005 splashed higher than 4 feet for a duration longer than 5 minutes.

SYLVAN SPRINGS GROUP

From the highway, Sylvan Springs is the most visible hot spring group in the Gibbon Geyser Basin. It is the large barren area at the far western end of the Gibbon Meadows. The hot springs are almost entirely acid, relatively cool, and often muddy. The most important springs here have undergone frequent and rather drastic changes.

Evening Primrose Spring was once regarded as one of the most beautiful pools in the Park, and it was probably the one reason a maintained trail used to lead here. In shape and color it was comparable to Morning Glory Pool in the Upper Geyser Basin. At some point (just when does not seem to have been recorded) the water changed from alkaline to acid. The surface became covered with a thick froth of elemental sulfur of brilliant yellow color. Then, during 1972 Evening Primrose was invaded by *Sulfolobus*, an Archaea that metabolizes sulfur. Today's pool has become one of Yellowstone's ugliest, with a murky yellowish-green or brown color. The temperature has been recorded as low as 106°F (41°C), and the pH was once measured at 0.95, perhaps the most strongly acid value ever determined in a Yellowstone spring.

Another important Sylvan Spring is Dante's Inferno, a name probably originally intended for this entire group. Although it is not a geyser as such, the shocks of the 1959 earthquake caused it to erupt violently to more than 100 feet high. The spring churns and boils vigorously while building extensive geyserite terraces.

Within the small gorge at the center of the group are several perpetual spouters, some of which are gas powered with temperatures below boiling. Just over a small ridge west of these spouters is Sylvan Spring, a vigorously boiling pool named after the group as a whole. A dark brown mud pot near the area's northeastern limit is called Coffin Spring, near which a small pool was seen to erupt as a true geyser, bursting 3 to 4 feet high every few minutes during the early 1980s.

What is sometimes called the "South Group of Sylvan Springs" lies about ¼ mile through the forest to the south. These springs are much more fitting of the term "sylvan," and they probably comprise the original Sylvan Springs. One of these pools is a geyser.

22. UNNG-GIB-13 has been active on all reported visits to it since 1974. It is a large, oval pool that measures 30 by 7 feet. It evidently developed along a fracture, because there are several vents along the length of the bottom. Eruption intervals of 10 to 12 minutes lead to durations of 8 to 9 minutes. The splashing play is from the northernmost vent and reaches as high as 5 feet.

THE CHOCOLATE POTS

The Chocolate Pots (possibly the "Red Geyser Basin" of 1877) are near the highway, scattered along both sides of the river's Gibbon Cascades in the small canyon between Gibbon Meadows and Elk Park. The spouting action from the large cone on the west side of the river can reach up to 2 feet high, but this is a result of hydrostatic (artesian) pressure and not boiling. The highest temperature in these small springs is only around 130°F (54°C), but the deposits being formed are unique. Rich yellow- and red-brown, they are more than 50 percent iron oxide, 5 percent aluminum oxide, and 2 percent manganese oxide; their silica content is only 17 percent. A sample of the clear water allowed to sit for a few minutes will develop a cloudy brown cast as iron oxide compounds spontaneously precipitate from the water.

CHAPTER TEN

Third (or "Lone Star") Geyser Basin

The Third Geyser Basin (Map 10.1, Table 10.1, numbers 1 through 10) lies along Firehole River, about 5 miles upstream from Old Faithful. Nearly everybody refers to it as the "Lone Star Geyser Basin," but that name has never had official status. The name "Third Geyser Basin" was first used in 1872 and was formally adopted by the geological survey of 1878. In terms of the number of geysers (only ten), this is Yellowstone's smallest geyser basin. Its most prominent feature is Lone Star Geyser, named because of its isolated location with respect to the Old Faithful area.

The Lone Star Basin naturally divides into five parts, each of which contains at least one geyser. These groups, with their informal names, are shown on Map 10.1.

The Lone Star Basin is reached via a wide trail beginning from the main highway near the Kepler Cascades of Firehole River. An old road open to traffic until 1971, it is mostly paved and is recommended as a bike trail. The distance from the highway to Lone Star

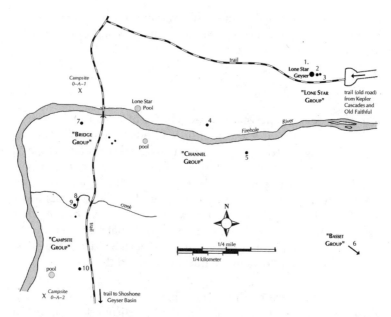

Map 10.1. *Third Geyser Basin*

Table 10.1. Geysers of the Third (or Lone Star) Geyser Basin

Name	Map No.	Interval	Duration	Height (ft)
Black Hole Geyser	2	10 min–hrs	3–15 min	3–25
Buried Geyser	6	7–15 min*	3½–5 min	3–20
Lone Star Geyser, minor	1	see text	5 min	45
Lone Star Geyser, major	1	2½–3½ hrs	30 min	45
Perforated Cone Geyser	3	infrequent	minutes	inches
UNNG-LST-3	4	frequent*	minutes	2–3
UNNG-LST-4	5	frequent*	seconds	1
UNNG-LST-5	8	hours*	30 min	1
UNNG-LST-6	9	hours*	hours	1
UNNG-LST-7	10	rare	10 min	3
UNNG-LST-8 ("Clam Shell")	7	minutes*	seconds	1–8

* When active.

Geyser is about 2½ miles. Partway along this trail, not far beyond the bridge over the river, is a small cluster of hot springs. The largest feature here, playing from a sinter crater on the hillside, is

Halfway Spring. Usually active as a variable perpetual spouter, it has been known to act as a geyser with intervals and durations of a few seconds. Its height is up to 3 feet.

LONE STAR GROUP

The Lone Star Group is a compact cluster of hot springs, most of which are small, acid features of little importance on the hillside. Separated slightly from them, though, is a geyserite mound topped by three alkaline springs, each of which is a geyser. The star attraction is Lone Star Geyser itself. To help the visitor better anticipate when Lone Star might erupt, the National Park Service maintains a logbook. It is hoped that visitors who see Lone Star or the other geysers erupt will write the time in the book. That way, later observers will know when it last played, and the Park Service will be able to gather the data needed to better understand the area.

1. LONE STAR GEYSER gained early fame because of its very large geyserite cone. Over 9 feet tall with nearly vertical sides, it is one of the biggest cones in Yellowstone. Lone Star erupts from one main and several minor vents at the summit of the cone. Collectively, the minor vents have their own official name: The Pepper Box.

The cone was built to its present height by the frequent splashing of the geyser during its quiet phase. This activity generally starts 1 to 1½ hours after the previous eruption. With gradually increasing force, the pre-play leads into a minor eruption. Lone Star often has just one minor preceding the major eruption, but there can be two or even three minors before the culminating major eruption. The usual single minor play lasts about 5 minutes and reaches as high as 45 feet. Then, after a quiet pause of 25 to 35 minutes, renewed splashing builds into the full, major eruption. It also jets water to 45 feet, but the eruption lasts fully 30 minutes. In the waning stages, the water gives out, and the final part of the play is a powerful steam phase, loud enough to be heard a mile away. No matter what the nature of the pre-play and minor eruptions, Lone Star's intervals are always close to 3 hours. No dormant period has ever been recorded.

2. BLACK HOLE GEYSER is just a few feet from the base of Lone Star's (1) cone. Its vent is a symmetrical funnel that penetrates the geyser-

Lone Star Geyser is one of the most regular geysers anywhere, repeating its major eruptions every three hours with little variation. Its geyserite cone is also one of Yellowstone's largest.

ite mound. This geyser probably had some activity during the early years of Yellowstone, but it was not reported in modern times until 1973. Black Hole's activity is variable. Regular intervals as short as 10 minutes with durations of 3 to 4 minutes are known, but so are periods of several hours and durations longer than 15 minutes.

The better eruptions jet rockets of water as high as 25 feet, but splashing only 3 to 4 feet high is more typical. During the quiet interval, water may be visible within the vent, and this will start to burst outward and produce runoff before the eruption begins. This preliminary bubbling can continue for several hours. The initial eruption begins abruptly, with no other warning.

3. **PERFORATED CONE GEYSER** is located within a low geyserite mound directly on the opposite side of Black Hole (2) from Lone Star's (1) cone. The mound is punctured by numerous small holes. The eruptions are erratic and infrequent and consist of nothing more than steamy sputtering out of the holes. This is barely enough to keep the mound decorated with spiny sinter, and there is never any runoff.

"CHANNEL GROUP"

Along both banks of Firehole River, starting roughly ¼ mile upstream from Lone Star Geyser, are numerous hot springs. They make up the "Channel Group." The majority are in the grassy area on the south side of the river. Most of the springs are small and of little importance, but a few are geysers. At the west end of this group, north of the river and near the trail's bridge across the river, is Lone Star Pool, which behaves as a long-cycle intermittent spring and has one report of small eruptions. Another pool and a perpetual spouter lie in the woods across the river from Lone Star Pool.

4. **UNNG-LST-3** is an assortment of small springs surrounded by meadow on the north side of the river. They include one pool about 3 feet in diameter plus several geyserite cones a few inches high. The springs of LST-3 are closely related to one another, as they are all active at the same time. Usually quiet or with tiny perpetual sputtering, the springs infrequently act as geysers, with some of the play reaching 2 to 3 feet high. The action may be cyclic, as times with intervals only a few minutes long have been recorded. A large runoff channel drains this cluster, implying that greater discharge takes place on rare occasions.

5. **UNNG-LST-4** is across the river and slightly downstream from LST-3 (4). The small pool, which is only a few inches above the

stream level, constantly surges and bubbles. Its eruptive activity may be cyclic, since it is often seen to play with short intervals but sometimes goes for long periods without activity. The eruptions last only a few seconds and are just 1 foot high.

All along the river on both sides of LST-4 and up the hillside to the south are many other hot springs. Most of them possess runoff channels, yet few discharge more than a trickle of water, and the channels are filled with vegetation. Evidently, these springs were once active as geysers, but it has been at least several decades since they last played.

"BASSET GROUP"

The "Basset Group" was named for the large rock faces exposed in the cliffs near its largest geyser; *basset* is French for a jagged, rocky outcrop. Well separated from the other Lone Star Basin springs, the group is easiest to locate from Lone Star Geyser, from where the steam of the one large geyser can be seen. Hike in that direction, either by fording Firehole River near Lone Star (not recommended) or by crossing on the Shoshone Lake trail bridge about ½ mile upstream and backtracking through the "Channel Group." The geyser is the first hot spring encountered in this group, at the base of the cliff at the head of a wide, barren drainage area. Around the ridge east of the geyser are several additional springs, some of which erupt as perpetual spouters, and up the steep slope above them is a fine mud pot.

6. BURIED GEYSER is probably a more significant geyser than Lone Star (1). Although not as high, its eruptions are frequent and powerful and often have great water discharge. Buried was so named because its crater was once partially covered by a sinter ledge; that is now gone, removed by the bursting activity of the past thirty years. Buried was apparently a quiet spring until perpetual spouting was reported in 1973. It became a geyser in 1983 and now has minor, intermediate, and major eruptions. Usually about two-thirds of the eruptions are minors. Their duration is about 3½ minutes, followed by intervals of 7 to 9 minutes. The water is splashed 3 to 9 feet high. This play ends abruptly just as the first overflow is reached. The water level then quickly drains down as far as 2 feet, only to immediately begin to rise again.

Intermediate eruptions are distinct from the minors because the water level rises more rapidly during the eruption, so there is some overflow before the play stops, again after a duration of about 3½ minutes. This play is also somewhat more vigorous, and the bursts may reach as high as 12 feet. The major eruptions begin in the same fashion as the others, but rather than stopping before or with little runoff, the play becomes violent. Bursts 20 feet high are accompanied by a gushing overflow that floods the formations down the slope below the crater. Most major eruptions have durations of 4 to 5 minutes and are followed by intervals of 11 to 15 minutes. In the long run, Buried is quite consistent in its performances. Times have been known when all the eruptions were minors, and for a while in 1990 nearly half were majors. Since 1973, only one short dormant period is known, when the water lay quietly about 1 foot below overflow.

"BRIDGE GROUP"

The "Bridge Group" includes a compact cluster of hot springs next to the Shoshone Lake trail, just south of the footbridge that crosses Firehole River. Mostly of no account, two of these springs were seen to have small eruptions in 1989. Along the fringe of forest in the upstream direction is a scattering of larger springs, at least one of which can be a significant geyser.

7. UNNG-LST-8 ("CLAM SHELL GEYSER") is a strange feature. It plays out of a small, cone-like buildup of geyserite within a more open, saucer-like crater. Most of the time the play is perpetual spouting only 1 foot high and confined to the cone, but LST-8 has been seen to fill the outer crater and burst as high as 8 feet. Such eruptions are brief but may occur as a series of plays every few minutes. There is never any overflow. The frequency of this action is unknown.

"CAMPSITE GROUP"

About 1 mile from Lone Star Geyser along the Shoshone Lake trail, starting several hundred feet beyond the "Bridge Group," is the "Campsite Group." A backcountry campsite is located next to this group. This is a fairly extensive cluster of hot springs. Most are small and many are muddy, but the area includes at least three geysers. The largest blue pool in the Lone Star Geyser Basin is also in this group. Surprisingly, it is not named.

8. UNNG-LST-5 plays from a shallow crater lined with spiny, yellow geyserite. It is just downstream along a small creek the trail crosses on a low boardwalk bridge. This geyser has long dormant periods during which the crater is only about half full of water. When it is active, the intervals are several hours long. The play lasts as long as 30 minutes but is only 1 foot high.

9. UNNG-LST-6 is almost identical to LST-5 (8) in every respect. It also has long dormant periods and, when active, undergoes small eruptions at intervals of several hours. The one difference is that the eruptions may have durations as long as a few hours.

Just a few feet west of LST-5 and LST-6 is an assortment of vents and fractures on a geyserite platform. Although no eruptions were witnessed, they apparently played in some fashion in 1988 and 1989, when washed areas and small runoff channels were formed. Just across the small creek from LST-6 is a muddy pool with a unique red-orange color.

10. UNNG-LST-7 is, by size at least, the most important geyser in the "Campsite Group." It is a small pool that constantly bubbles and splashes weakly. Active periods as a geyser are rare. During the mid-1970s a number of eruptions were seen. They lasted about 10 minutes and reached 3 feet high. Only a few eruptions have been recorded since 1980, but this is likely because of the remote location rather than inactivity.

Continuing another mile along the Shoshone Lake trail beyond the "Campsite Group," the hiker enters the large area of the Firehole Meadows. Between the trail and the hillside to the east are a number of hot pools collectively called the "Firehole Springs," and of difficult access up a small canyon farther east is the "Divide Group," where one geyser has been reported.

CHAPTER ELEVEN

Shoshone Geyser Basin

The Shoshone Geyser Basin (Map 11.1, Tables 11.1, 11.2, and 11.3, numbers 1 through 56) is one of the most important thermal areas in the world, even though its major portion measures only 1,600 by 800 feet. The basin contains over eighty geysers, more than any place on Earth other than the remainder of Yellowstone, the Valley of Geysers on Russia's Kamchatka Peninsula, and El Tatio, Chile. Some of the spouting at Shoshone can be of considerable size, and one geyser—Union—is of truly major proportions. On the other hand, many of these geysers are small, infrequent performers about which relatively little is known. Accordingly, only fifty-six geysers are described here. Extensive areas of mud pots, frying pans, and acid pools lie in and around the hills between the geyser basin proper and Shoshone Lake.

The Shoshone Geyser Basin in general resembles the Upper Basin. The hot springs form compact groups scattered along the course of Shoshone Creek. But Shoshone has its own special attri-

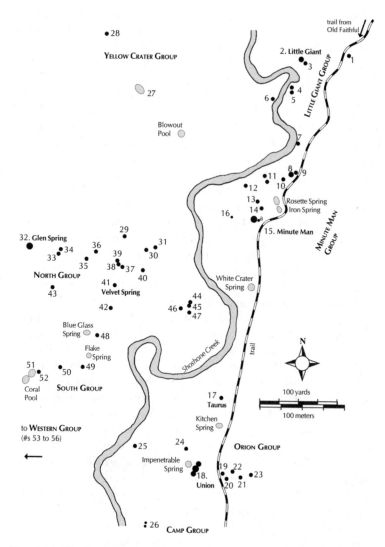

Map 11.1. *Shoshone Geyser Basin*

butes as well. Some deposits are brightly colored by iron oxide minerals, and substantial amounts of mercury have been reported in a few formations. Yellow crystalline sulfur forms at other springs, and small amounts of arsenic sulfide minerals spot the ground with brilliant orange-reds and yellows at a few places among the acid springs around the perimeter of the basin. As everywhere, the geyser basin's alkaline springs support colorful cyanobacteria along the runoff channels and in the cooler pools.

One of the earliest descriptions that can be ascribed to a particular geyser was written by trapper Osborne Russell in 1839. He called it "Hour Spring" because of its regular activity. Unfortunately, nothing now active at Shoshone closely resembles the geyser Russell described. Various researchers have tried to equate Hour Spring with Minute Man Geyser or its adjacent "Minute Man's Pool," or with either Union Geyser or Little Giant Geyser, but we will never know for sure. During the 1870s the Shoshone Geyser Basin was only slightly more remote than any other area of Yellowstone, and it was studied extensively by the geological surveys of 1871, 1872, 1878, and 1883–1887. Most of the recognized names were given to the springs in those years. From that time on, however, no detailed studies or maps were made of the area until the late 1960s.

The Shoshone Geyser Basin is reached most easily via the same trail that passes Lone Star Geyser (see Chapter 10). The total distance from the highway at Kepler Cascades is about 8½ miles; riding a bicycle to Lone Star Geyser reduces the hiking distance to a little less than 6 miles one-way. The route is well maintained and generally easy. Just beyond the high point on the trail at Grant Pass is a trail junction in Shoshone Meadows, where the left branch leads to the geyser basin. The basin can also be reached by way of other, longer trails that lead from the east along both shores of Shoshone Lake or by canoe. There are backcountry campsites nearby.

It is easy to imagine yourself as the first person ever to see Shoshone's geysers. The basin seems nearly untouched. Unfortunately, not everybody appreciates the natural world. Some of the formations have been vandalized, one even chopped out with an ax. When in the area, please do everything you can to avoid damaging any of the springs and geysers or their deposits. As always, remember the dangers of hot springs, too—even a small thermal burn is a serious injury in a place as remote as this.

LITTLE GIANT GROUP

The Little Giant Group (Table 11.1, numbers 1 through 6) is the first collection of hot springs encountered when entering the basin along the trail from the north. It is the smallest in area of Shoshone's

Table 11.1. Geysers of the Little Giant and Minute Man Groups, Shoshone Geyser Basin

Name	Map No.	Interval	Duration	Height (ft)
Black Sulphur Spring	7	steady	steady	3–6
Double Geyser	3	55 min–hrs	5–7 min	8–12
Five Crater Hot Spring	12	3–7 min*	1½–3 min	3
Gourd Spring	13	1–2 hrs	45–90 min	8–10
Little Bulger Geyser	10	see text	sec–min	2–10
Little Giant Geyser	2	see text	sec–min	1–50
Locomotive Geyser	6	[1989]	2 hours	15
Meander Geyser	5	hours*	hours	3
Minute Man Geyser, entire series	15	5–9 hrs	2–4 hrs	—
Minute Man Geyser, within series	15	minutes	10–50 sec	10–40
Minute Man's Pool, minor	15a	w/Minute Man	seconds	1–8
Minute Man's Pool, major	15a	[1978]	10–30 sec	35–75
Shield Geyser	14	1–2 hrs	45–90 min	2–10
Soap Kettle	8	6–27 min	30 sec–4 min	4–6
Trailside Geyser	1	6–25 min	20–50 sec	1–3
UNNG-SHO-10 ("Skylight")	9	[1986]	unknown	flood–1
UNNG-SHO-11	11	9–15 min*	1–2 min	boil–3
UNNG-SHO-12	16	[1997]	seconds	1
UNNG-SHO-17 ("Trio")	4	irregular	30–40 min	4

* When active.
[ ] Brackets enclose the year of most recent activity for rare or dormant geysers. See text.

hot spring groups. Historically, Little Giant Geyser was the second largest geyser in the basin, but it rarely, if ever, has major eruptions now. The group contains several small pools in addition to the geysers, most of which have been known to erupt rarely.

1. TRAILSIDE GEYSER is located at the base of the hill immediately to the left of the trail as you enter the basin. Eruptions were first recorded during the late 1970s. The sinter of the shallow crater is lightly stained by iron oxides. During the quiet intervals, which range from 6 to 25 minutes in length, the crater is about half full of water. Eruptions begin with a sudden filling of the basin. Water is splashed 1 to 3 feet high for durations of 20 to 60 seconds. Very brief minor eruptions sometimes occur about 1 minute before the full play; on some occasions, they seem to comprise the only activity.

At the base of the slope just behind Trailside is another geyser, first observed in 1981. Its eruption is usually little more than mild bubbling that is nearly constant, but small splashing eruptions are sometimes seen. In the opposite direction, within the geyserite drainage area from Trailside, are several tiny holes that sometimes have sputtering eruptions of their own.

2. LITTLE GIANT GEYSER plays from a heavily iron-stained vent in the middle of a round sinter platform. In the early days of Yellowstone it was a significant geyser, spouting as much as 70 feet high about twice per day. There is no record of when such eruptions ceased; indeed, there is no mention of Little Giant in any reference after 1886—other than to repeat the data of the 1800s—until after the 1959 earthquake. A survey of the Shoshone Geyser Basin two months after the shocks found Little Giant to be active, with eruptions 15 to 20 feet high. It is interesting that this report noted the eruptions as more frequent than they had been before the quake— puzzling, because it is not mentioned in any thermal reports of the previous few years.

In any case, 1959's active episode was evidently brief. Except for a single report of a 20-foot eruption seen in 1988 and some sort of large eruption in late 1991, when a long-unused runoff channel was found cleanly washed, Little Giant has had only minor eruptions since before 1970. This play is erratic bursts 1 to 6 feet high that usually take place while nearby Double Geyser (3) is in eruption; even these appeared to have ceased in 2006. Overall, Little Giant has largely had a mysterious history. Major eruptions clearly are possible, and more probably lie in its future.

3. DOUBLE GEYSER jets two columns of water into the air. There is actually only one vent, but a bridge of sinter above it splits the stream. Double has also been called "The Pirates" because it is likely the cause of Little Giant Geyser's (2) weak activity. Double apparently did not exist during the early years of Yellowstone, when Little Giant was a regular performer. Now a reliable performer with only a few minutes' variation at any given time, its average interval was near 1 hour until 2000, when it abruptly began to increase. By 2007, full eruptions sometimes were several hours apart, with numerous minor eruptions 1 to 2 feet high between them. The play begins with progressively more vigorous welling of water out of the vent. It takes several minutes to build to the full height. Of the twin water jets, the vertical one is the highest, reaching about 10 feet, while the other shoots at an angle to 6 feet high. The total overflow duration is 10 to 15 minutes, and the actual jetting lasts 5 minutes.

4. UNNG-SHO-17 ("TRIO GEYSER") erupts from three small vents at the upstream side of Meander Geyser's (5) sinter platform. Eruptions were unknown before 1998. The play usually begins about 30 minutes before and then ends within 10 minutes after Meander starts its own eruption, giving a duration of 35 to 40 minutes. Some of the jetting reaches as high as 4 feet. In 2007, minor action only inches high continued although Meander was dormant.

5. MEANDER GEYSER rises from a small cone just downstream from a meander in Shoshone Creek. It was first observed in 1974, when it was active as a perpetual spouter. More recently, it has generally acted as a highly variable geyser. Intervals as short as 3 hours leading to 12-minute durations were known in the early 1980s, but now both the interval and duration are apparently many hours long. The play reaches 1 to 3 feet high. Meander was dormant in 2007.

6. LOCOMOTIVE GEYSER was originally described in 1947, when its water jet was periodically cut off by a puff of steam. The resulting sound resembled that of a steam engine. Over the years, intervals were known to range from as little as 4 to more than 24 hours. The duration was as long as 2 hours, and the pulsating water jet squirted as high as 15 feet. Unfortunately, Locomotive has been completely dormant since 1989, and the sinter platform that surrounds the vent is severely decayed.

MINUTE MAN GROUP

The Minute Man Group (Table 11.1, numbers 7 through 16) contains many hot springs, and most of them have histories as geysers. The most important is Minute Man Geyser, which is cyclic and plays frequently during active phases only a few hours apart. Either the hillside above Minute Man or the rock outcrop south of Little Bulger Geyser is a wonderful place to sit and take in the view that encompasses most of the geyser basin. Rosette Spring, just south of the outcrop, was well publicized during Yellowstone's

Soap Kettle exhibits boiling eruptions from a massive cone next to the trail through Shoshone Geyser Basin. When dormant, it bubbles gently from a water level deep within the crater.

early days because of its delicate sea-green color. It sometimes acts as an intermittent spring. The smaller pool just south of Rosette is Iron Spring.

7. BLACK SULPHUR SPRING is a perpetual spouter that plays from a vent within the steep sinter embankment of Shoshone Creek. The water is jetted in a series of rapid pulses that reach 6 to 10 feet outward at an angle into the stream. The black color is probably a result of manganese oxide, not sulfur, incorporated into the geyserite.

8. SOAP KETTLE is a superheated spring within a massive sinter cone near the trail that in most years exhibits true geyser activity. The intervals generally range between 6 and 27 minutes and are very regular at any given time. Eruptions begin when the boiling water begins to rise within the crater. When it nears overflow the boiling increases in force, and some bursts throw water as high as 6 feet above the rim. The duration is 1 to 3 minutes. Severe erosion around the base of the cone indicated that some exceptionally voluminous eruptions took place during 1987. When Soap Kettle is not active, the water level stands low within the crater, and the boiling is gentle and more constant, with no overflow. This has been the case for most of the 2000s.

9. UNNG-SHO-10 ("SKYLIGHT GEYSER") lies within a cavernous crater next to the trail near the northeast base of Soap Kettle's (8) sinter mound. It underwent some powerful eruptions during 1986. None was witnessed, but the overflow washed out a wide and deep runoff channel, while muddy spray coated vegetation on the slope as far as 15 feet away. Weaker action, only 1 foot high, is nearly constant and is confined to within the cave.

10. LITTLE BULGER GEYSER used to be one of the nicest geysers at Shoshone, with eruptions 10 feet high recurring every 8 to 13 minutes. But in 1985 a subsidiary vent known as "Little Bulger's Parasite" developed near the east edge of the crater. Active with about the same interval as Little Bulger used to show, it has brief erratic bursts up to 3 feet high. This new vent is between Little Bulger's vent and Soap Kettle (8). Some have surmised that its evolution was related to the general increase in Soap Kettle's action and the blowout of SHO-10 (9) in the mid-1980s. If so, Little Bulger may have met its demise. Although the old vent still splashes on occasion, no regular eruptions have been seen since the other developments.

11. UNNG-SHO-11, known to most geyser gazers as "USGS #11," is a very deep, oval spring along the runoff channel leading from Shield Geyser (14). Its activity depends on the volume of flow from Shield. When Shield is active, which is a great deal of the time, its water flows directly into the crater of SHO-11. The play then consists only of a gushing, bubbling overflow. However, when Shield is inactive and there is no runoff, SHO-11 can burst as high as 3 feet. Either way, the intermittent action recurs every 9 to 15 minutes and lasts 1 to 2 minutes. Perched as it is within a massive formation right at the brink of a steep slope, it is a fascinating feature.

At the base of the slope below SHO-11 is another pool, mapped by the U.S. Geological Survey as #12. It normally acts as an intermittent spring, but geyser eruptions as much as 5 feet high are known to occur infrequently.

12. FIVE CRATER HOT SPRING lies near the level of Shoshone Creek down the slope from Gourd Spring (13). The crater contains numerous highly convoluted and decorated ridges and projections of geyserite. They are what produce the "five craters," which are really just separate openings above one vent below. Five Crater is mostly active as a pulsating intermittent spring. The action rocks the water about within the crater, occasionally causing squirts to rise as much as 2 feet above the openings. More vigorous action often takes place during exceptionally long cycle intervals of Gourd Spring and Shield Geyser (14), when no runoff has entered the crater for an hour or more. Then the play is regular, repeating every 3 to 7 minutes and lasting 1½ to 3 minutes with actual jetting as high as 3 feet.

13. GOURD SPRING is the first of the four members of the Minute Man Complex. Its low sinter cone is occupied by a crudely gourd-shaped crater, and the massive geyserite shoulder is punctured by numerous minor vents. As is nearby Shield Geyser (14), Gourd is cyclic in its action. The intervals between active phases range from a few minutes to several hours. Their length depends on the previous duration, which is known to vary from 15 to 70 minutes. When active, the play is almost continuous splashing 2 to 3 feet high.

14. SHIELD GEYSER erupts from a somewhat square cone a few feet from Gourd Spring (13). An open, flat-bottom crater centered by a small vent occupies the top of the cone. Like Gourd, Shield is cyclic. It is usually active when Gourd is in eruption, but it is also known to undergo independent, albeit short, active phases of its own. Shield erupts only when its crater is full and overflowing. The bursting play reaches 2 to 4 (rarely 10) feet high, lasts a few seconds, and repeats every minute or two until the crater drains at the end of the active phase.

15. MINUTE MAN GEYSER erupts from a prominent cone 5 feet high and 12 feet long. Most of its sinter is exquisitely beaded, unlike any formation that can be seen close at hand elsewhere in Yellowstone. Minute Man is cyclic, but its cycles bear no discernible relationship to those of Gourd Spring (13) or Shield Geyser (14). Detailed studies conducted during the 1990s revealed Minute Man to be a very complex geyser.

Several hours—4½ to 9 and usually about 7 hours—pass between active cycles. Often there will be 1, 2, or 3 preliminary

Minute Man Geyser is the largest frequent performer in the Shoshone Geyser Basin.

eruptions before the full cycle starts. The full series begins 30 to 100 minutes after the preliminary activity. The total duration is 2 to 4 hours, and eruptions take place as often as every 1 to 3 minutes, typically adding up to more than 100 individual eruptions per cycle. The durations are from 10 to 30 seconds, and most play is about 20 feet high.

Late in a cycle, Minute Man *may* be joined by Minute Man's Pool (15a), and the jet then reaches as high as 40 feet. However, if the Pool is not active, the height drops to as little as 10 feet as the intervals grow longer near the end of the action.

15a. MINUTE MAN'S POOL. During the 1870s, the pool between Minute Man Geyser's (15) cone and the hill was reported to "sometimes spout," but no details as to the frequency or height were given. A photograph taken in 1930 shows a small eruption, perhaps 4 feet high. Similar eruptions were seen in 1974, when they took place several times during each of Minute Man's active phases throughout that summer.

"Minute Man's Pool," a vent between the cone of Minute Man Geyser and the hillside, underwent eruptions of unprecedented power between 1975 and 1978. Of all the geysers in the Shoshone Basin, only Union Geyser has been known to play to greater heights.

Beginning in 1975 and increasing in power and frequency through 1977, the Pool began exhibiting unprecedented major action. The eruptions were so strong and persistent that they became the major part of Minute Man's activity. At times, Minute Man itself was practically dormant. During an active period the bursts would recur as often as every 6 minutes. They were brief, generally lasting less than 20 seconds, but throughout the play jets of water were propelled at least as high as 35 feet. Many eruptions had some bursts of more than 50 feet, and heights of 75 feet were estimated by some observers. The explosive concussions could be felt and heard throughout Shoshone Geyser Basin. The eruptions washed out a section of the trail through the basin, eroded away some of the adjacent hillside, and considerably widened the runoff channels, making it all but certain that such eruptions had never happened before.

Unfortunately, the major activity ended during 1978. The Pool's water level now rises and falls in synchrony with Minute Man's eruptions, and sometimes, usually near the end of Minute Man's cycle, it will join in with bursts 4 to 10 feet high.

16. UNNG-SHO-12 was first observed in 1992, initially as a small hole that frequently sputtered about 1 foot high within Minute Man's (15) westward runoff channel. Through time it enlarged its crater. At about the same time it developed a small, open pool in 1997, it stopped erupting.

ORION GROUP

The Orion Group (Table 11.2, numbers 17 through 25) contains at least seven geysers of note plus several perpetual spouters, small flowing springs, and pools. Union Geyser is the largest of all the geysers at Shoshone. It is believed that its three cones, which have a fancied resemblance to the three stars in the belt of the constellation Orion, are what gave the group its name. The fact that the group also includes Taurus Spring supports the astronomical connection.

For reasons that are entirely unclear, the Orion Group's water levels drastically declined during the late 1970s, and since then there has been little activity in the area. This might be the result of an exchange of function, but none of the neighboring spring

Table 11.2. Geysers of the Orion and Camp Groups, Shoshone Geyser Basin

Name	Map No.	Interval	Duration	Height (ft)
Fifty Geyser	21	[1980]	5 min	1–2
Geyser Cone	26	[1996]	seconds	3
Sea Green Pool	19	[1976]	seconds	10
Taurus Spring	17	[2003]	seconds	50
Union Geyser	18	[1977]	40–60 min	100–125
UNNG-SHO-3	22	unrecorded	minutes	4
UNNG-SHO-4	23	15–30 min*	2 min	6–12
UNNG-SHO-13 ("USGS #86a")	24	1–4 min	1–2 min	1–2
UNNG-SHO-14	26	minutes*	seconds	1–4
UNNG-SHO-24	25	[1997]	seconds	3–6
White Hot Spring	20	frequent*	minutes	1–2

* When active.
[ ] Brackets enclose the year of most recent activity for rare or dormant geysers. See text.

groups has shown a corresponding increase in activity as a result. For a time in 1997, the water levels rose within the group and there were some geyser rejuvenations, but that action was temporary.

The pool west of the trail about midway between Taurus Spring and Union Geyser is Kitchen Spring, named in 1878 when it was used daily by the Hayden Survey's camp cook. Right next to Union Geyser is a dangerously deep crater occupied by a hot, sometimes boiling pool. Please keep away from the edge, for there would be no escape from Impenetrable Spring.

17. TAURUS SPRING is a small but very deep pool located just at the point where the basin begins to open out into north end of the Orion Group. Superheated, Taurus boils constantly and vigorously, and this is usually the only activity noted. Taurus erupted as high as 50 feet for an unknown but probably short time following the 1959 earthquake. Other eruptions occurred during the early 1970s. Those 4-foot splashes were always seen at times when Union Geyser (18) was in actual eruption. Although this was taken at the time as proof of an underground connection between the two geysers, similar action also occurred in 1991 with Union completely dormant. No further eruptions took place until shortly after a nearby magnitude 4.2 earthquake in late June 1997. One eruption, on July 4, was seen; it reached 50 feet high but lasted much less than 1 minute. Another major eruption, known only because of extensively washed surroundings, happened in July 2003, and another (again following an earthquake) was suspected in January 2004.

18. UNION GEYSER is a spouter of the first rank. Unfortunately, it alternates between periods of extreme activity and others of very little. It was first seen in 1872 and was reported by every geological survey and numerous tourist guidebooks until 1911. Whether Union was continuously active during all those years is doubtful, though, as the guidebooks tended to repeat the same data year after year. Union was definitely dormant from 1911 until 1949, with the sole exception of two eruptions noted in 1934. Rejuvenated in 1949, it apparently survived the 1959 earthquake and continued playing into 1977. It has been dormant since then.

Union erupts from three distinct geyserite cones. About 4 feet tall, the center cone is the largest and shoots the highest water jet. When Union is active, during the quiet period all three cones are nearly full of water. On occasion, an increase in the superheated boiling will splash some water out of the cones. An eruption begins with a series of heavy surges. It is a sight to behold. The water column rapidly climbs to its maximum height. The jet from the central cone is well over 100 feet high, and its play lasts for 12 minutes, with diminishing power. The northern, second largest cone begins spouting within seconds after the first. It plays its own water jet to at least 60 feet. The longest lasting of the three cones, it persists for over 20 minutes. The southern, smallest cone delays its start for as long as 2 minutes after the others and then briefly shoots a double stream of water to about 30 feet. As each jet gives out, a steam phase takes over. When the third spout of water dies, the entire eruption briefly becomes violent, and its roar can be heard throughout the Shoshone Geyser Basin. The complete display, including the steam phase, lasts about 1 hour.

Union has been known to have infrequent single eruptions, but usually it is cyclic, with active periods separated by about 5 days of quiet. Once an active phase begins, there will normally be three, but sometimes four or only two, eruptions in the next few hours. The second follows the first by about 3 hours; the next interval is near 7 hours, and the last, if any, is around 10 to 12 hours. All of the eruptions are of equal force, but each in succession has a shorter total duration.

When Union is inactive for long periods, as it has been since 1977, the water level remains well below overflow. It is always super-

Union Geyser, playing powerful jets of water from three separate cones, is one of the tallest geysers in Yellowstone and is the star performer in the Shoshone Geyser Basin. During its recorded history, it has been dormant much more than it has been active, and it has been completely quiet since 1978.

heated, so the inactive periods are apparently the result of an exchange of function that draws water volume away from Union. Where it goes is unknown; many other springs in the Orion Group are similarly affected at the same time as Union, but a corresponding increase in the activity of other groups is not seen. Since dormant periods have been known to last for decades, the eruptions of the 1970s are apt to have been Union's last for a long time.

19. SEA GREEN POOL was observed as an active geyser during 1976, at a time when Union Geyser (18) was in an active phase. The eruptions lasted only a few seconds but recurred as often as every 2 or 3 minutes. The height was about 10 feet. This was the first known active period of this geyser, and since 1976 it has never been known to fill as it did before. What had been a beautifully beaded geyserite crater is now severely weathered, and the remaining small pool stands several feet below ground level.

20. WHITE HOT SPRING has a cavernous vent that opens out into a broad, shallow pool. Records of its geyser activity date to 1872, but apparently at no time was it a frequent or regular performer. During the last active episode of Union Geyser (18), White Hot erupted fairly often for long durations. The water burst from the vent at a low angle, reaching well out into the pool. Since the 1977 dormancy of Union, White Hot has had a low water level. The pool area is dry except for the slight spray of erratic eruptions largely confined to the vent. The only exception was during the brief episode of high water levels in 1997, when White Hot had a few eruptions that sprayed outward about 3 feet into the basin of the old pool.

21. FIFTY GEYSER was located just a few feet east of White Hot Spring (20). First observed during the early 1970s when its interval was extremely regular at 50 minutes, it actually proved to be an erratic performer. However, whatever the interval, the eruptions always lasted 5 minutes and splashed water about 2 feet high. With declining frequency through the 1970s, Fifty entered dormancy before 1980. Brief episodes of sputtering action only inches high were seen in 1995 and 1997. Fifty's site is now only a slight depression filled with gravel.

22. UNNG-SHO-3 was active only for a short time during the late 1970s. The eruptions came from a symmetrical vent at the northern apex of a triangle formed with White Hot Spring (20) and Fifty Geyser (21). Few details about SHO-3's eruptions are known, except that they reached 4 feet high. It was active as a 1-foot perpetual spouter during part of 2002.

23. UNNG-SHO-4 might be a recipient of some of the energy lost by the other members of the Orion Group. Prior to the first known activity in 1978, the site was a depression sometimes containing a bit of tepid water. Now there is a large crater lined with spiny geyserite. It is clear that SHO-4 is an old spring that had been buried by erosion during a long dormancy and that its crater has simply been reopened by the current activity. The geyser played murky water in bursts as high as 12 feet during the 1980s, but only weak, low-level splashing was observed in 2007.

24. UNNG-SHO-13 is a small spring beyond the weathered geyserite mound northwest of Union Geyser (18) and Impenetrable Spring.

It is also known as "USGS #86a." It apparently formed after the 1968 mapping by the U.S. Geological Survey. Although usually less than 1 foot high, SHO-13 is a regular and persistent geyser. The intervals are 1 to 2 minutes and the durations generally 20 to 50 seconds. Three other small geysers lie near SHO-13.

25. UNNG-SHO-24 erupted from a narrow vent atop a small geyserite platform adjacent to Shoshone Creek. Its only known activity was during 1997. The eruptions lasted only a few seconds each but splashed as high as 6 feet. A similar pool next to SHO-24 might have had some eruptions at about the same time.

CAMP GROUP

The Camp Group (Table 11.2, number 26) is a cluster of small springs near the base of the hill south of Union Geyser. Rather decayed in appearance, most of the springs have little or no overflow and are at relatively low temperatures. Two or three besides those described here have had geyser eruptions, but their active episodes have been brief and scattered. One of them is Lavender Spring. Across Shoshone Creek in an area studded by numerous small cones and one good-sized pool is the Island Group. Several of these cones have recorded histories as irregularly active geysers never more than a foot or two high.

26. GEYSER CONE was named in 1878, probably because of its appearance since no actual eruptions were described. It reactivated in 1974 and was a frequent performer until the winter of 1995–1996, when an unseen eruption blew debris from the vent. Geyser Cone has not been known to play since that time, and the entire formation is now severely weathered.

UNNG-SHO-14 is a small pool near Geyser Cone (26). Apparently independent of Geyser Cone, it has been known to have infrequent and brief splashes as high as 4 feet.

YELLOW CRATER GROUP

Until the early 1970s, the National Park Service maintained a backcountry patrol cabin near the west edge of Shoshone Geyser Basin. Access to it was via a trail with a footbridge that crossed Shoshone

Creek below Minute Man Geyser. Unfortunately, the cabin burned and was replaced with a new patrol station outside the geyser basin, and the bridge was removed in 1976. The old trail passed through the Yellow Crater Group (Table 11.3, numbers 27 and 28). The large spring closest to Shoshone Creek is Blowout Pool. It apparently formed as the result of a steam explosion in the winter of 1928–1929 and had some early geyser action. As it quit playing, another nearby spring (possibly #27) began erupting during 1930, but it soon quit, too. The modern identity of Yellow Crater Spring itself is uncertain, as no existing feature matches the original description. It might be a nearly extinct pool atop a dilapidated geyserite mound near the old patrol cabin site.

27. UNNG-SHO-22 ("WEDGE SPRING") is a large triangular pool, measuring 30 by 15 feet, at the north side of the old trail west of Blowout Pool. It may be the same as Boiling Pond, named by the Hayden Survey of 1872, and it might have been briefly active during 1930, shortly after the creation of Blowout Pool. Eruptions during the winter of early 2000 cleanly washed wide areas of the surroundings, but details of that action are completely unknown.

28. UNNG-SHO-16 ("USGS #110") lies well to the northwest of Blowout Pool, where it is isolated from other hot springs by a surrounding meadow. Known as "USGS #110" because of a map designation, it has also (probably incorrectly) been identified as Yellow Crater Spring. It generally acts as a perpetual spouter. Otherwise, it plays as a geyser every 2 to 5 minutes, with durations of about 3 minutes. The play is 2 to 4 feet high.

NORTH GROUP

The concentrations of springs and geysers on the west side of Shoshone Creek were separated into the North Group and South Group by the researchers of 1878. There is only a slight natural divide between the two on the surface, but they clearly are separate units at depth.

The North Group (Table 11.3, numbers 29 through 47) contains the greatest number of hot springs at Shoshone. Seventeen are geysers important enough to have been given names. Numerous other springs erupt as perpetual spouters.

Table 11.3. Geysers of the Yellow Crater, North, South, and Western Groups, Shoshone Geyser Basin

Name	Map No.	Interval	Duration	Height (ft)
Bead Geyser	38	1–2 hrs	1½–2 min	10–25
Boiling Cauldron	56	near steady	near steady	6
Bronze Geyser	46	rare	seconds	3
Brown Sponge Spring	33	near steady*	near steady	1
Fissure Spring	43	hours	6–10 min	3–6
Frill Spring	30	8–25 min*	20–30 sec	10–15
Glen Spring	32	1½–3 hrs	minutes	1–2
Iron Conch Geyser	47	3½–30 min	1–2 min	2–3
Knobby Geyser	37	2½–15 min	sec–min	5–25
Lion Geyser	44	50–100 min*	4–6 min	20–40
Mangled Crater Spring	29	90–100 min	9–12 min	8–10
"Old Lion Geyser"	45	unknown	minutes	10–25
Outbreak Geyser	50	[2002]	1½–2½ min	6–15
Pearl Spring	31	[1996]	2 hrs	1–2
Pectin Geyser	54	steady	steady	1–6
Small Geyser	36	minutes	minutes	2–8
Three Crater Spring	51	irregular	seconds	1–2
UNNG-SHO-7	49	minutes*	seconds	3–6
UNNG-SHO-15 ("Not-Pectin")	55	11–25 min	2 min	3
UNNG-SHO-16 ("USGS #110")	28	2–5 min	sec–3 min	2–4
UNNG-SHO-18 ("Chocolate")	34	[1997]	sec–min	1
UNNG-SHO-19 ("Terracette")	39	[2003]	1–1½ min	1–2
UNNG-SHO-20 ("Slosh")	40	hrs–days	30 min	20
UNNG-SHO-21 ("The Hydra")	42	1–4 days*	minutes	1–15
UNNG-SHO-22 ("Wedge Spring")	27	[2000]	unknown	unknown
UNNG-SHO-23 ("Diverted")	52	5–12 min*	1–3 min	2–5
UNNG-SHO-24	25	[1997]	seconds	6
UNNG-SHO-25 ("Rototiller")	48	5½–7 min	20 sec–2 min	5–10
UNNG-SHO-26 ("USGS #132")	53	unknown	minutes	2
Velvet Spring	41	11–13 min*	2 min	10
Yellow Sponge Spring	35	[1997]	near steady	10

* When active.
[ ] Brackets enclose the year of most recent activity for rare or dormant geysers. See text.

A branch of the trail used to cross Shoshone Creek by way of a bridge near Minute Man Geyser, but the bridge was removed in 1976. Now, to reach this area one must either ford the stream or cross it via downed logs at the south end of the basin near the Camp Group.

Most of the spring names in this group were applied during the 1800s, but in the intervening years many were shifted to other springs. Corrections have been made, and the descriptions here place the names where they belong.

29. MANGLED CRATER SPRING was named in 1872 because of the complexity of numerous vents and geyserite projections within the large crater. Mangled Crater itself does not discharge any water during its eruptions, but a related pool nearby does. The activity is erratic. Intervals are known to range from just 30 minutes to as long as several hours. The duration varies accordingly, from as short as 5 minutes on some occasions to as long as 30 minutes when the intervals are hours long. Bursts from the main vent can reach over 10 feet high.

30. FRILL SPRING as a name was long thought to belong to a wide, shallow orange pool near this geyser. Frill apparently got its name because of the decorative sinter about the edge of the teardrop-shaped crater. The vent, at the narrow end of the pool, extends vertically downward to a considerable depth. Frill's action is cyclic. Several days pass between active phases that last a few hours, during which eruptions may recur as frequently as every 8 to 25 minutes. The play lasts as long as 4 minutes and jets water up to 20 feet high. At the end of an active phase, the water drops several feet within the crater, which slowly refills. When full but not active, the geyser acts as a nondescript intermittent spring whose overflow is occasionally punctuated by splashes 1 foot high.

31. PEARL SPRING is little more than an intermittent spring, but its periods of overflow may include a few bursts 2 feet high. These eruptions last as long as 2 hours. Following the play, the water level of the pool drops very slowly, requiring about 2 hours to fall 12 inches. Refilling takes another 2 hours, and a long period of intermittent overflow is necessary before another bit of splashing occurs. Eruptions have not been recorded since 1996.

32. GLEN SPRING is situated in a deep alcove in the hill, largely out of sight from the rest of the basin. It is a geyser, but little is known about the activity. The wide, shallow pool is colored a strange mixture of oranges and yellow-greens everywhere except at the vent. Along the front of the pool is a series of logs that appear to have been placed there purposely long ago. The erratic eruptions are a series of individual bursts of water, separated from one another by as much as 30 seconds. Some reach 5 feet high. The duration of such action is believed to be several hours.

33. BROWN SPONGE SPRING was named because of a brown mineral stain on the inside of the crater, which is composed of a porous-looking geyserite. Brown Sponge probably plays rather frequently, but little attention has been paid to it because of the small size of the play and lack of significant runoff. The height of its boiling eruption is less than 1 foot.

34. UNNG-SHO-18 ("CHOCOLATE GEYSER") usually operates as a quiet intermittent spring in a small pool a few feet from Brown Sponge (33), but it can play up to 1 foot high. It was active as a geyser in 1994 and 1997.

35. YELLOW SPONGE SPRING erupts from a water level well below the surrounding ground surface. The geyserite is tinted a pale, pure yellow by a trace of iron oxide. One of the most vigorous geysers in the area, Yellow Sponge was once very close to being a perpetual spouter, with intervals never more than a few seconds long. Water jets sprayed 4 to 8 feet high. However, with the reactivation of nearby Small Geyser (36) in 1997, Yellow Sponge's play weakened considerably to little more than vigorous boiling.

36. SMALL GEYSER has changed its behavior substantially since the 1870s. Back then, it erupted frequently and as high as 20 feet. No geyser of that sort exists in the vicinity now, where there are several craters with water levels well below the surface. One among these did erupt, so the name was applied to it—justified, it seems, when it filled to overflow and began stronger eruptions in 1997. The play by the renewed Small Geyser is mostly less than 2 feet high, but exceptional bursts can reach as high as 8 feet. The activity recurs every few minutes.

37. KNOBBY GEYSER is cyclic in its action and may be affected by activity in Velvet Spring (41). Although mapped in 1878, no eruptions were observed until the mid-1970s. Knobby's crater is square in general outline and entirely decorated with exceptionally ornate geyserite. The vent is at the uphill corner of the crater. When undergoing major eruptions, Knobby can be the tallest active geyser at Shoshone.

Three varieties of eruptions are known. Most are minor in scale and duration, reaching less than 5 feet high for a few seconds. Action of intermediate size and duration reaches as high as 10 feet and lasts as long as 2 minutes. Major eruptions are generally uncommon but have been known to reach 35 feet high for durations in excess of 7 minutes. At times, there is a clear progression of gradually stronger minor and intermediate eruptions leading to a major play that ends a cycle. Often, however, there is no clear sequence. Quiet periods between active cycles can last from 1 to 4 hours, the action then lasting from a few minutes to several hours. Indeed, since 2001, Knobby has been active almost all the time.

Before the 1982 dormancy of Velvet Spring, Knobby was comparatively weak, and its cycle intervals were sometimes longer than 24 hours. It only became frequent when Velvet began its dormancy. However, subsequent active phases by Velvet have not produced observable changes in Knobby, so a relationship between the two geysers is only hypothetical.

38. BEAD GEYSER was an impressive spouter during the 1870s, but then it apparently was dormant for more than 100 years. Indeed, exactly which spring was Bead was not certain until June 2003, when a quiet intermittent spring within an irregular vent upslope from Knobby Geyser (37) began to erupt. Bead now plays every 1 to 2 hours. Water is jetted 10 to 25 feet high throughout durations of 1½ to 2 minutes.

39. UNNG-SHO-19 ("TERRACETTE SPRING") is a small pool long known to vary its water level in synchrony with that in the crater now known to be Bead Geyser (38), rising a few inches as Bead fell but never reaching overflow. In August 2000, "Terracette" began to erupt. At first, the intervals were 8 to 11 minutes, but they gradually grew longer so that some exceeded 30 minutes by late 2002. There was always a pair of eruptions, each lasting 1 to 1½ minutes separated by a pause of about 1 minute. The splashing play was 1

Bead Geyser was apparently inactive for more than 100 years until it reactivated during 2003. It is now one of the largest and most reliable geysers in the Shoshone Geyser Basin.

to 2 feet high. "Terracette" has not erupted since the start of Bead's activity, and its water level drops as far as 4 feet, then rapidly refills with every eruption by Bead.

40. UNNG-SHO-20 ("SLOSH GEYSER") was first seen in 1999, but well-developed runoff channels showed it to have been active for some time before then. Most activity is minor in scale, splashing briefly a few feet high. This occurs only when runoff from Knobby Geyser (37) has *not* flowed into the jagged crater for a considerable time—something uncommon since 2001. Major eruptions are rare. They take place only during extraordinarily long intervals by Knobby, and even then the major intervals are usually several days long. However, the eruption is most impressive. An angled jet of water reaches over 20 feet high for several minutes and is followed by a noisy steam phase that lasts more than half an hour.

41. VELVET SPRING was not described as a geyser until 1886, but it was (apparently) continuously active from then until 1982. Based primarily on the size of its eruptions, it was long believed to be Bead

Geyser (38) until a comparison of an 1878 drawing with the modern crater revealed the error. The eruptions by Velvet were remarkably regular, with only a few seconds variation—making it one of the most regular geysers in Yellowstone. The play recurred every 12 to 14 minutes and lasted 2½ minutes. Velvet has two vents. The main crater contains a deep, blue pool. Its bursting eruption would spray water up to 20 feet high. The smaller vent to the west would play with more of a jetting action, at an uphill angle to as high as 25 feet. Velvet fell dormant in 1982, and since then it has had only a few brief active phases during 1987, 1994, 1996, and 1998–2001. When not active, Velvet behaves as an intermittent spring, and the times of overflow correspond to a few seconds of increased, superheated boiling above the main vent. The decline in activity may be a result of an exchange of function with Knobby Geyser (37), as that geyser has been active most of the time since 1982 and almost constantly since 2001.

42. UNNG-SHO-21 ("THE HYDRA") comprises a curious group of small cones and vents long known to sputter frequently enough to keep their geyserite freshly beaded. However, true eruptions were not known until 2002, when "The Hydra" became a significant geyser. A small pool bursts as high as 15 feet, the largest cone jets up to 10 feet, and numerous other holes spray between 1 and 5 feet high, all playing simultaneously for durations of several minutes. Active phases take place erratically, varying between 8 hours and 4 days apart in an entirely unpredictable pattern; the average interval is around 20 hours. During an active series, there can be several eruptions that recur every 1½ to 2 hours and last about 1½ minutes each.

43. FISSURE SPRING and **SNAIL GEYSER** always erupt together. Fissure consists of an elongated vent plus three other openings a few feet to its east. Directly associated with it are two pools just to the west, one of which is Snail. Fissure and Snail have become much more active during the last few years. Eruptions recur on intervals of 1½ to 7 hours and last 6 to 10 minutes. Fissure is about 3 feet high, while Snail bursts up to 6 feet.

44. LION GEYSER is one of the most regular geysers in the Shoshone Geyser Basin, except that it is subject to rather frequent and long dormant spells. When active, which might be as little as one year out of three, the intervals usually range from 50 to 70 minutes, although some as great as 2 hours were recorded in 1993 and others up to 4 hours in 2003. The eruptions last 4 minutes. Most of the bursts are only 2 to 4 feet high, but Lion's vent is slit-like and penetrates the ground at an angle, so some bursts squirt water at a sharp angle. In exceptional cases, these bursts may reach as much as 20 feet high and 40 feet outward, the water falling well beyond the crater rim. At the end of the eruption the crater drains, forming a sucking whirlpool over the vent. Very little refilling takes place until immediately before the next eruption. The name "Lion" is entirely mysterious, since nothing of either the geyser's behavior or its crater bears any resemblance to the animal.

45. "OLD LION GEYSER." In 1991 an inactive vent a few feet south of Lion (44) began to play. In time it revealed an old vent that penetrated the ground at an angle so that, like Lion, the geyser played at an angle to a distance as great as 25 feet. One time, this geyser and Lion were observed in eruption at the same time, and their water jets actually collided in midair. "Old Lion" was not seen again until late 2007. The intervals were hours long, but the play lasted at least 30 minutes. The splashing was only 2 to 4 feet high, but the discharge was heavy and flowed into adjacent Lion Geyser. "Old Lion Geyser" may have been the original Lion of 1878, but its appearance does nothing to answer the question about the name.

46. BRONZE GEYSER was evidently a frequent performer several feet high during the 1870s and 1880s, but it rarely erupts now. Located a short distance south of Lion Geyser (44), the most notable thing about it is the color of its geyserite rim. The sinter has incorporated iron oxide in such a way as to give it an almost perfect metallic bronze luster. When Bronze does erupt, the play lasts only a few seconds and reaches 2 to 3 feet high. In August 2007, extensive runoff channels implied that voluminous overflow and potentially significant eruptions had recently occurred, but such activity was never actually observed.

47. IRON CONCH SPRING forms the third point of a triangle, along with Lion (44) and Bronze (46) geysers. The crater is even more heavily stained with iron oxide than Bronze's is; the result is a brilliant red-orange vent surrounded by bumpy, bronzy mounds of geyserite. Iron Conch is a regular geyser. Intervals range from 3½ to 30 minutes but usually average near 9 minutes. The play lasts 1 to 2 minutes and reaches 2 to 3 feet high.

SOUTH GROUP

The South Group (Table 11.3, numbers 48 through 52) springs are located south of the North Group, just beyond a slight ridge of old geyserite. Eight of its fifteen springs have been known as geysers, but most of them have had only very brief episodes of small eruptions. Among them are well-named Blue Glass Spring (also known as Ornamental Spring), named for its color; Flake Spring, named after the nature of its geyserite deposits; and Coral Spring, farthest to the southwest against the hillside.

48. UNNG-SHO-24 ("ROTOTILLER GEYSER") was first observed when a crack in the ground began erupting in September 2006, when the action resembled the gardening machine of that name. By mid-2007 it had developed a large, jagged crater. Eruptions recurred every 5½ to 7 minutes except when a "second burst" took place no more than 1 minute after the initial eruption. Most of the play lasted less than half a minute. Splashes 5 feet high were punctuated by strong angled jets that reached as far as 10 feet upward and outward to the south.

49. UNNG-SHO-7 is a narrow spring surrounded by smooth geyserite deposits. Although dormancies are known, it is usually a frequent performer, with intervals and durations of a few seconds. The height is 2 to 4 feet.

50. OUTBREAK GEYSER might be an old, rejuvenated feature that reappeared following a small steam explosion in 1974. At first, the eruptions were extremely erratic, but Outbreak soon settled down to reliable behavior. At times almost stopwatch regular, the intervals ranged between 28 and 36 minutes. The play lasted 2 to 2½ minutes and burst water 5 to 15 feet high. There were often a number of "false start" minor eruptions in the last few minutes before the full play. Outbreak was dormant in 1992. It was active again in 1993, but the renewed play sometimes regressed into weak perpetual spouting. Even that slowly declined into complete dormancy by

2002. The vent is now filled with and surrounded by a wide area of orange cyanobacteria.

51. THREE CRATER SPRING actually contains several vents, but there are three large ones within the wide, shallow basin. One of these vents has had several short episodes of small but frequent eruptions since the 1970s. The action is always irregular.

52. UNNG-SHO-23 ("DIVERTED GEYSER") began erupting in 1995, after bison trampled the rim of Three Crater Spring (51), diverting its overflow away from this geyser. Although there have been some brief dormancies, "Diverted" typically plays every 8 to 10 minutes, splashing as high as 5 feet for several minutes. It was dormant in 2006–2007.

WESTERN GROUP

The Western Group (Table 11.3, numbers 53 to 56), also known as the "Fall Creek Group," is isolated from the rest of the Shoshone Geyser Basin, through the woods and around the hill southwest of the South Group. The springs are near Fall Creek, which flows into Shoshone Creek just downstream from the Camp and Island groups. The Western Group is an area of many large and very active hot springs. Aside from the three geysers and one large perpetual spouter described here, it includes numerous small spouters, beautiful pools such as Cream Spring and Great Crater, and several mud pots.

53. UNNG-SHO-25 ("USGS #132") is a wide and very shallow pool that was never known to erupt until 2007. Then, two of the several small vents at the northwest end of the crater played vigorously, some splashes reaching as high as 2 feet. Many of the rising steam bubbles collapsed within the pool, causing a thumping of the ground that could be both heard and felt.

54. PECTIN GEYSER is an interesting geyser in that most of the time, although unquestionably a true geyser, its surface water temperature is well below boiling. The eruptions may be triggered by the evolution of gas other than steam (probably carbon dioxide), but there is still boiling at depth once the play has started. Pectin plays frequently. The intervals usually fall within the range of 3 to 7 minutes. Over the course of a few seconds to 2 minutes, water is splashed 1 to 6 feet high. The name came about because of a gelatinous bacterial growth along the runoff that has since disappeared. Since 2000, Pectin has acted as a weak perpetual spouter.

55. UNNG-SHO-15 ("NOT-PECTIN GEYSER") is a new geyser. The U.S. Geological Survey mapping of the Shoshone Geyser Basin was completed in 1968, and this spring was not shown. Its vent is a jagged opening in the sinter that indicates an explosive origin. Eruptions were first seen in 1987. During active phases, they recur every 10 to 25 minutes but show a wide variation in the durations, sometimes lasting only a few seconds but usually 1½ to 2 minutes. The splashing play of 1 to 3 feet is largely confined to the crater, and eruptions often end just as the pool reaches overflow.

56. BOILING CAULDRON produces most of the large stream of scalding water that flows past Pectin Geyser (54) and SHO-15 (55). Boiling Cauldron is a perpetual spouter with several vents. One vent plays to about 6 feet and has been known to have brief pauses. Another opening jets high-pressure steam, producing a hissing roar that can be heard several hundred feet away.

CHAPTER TWELVE

Heart Lake Geyser Basin

The Heart Lake Geyser Basin (Index Map 12.1) lies in a setting rather different from those of the other geyser basins. Instead of the hot springs being found in nearly contiguous groups, they occur as a series of distinct, widely spaced clusters. Hiking into the area after having walked through miles of lodgepole pine forest (now largely open since the forest fires of 1988), the hiker emerges from the trees at Paycheck Pass on the edge of a deep valley. With the muddy hot springs of White Gulch just below, the view from the overlook is down Witch Creek toward Heart Lake. Spotting the length of the green valley are white patches, the sinter deposits of the hot spring groups. Beyond is the lake and, farther still, the snowy peaks of the Absaroka Range. The view is of one of the prettiest in Yellowstone.

The Heart Lake Geyser Basin is divided into five groups of importance, each separated from the others by extensive nonthermal ground. Geysers have been known in each of them, although

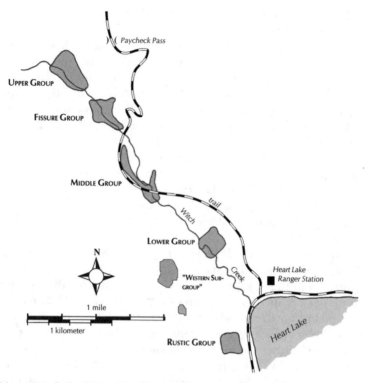

Map 12.1. *Index Map of the Heart Lake Geyser Basin*

those of the Middle Group were active only during the late 1980s. Most of the geysers are of small size; even the largest, at a maximum of 60 feet high, is quite small compared to what can be seen in most other geyser basins. Overall, though, the activity at Heart Lake is intense.

Perhaps because its geysers tend to small sizes, the Heart Lake Geyser Basin has seldom attracted much attention. A detailed but ridiculously small-scale map was published in 1883, thorough descriptions without a map appeared in 1935, and a cursory, unpublished study by the U.S. Geological Survey was compiled in 1973. It was not until 1988 that truly accurate maps, tables, and descriptions were completed through the extraordinary personal efforts and research of geyser gazer Rocco Paperiello. Additional work has been accomplished by other volunteers since then. Their reports are published by the Geyser Observation and Study Association, and much of the information in the descriptions here has been extracted from them.

Few of the Heart Lake geysers have been given official names, so most of the names used here are unofficial. Also, these springs are identified by a numbering system slightly different from that used elsewhere in this book. It is based on Paperiello's 1988 maps, which are incomparably the best available. For example, UNNG-HRG-P7 is the spring labeled number 7 on Paperiello's map of the Rustic Group.

As a historical aside, the name of the lake and geyser basin really should be "Hart." It was named by explorers before 1871 after an early prospector, Hart Hunney, who prospected throughout the Montana-Wyoming region. The spelling was inadvertently changed by the geological survey of 1871 and became official in the incorrect form.

Heart Lake is one of the least impacted areas of Yellowstone. The foot traffic into the lake is mostly that of fishermen with little interest in the hot springs. The trail leaves the South Entrance highway at a point about 5½ miles south of the Grant Village junction and just north of Lewis Lake. From the trailhead it is 5½ miles to the Paycheck Pass overlook at the head of the valley. It is another 2½ miles from there to Heart Lake, where there is a series of backcountry campsites along the shore. The Heart Lake Ranger Station is occasionally staffed during the summer months and, if so, may be able to provide assistance and outside communication in case of an emergency. Such help cannot be counted on, however. In this backcountry area, as in all others, your safety is your own responsibility. Because of bear management, the area is closed to entry between April 1 and July 1 every year.

RUSTIC GROUP

The Rustic Group (Map 12.2, Table 12.1, numbers 1 through 8), named after Rustic Geyser, is the closest hot spring group to Heart Lake. Set against the base of Mt. Sheridan, it is about ½ mile south of the Heart Lake Ranger Station and 1,000 feet from the lakeshore. The best approach is to take the shortest possible route from the lake. The meadow outside the tree line is normally very soggy, and it is often impossible to find a dry route to the springs. The group contains at least eight geysers. Columbia Spring has no

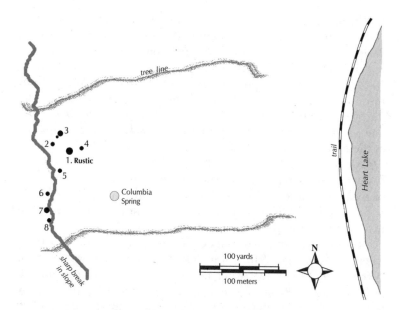

Map 12.2. *Rustic Group, Heart Lake Geyser Basin*

Table 12.1. Geysers of the Rustic Group, Heart Lake Geyser Basin

Name	Map No.	Interval	Duration	Height (ft)
"Composite Geyser," minor	3	minutes	seconds	2–6
"Composite Geyser," major	3	1–3 hrs	30 sec–5 min	20–30
Prometheus Spring	6	[1978]	seconds	6–15
Rustic Geyser	1	20–36 min	40–60 sec	10–50
UNNG-HRG-P3	4	[2000]	40 sec	10–15
UNNG-HRG-P7	2	1–3 hrs	1–4 min	10
UNNG-HRG-P9 ("Trapezoidal")	5	*not a geyser—see text*		
UNNG-HRG-P12	7	[1997]	sec–min	4–25
UNNG-HRG-P15 ("Threaded")	8	7–11 min*	2–4 min	1

* When active.

[ ] Brackets enclose the year of most recent activity for rare or dormant geysers. See text.

history of eruptions, but it is the largest hot spring at Heart Lake and is among the most beautiful pools in Yellowstone. It cannot be approached closely because of extensive overhanging geyserite rims.

1. RUSTIC GEYSER is spectacular. For most of its known history, it has been the only large and frequent geyser in the entire Heart Lake Basin. Rustic was named in 1878, and from then until 1984 there was little change in its performances. The intervals ranged from 10 to 90 minutes. The play began with little warning. Water slowly rose in the crater. When it was still about 6 inches below the rim, the water suddenly surged, filling the crater to overflowing within a few seconds. Bursts came in rapid succession, with perhaps 100 of them during the usual 40- to 60-second eruption. The water was thrown into the air as steam bubbles exploded within the shallow pool, and the pop could be heard with each one. The forcefulness of the bursts varied. Although most were only 10 to 20 feet high, every eruption also included a few that reached 35 to 50 feet. The play ended as abruptly as it began.

Rustic's only known dormancy began in late 1984 or early 1985, when an exchange of function shifted the energy flow to nearby "Composite" Geyser (3), which became a much more frequent and powerful geyser. Rustic rejuvenated in 1996, and its current activity is nearly identical to that seen before the dormancy: intervals of 20 to 36 minutes lead to 40- to 60-second eruptions that burst as high as 50 feet.

There is evidence that the length of the interval is dependent on the level and temperature of the subsurface groundwater table—rather persistently, as the summer season progresses and the surrounding environment dries out, Rustic develops progressively shorter intervals. For example, during June 1974, when consistent records were kept by a backcountry ranger, the average interval was near 25 minutes; by August it had decreased to just 14 minutes, and it lowered still further in September to only 10 minutes (the shortest running average ever recorded). Whatever the current average, the intervals are fairly regular at any given time, and 26 minutes seems typical in the long run. Some of the other Rustic Group features undergo similar changes, always gradually rather than abruptly, as with the "disturbance" and "energy surge" changes known elsewhere.

At some time in the distant past, Rustic was altered by humans. Logs were placed around the crater, giving it a square outline. Completely covered by pale-brown sinter, the logs appear virtually unchanged from the early days of the Park, when they were shown in a woodcut published in 1883. Given that, the logs were probably placed by Indians as long as a few hundred years ago. Why they

The unnamed geyser HRG-P7 undergoes its most vigorous eruptions shortly after nearby "Composite" Geyser has ended its activity.

would do such a thing is a mystery, but it indicates that they did not shy away from the geysers. Rustic is one of only three places where such early "vandalism" is known in Yellowstone, the others being at Old Bath Lake in the Lower Geyser Basin and Glen Spring in the Shoshone Geyser Basin.

2. UNNG-HRG-P7 lies at the base of the grassy slope a few feet from Rustic Geyser (1). About 10 feet across, the crater is coated with a pale-brown geyserite identical to that at Rustic, and there clearly are direct subsurface connections among HRG-P7, Rustic, and "Composite" Geyser (3). When Rustic was active prior to 1984, HRG-P7 was very regular in its activity. Most of the time, the average interval was near 25 minutes, but some as short as 5 minutes were seen. The bursting play reached 5 to 15 feet high for durations of 1½ to 5 minutes. During Rustic's dormancy between 1984 and 1996, HRG-P7 operated on a different, cyclic pattern directly related to "Composite." A series of weak eruptions, each somewhat more vigorous than the one before, took place during "Composite's" quiet interval. A concluding major eruption then occurred about 2 minutes following the end of "Composite's" major eruption. Although less frequent since Rustic reactivated, these majors still occur. They reach as high as 10 feet, last 1 to 4 minutes, and end with a brief but impressively loud steam phase that does not occur during the minor eruptions.

3. "COMPOSITE GEYSER" is a few feet from Rustic (1) and HRG-P7 (2), forming the third apex of a triangle. It received its informal name because the larger of the two vents acts as a bursting fountain-type geyser, while the other jets with cone-type play. "Composite" was a rare performer prior to the 1984 dormancy of Rustic. The lack of well-defined runoff channels leading away from "Composite" implied that it had seldom, if ever, erupted prior to that year. It is interesting that it has continued with little change since Rustic's rejuvenation. "Composite's" modern activity is cyclic. Frequent minor eruptions, usually lasting only a few seconds and reaching 2 to 6 feet high, take place as the "quiet" interval progresses. The major eruptions occur at intervals of 1 to 3 hours. The play from the main crater is massive bursting as high as 20 feet that persists throughout the eruption's duration. It is only during these major eruptions that the small cone-type hole, in the geyserite a few feet toward HRG-P7, joins in with a steady jet of water as high

"Composite" Geyser bursts as a fountain-type geyser from a crater and jets as a cone-type geyser from an adjacent vent; as such, it is Yellowstone's only geyser to exhibit both kinds of eruptions from a single feature.

as 30 feet. The only significant change in "Composite's" behavior is in the durations of the majors: they lasted about 5 minutes while Rustic was dormant but only 30 to 90 seconds since then. It is still an amazing spectacle, especially when followed by the major steam-phase eruption of HRG-P7.

4. UNNG-HRG-P3 is a rare performer. The eruptions rise from within a deep hole about 50 feet from Rustic Geyser (1). The first record of major eruptions was in 1973, when jetting play reached as far as 10 feet above the ground at intervals about 1½ hours long. The only other known major activity occurred in 2000, when intervals of 8 to 27 minutes led to bursting jets over 15 feet high. At other times, HRG-P3 acts as a subterranean perpetual spouter, splashing 1 to 3 feet high from a boiling pool as far as 10 feet below the surface.

5. UNNG-HRG-P9 ("TRAPEZOIDAL SPRING") was incorrectly included as a geyser in an early edition of this book on the basis of erroneous reports that it had undergone an explosion. The new activity actually took place in another, previously quiet spring (HRG-P12 [7]). "Trapezoidal" is a pretty, gently bubbling pool that has never been known as a geyser.

6. PROMETHEUS SPRING has behaved as a non-erupting flowing spring, as a perpetual spouter, and as a geyser during its known history. Just how dominant each of these kinds of action was is unknown. When it was a perpetual spouter, the play was usually about 6 feet high. When acting as a true, periodic geyser, Prometheus was more spectacular. Intervals ranged from 8 to 15 minutes. Although the jetting eruption lasted less than 1 minute, it sent fan-shaped spray as high as 15 feet at an angle away from the hillside. Sometime during late 1977 or early 1978, Prometheus lost its water supply to nearby features, especially HRG-P12 (7). It is now so completely dormant that the once beautiful geyserite formations have almost entirely weathered away and are covered with grasses. In 2000, people familiar with its location had to use probes to find the old crater.

7. UNNG-HRG-P12 formed in 1977–1978, apparently because of an exchange of function from Prometheus Spring (6). This exchange is completely unrelated to the exchange between Rustic Geyser (1) and "Composite Geyser" (3) that occurred about seven years later. HRG-P12 had previously been a pair of small murky pools that barely overflowed and had occasional periods of weak spouting. The new activity was initially reported as resulting from a steam explosion at "Trapezoidal Spring" (5). In fact, there was no explosion. Instead, the new eruptions, clearly the first ever by these springs, quickly eroded a large crater complex in the hillside. Eruptions recurred on intervals of 2 to 12 minutes. Most of the splashing was from the rear of the two vents. It reached 4 to 10 feet high. The front vent was less frequent, but it then burst violently as high as 25 feet. Almost all durations were shorter than 20 seconds. Major activity ceased during 1997, when the action regressed to weak spouting from the back vent. Given the vigor and volume of the major eruptions, it seems likely that HRG-P12 will have major activity in the future and that Prometheus Spring is unlikely to ever resume its historical form.

8. UNNG-HRG-P15 ("THREADED GEYSER"), named because of the numerous small runoff streams that lead into it from a spring above, is usually a perpetual spouter, but during 2000 it acted as a true geyser. Intervals of 7 to 11 minutes led to durations that ranged between 2 and 4 minutes. The eruption is only about 1 foot high. Just in front of the crater is an old geyserite cone within which the water level slowly rises and falls. It may also have small eruptions at times.

LOWER GROUP

Until the early 1970s, the Heart Lake trail passed through the center of the Lower Group. Now it runs across the hillside in the forest nearly 400 feet from the nearest of the springs, so many hikers simply pass by the group. The Lower Group (Map 12.3, Table 12.2, numbers 9 through 16) is a compact cluster that contains at least nine geysers and several perpetual spouters among its thirty hot springs.

About ¼ mile southwest of the Lower Group is another cluster of hot springs, sometimes called the "Western Subgroup." It contains several pools, one of which had some eruptions 1 to 3 feet high during the 1980s; it has not been seen more recently. Another

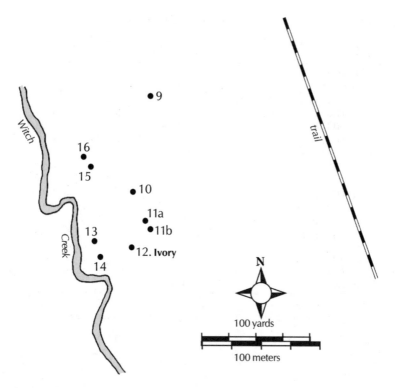

Map 12.3. *Lower Group, Heart Lake Geyser Basin*

¼ mile south of the "Western Subgroup" is a small, unnamed thermal area that includes one perpetual spouter. Access to both of these areas is difficult because they are surrounded by extensive wet meadowlands.

9. UNNG-HLG-P1 ("TURBINE GEYSER") splashes out of a 2-foot-high cone. The geyser's activity is weakly cyclic. Most of the time, the eruption is nearly perpetual and confined to within the cone. Some spray rises above the rim, but it nearly all falls back inside. During such periods there are infrequent surges when the splashes reach 3 feet above the top of the cone, and then there can be some runoff. Very rarely, "Turbine" has more powerful eruptions. For a few seconds water seems to spin turbine-like as it is thrown over 10 feet high with enough discharge to create a considerable runoff stream. Such eruptions have not actually been seen since the 1970s, but the

Table 12.2. Geysers of the Lower and Middle Groups, Heart Lake Geyser Basin

Name	Map No.	Interval	Duration	Height (ft)
Lower Group				
Ivory Geyser, minor	12	seconds	seconds	1–3
Ivory Geyser, major	12	11–18 min	2 min	4–7
UNNG-HLG-P1 ("Turbine")	9	infrequent	seconds	3–10
UNNG-HLG-P7	16	35–75 sec*	10–50 sec	1–2
UNNG-HLG-P8	15	steady	steady	1
UNNG-HLG-P19 ("Calyx")	10	30–40 min	seconds	5
UNNG-HLG-P32 ("North Reciprocal")	11a	4–8 min	40–60 sec	1–2
UNNG-HLG-P34 ("South Reciprocal")	11b	5 min	1 min	2–4
UNNG-HLG-P53	13	[1989]	2 min	2–6
UNNG-HLG-P54	14	steady	steady	1
Middle Group				
UNNG-HMG-P42	17	[1988]	3 min	1–2
UNNG-HMG-P56	18	[1988]	seconds	1

* When active.
[ ] Brackets enclose the year of most recent activity for rare or dormant geysers. See text.

fact that the runoff channel is clear of debris implies that they still take place.

10. UNNG-HLG-P19 ("CALYX GEYSER") used to be one of the smallest true geysers anywhere. Centered in a small splash zone covered with beaded sinter, water was spattered only a few inches high from the 1-inch-wide vent. "Calyx" has been a very different geyser since 1993. Regular eruptions now recur every 30 to 40 minutes. The play lasts only a few seconds, but water is sprayed as high as 5 feet and produces substantial runoff.

11a. UNNG-HLG-P32 ("NORTH RECIPROCAL SPRING") is a very pretty little pool. The rim of the crater is beautifully beaded and scalloped with pearly gray sinter. Near its edge are two tiny vents that began erupting as true geysers in 1986. HLG-P32 and nearby HLG-P34 (11b) are called "Reciprocal" because their eruptions alternate with one another, the water rising and erupting in one as it drops a few inches in the other. In "North Reciprocal," the eruptions generally recur every 4 to 8 minutes, last 40 to 60 seconds, and reach 1½ feet high.

11B. UNNG-HLG-P34 ("SOUTH RECIPROCAL SPRING") is another small pool with an ornamental geyserite rim. Its play recurs on average about every 5 minutes. More vigorous than its "North Reciprocal Spring" (11a) neighbor, the eruption lasts about 1 minute and reaches 2 to 4 feet high.

12. IVORY GEYSER, named for the smooth, creamy geyserite within the crater, is the premier geyser of the Lower Group. It was first seen during 1985 and has been continuously active, with little change, ever since. The cyclic play consists of both minor and major eruptions. A few minutes of quiet follow a major. Then a series of brief minor eruptions, only 1 to 3 feet high but each a bit stronger than the previous one, leads to the next major eruption. It reaches 4 to 7 feet high for durations as long as 2 minutes. The major intervals depend on how many minor eruptions take place, but they are usually between 11 and 18 minutes long.

13. UNNG-HLG-P53 is nearly a perpetual spouter. The pale-blue pool sits within a 2-foot by 5-foot crater with a well-established runoff channel. Much of the crater and the channel are brilliantly colored by cyanobacteria, but the vent itself is nearly black because of manganese oxide. Boiling above the vent is nearly constant, but it frequently increases in vigor so the water is thrown 2 to 6 feet high. Such eruptions can last as long as 2 minutes but have not been observed since 1989.

14. UNNG-HLG-P54 is a perpetual spouter. It might have had some unobserved eruptions during 1986, when intermittent overflow was observed. The normal steady play is 1 foot high.

15. UNNG-HLG-P8 is a small but vigorous perpetual spouter that plays out of a small cone. Significant runoff channels lead away from the spring, but eruptions large enough to have carved them have never been observed.

16. UNNG-HLG-P7 was a stable perpetual spouter for all observers until 1995. Since then, it has been consistently active as a true geyser, with intervals of 35 to 75 seconds, durations of 10 to 50 seconds, and heights of 1 to 2 feet.

MIDDLE GROUP

The Middle Group (not separately mapped, Table 12.2, numbers 17 and 18) was not known to contain any geyser until 1986, when two springs were active well away from the trail. Indications are that none of the other hot springs are geysers. Among them are a few perpetual spouters and some small but pretty pools, including Double Spring immediately next to the trail.

17. UNNG-HMG-P42 lies at one end of a fracture within a shallow depression; vents at the other end of the crack do not erupt. It was first seen as a geyser in 1986, when the play recurred with a high degree of regularity. Intervals were 6 to 6½ minutes long, and all recorded durations were within a few seconds of 3 minutes. The play was 1½ feet high. HMG-P42 was dormant by late 1988.

18. UNNG-HMG-P56 was first seen in eruption during 1987. The play was brief but frequent and 1 foot high. Like HMG-P42 (17) a few feet away, this geyser had returned to dormancy by the end of 1988 and has not been seen since.

FISSURE GROUP

The Fissure Group (Map 12.4, Table 12.3, numbers 20 through 41) is the most extensive group in the Heart Lake Geyser Basin. The continuous bubbling, splashing, roaring, and steaming create a scene like that of an old steam works. Thus, the mountain behind the springs is known as Factory Hill. Geyser activity is most intense on and near the sinter-covered rise at the upper end of the group, where a long crack in the geyserite is the site of several geysers and numerous other hot springs jointly known as the Fissure Springs. This fissure is part of a fault that can be seen as a sharp change in the slope of the hillside to the south. During the 2000s, the geyser action in this area decreased to the point that many of the springs underwent only weak, infrequent eruptions.

Witch Creek flows through the center of the Fissure Group, and much of its volume actually runs belowground in the fissure area. The stream is almost entirely hot spring runoff and has a temperature of about 85°F (30°C). The Heart Lake trail crosses Witch Creek at the lower end of the Fissure Group.

Table 12.3. Geysers of the Fissure Group, Heart Lake Geyser Basin

Name	Map No.	Interval	Duration	Height (ft)
"Fissure Springs Geyser"	41	[2000]	6–36 sec	12–20
Glade Geyser	25	hrs–days	1½–2 min	30–60
Hooded Spring	32	steady	steady	1–4
Pit Geyser	23	6–35 min*	20–35 sec	8–12
Puffing Spring	29	erratic	seconds	2–4
Shelf Spring	33	[1992]	unknown	unknown
Shell Geyser	31	2–3 min*	seconds	10
Splurger Geyser	28	minutes*	minutes	3–6
UNNG-HFG-P3	21	steady	steady	1
UNNG-HFG-P5	20	[1980]	steady	1–2
UNNG-HFG-P6 ("Black Velvet")	20	2–4½ min	1½–4 min	2
UNNG-HFG-P7	22	1–2 min	20 sec	1–8
UNNG-HFG-P36	24	30 sec	seconds	1–4
UNNG-HFG-P41	30	steady	steady	2
UNNG-HFG-P51	34	steady	steady	inches
UNNG-HFG-P52 ("Siphon")	35	hours*	2–3 min	1–2
UNNG-HFG-P67	40	[1999]	15 sec	4–8
UNNG-HFG-P68	39	seconds*	seconds	3
UNNG-HFG-P69	38	1 min	5–10 sec	1–4
UNNG-HFG-P70	37	5–10 sec	sec–2 min	12
UNNG-HFG-P91	36	4–8 min*	4–8 min	1–2
UNNG-HFG-P116	27	hrs–days*	hrs–days	1–2
UNNG-HFG-P138 ("Wisp")	26	[2003?]	10 sec	15

* When active.
[ ] Brackets enclose the year of most recent activity for rare or dormant geysers. See text.

20. UNNG-HFG-P6 ("BLACK VELVET GEYSER") sits well apart from the main portion of the Fissure Group. It and adjacent UNNG-HFG-P5 were not mapped by the 1973 USGS study. HFG-P5 was active as a perpetual spouter 1 to 2 feet high until it entered dormancy in 1980. "Black Velvet" evidently came to life in 1986. Erupting every 2 to 4½ minutes, the 2-foot play lasts 1½ to 4 minutes, so the geyser is in eruption nearly half the time.

21. UNNG-HFG-P3 is a pair of perpetual spouters that developed along an old crack. Both are decorated with massive, convoluted geyserite rims and are colored black and orange-brown by mineral deposits. The play is seldom more than 1 foot high.

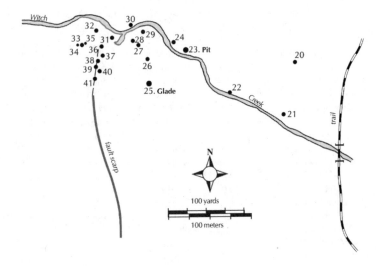

Map 12.4. *Fissure Group, Heart Lake Geyser Basin*

22. UNNG-HFG-P7 is a very interesting spring. The rectangular vent lies within a shallow basin several feet across. As with many springs in this area, the vent is colored black by mineral deposits (probably manganese oxide). As large steam bubbles rise into the bottom of the pool, they create a rich blue, flame-like flashing—perhaps the best seen anywhere in Yellowstone. The eruptive activity is frequent but erratic. It consists of vigorous bursting 1 to 8 feet high over durations near 20 seconds. Most intervals are less than 2 minutes long.

23. PIT GEYSER is the modern successor to a large pool or geyser. The crater is over 12 feet wide at the top and about 8 feet deep. Its sinter walls are deeply weathered, indicating a long period of inactivity. The modern geyser plays through a long, narrow crack at the bottom of the western crater wall. The play is nearly horizontal, reaching completely across the bottom of the crater—a span of 7 or 8 feet—and occasionally it is strong enough to clear the opposite crater wall entirely. Although some dormant periods are known and there have been times when the intervals were several hours long, most observers find the play recurring every 6 to 35 minutes. The typical duration is 20 to 35 seconds.

24. UNNG-HFG-P36 plays from a shallow pool within an eroded crater 10 feet wide and 6 feet deep. There are two vents in the pool. One acts as a vigorous perpetual spouter that jets at a low angle. The other vent is periodic, briefly bursting 1 to 4 feet high about every 30 seconds.

25. GLADE GEYSER is Heart Lake's tallest and most spectacular geyser. Unfortunately, its eruptions are brief and happen at intervals that are usually many hours or days long. Glade is also somewhat controversial. Some people have equated Glade with the "Hissing Spring" discussed by Professor T. B. Comstock in 1873, but his description does not match Glade's location. Reports subsequent to 1873 never described Glade, and even a special post–1959 earthquake survey of the Heart Lake Geyser Basin showed no spring at the site—this in spite of a photograph dated 1957 that unquestionably shows Glade in eruption. Glade's small cone is hidden within an erosional alcove, but its obvious runoff channel should have revealed its existence even if the action was infrequent. The oval vent at the top of the cone measures about 13 by 21 inches. The fact that it was only 8 by 9 inches in 1964 shows that the modern activity has caused substantial erosion. The play is a steady stream of water 30 to 60 feet high at a slight angle into the alcove.

Glade's known activity has always been highly variable. It might also be a long-term cyclic geyser. On most of the rare occasions when people have spent extended time in the Fissure Group, and as more recently shown by electronic monitoring, the intervals usually exceed 20 hours, and some as long as 60 hours are known. The small pool intermittently boils and overflows for many hours before an eruption. It is believed that these periodic episodes become stronger and longer shortly before the play and that it is one of these stronger episodes that finally triggers the eruption. A typical duration is just 1½ to 2 minutes. In 1973 and 1974 there were always two eruptions in sequence, the second following the first on intervals of only 7 to 10 minutes *or* a few hours. At that time, most full-cycle intervals were longer than 40 hours. But in early 1984 Glade was having preliminary series of so-called minor eruptions. Intervals of 1 to 3 hours were seen, and the play could last as long as 10 minutes and reach 30 feet high. The major eruptions, 45 to 60 feet tall, recurred every 18 hours and lasted as long as 20 minutes, most of which was steam phase. There was then a span of

6 to 8 hours of quiet before the minor eruptions resumed. The best years on record were 1997, when an electronic monitor showed average intervals of only 67 minutes, and 1998, with an average of 90 minutes. In the early 2000s it ranged from less than 12 hours to 30 hours, but, in keeping with Glade's erratic nature, some intervals during 2007 were longer than 9 days. Glade was dormant during 1985, 1987, and 1993–1994.

26. UNNG-HFG-P138 ("WISP GEYSER") is a small cone amid numerous hot springs between Glade Geyser (25) and Splurger Geyser (28). The fact that it showed geyser activity in 1878 is implied by a comment about quiet intervals in that year's Geological Survey report, but no eruption was described until September 1993, when "Wisp" was seen in full eruption three times in one day. The first interval was slightly more than 5 hours in length, and the next exceeded 6 hours. The eruptions consisted of about 10 seconds of water jetting 15 feet high, followed by another 35 to 40 seconds of strong steam phase. Washed areas around the cone implied eruptions during 2000, 2002, and 2003, but none is known to have been witnessed since that one day in 1993.

27. UNNG-HFG-P116 is located in a sandy basin about 10 feet upslope from Splurger Geyser (28). The eruption is 1 to 2 feet high. HFG-P116 is known to have both intervals and durations hours to days long. It is often dormant, and then the unstable crater virtually disappears because of erosion from the sandy slope above.

28. SPLURGER GEYSER (called **TRIPLE BULGER GEYSER** in the U.S. Geological Survey study of 1973) is one of the most impressive springs in the Fissure Group. The oval pool is colored aquamarine blue over each of the three vents. Splurger has shown much variation over the years. Although the name dates to the early years of Yellowstone, no eruption was actually described until 1973, and additional geyser eruptions were seen only in 1993, 1998, 2000–2003, and 2007; it also acted as a perpetual spouter during 1987 and 1996. When active, Splurger is in eruption nearly half the time, with both intervals and durations 30 to 40 minutes long. The play surges up to 3 feet high over each vent, with occasional bursts of 5 to 6 feet. In addition to the main vents, there are four other spouting holes around the sides of the crater. When dormant, Splurger is a quiet pool, but it still discharges more water than any other single spring in the Fissure Group.

29. PUFFING SPRING was described in passing by Professor T. B. Comstock in 1873. Although that spring was probably in the Upper Group, the name was transferred here in 1878. Puffing bears a massive cone 2½ feet tall that stands alone atop a geyserite mound. It was probably once a more significant geyser than it is now, but the only modern eruption of size ever witnessed occurred during 1992, when a steady jet 4 feet high produced a flood of water for more than 10 minutes. All other activity has been constant boiling within the crater. Some splashes throw out a little water, and exceptional surges may reach 2 feet above the rim, but usually the action is completely confined within the vent and the net discharge is zero.

30. UNNG-HFG-P41 plays from a beautiful small crater right next to Witch Creek. The delicately scalloped sinter rim is lightly stained by iron oxides, giving it a pink and orange cast. The spring is a perpetual spouter about 2 feet high.

31. SHELL GEYSER is almost impossible to see clearly. The deep crater is carved into a nearly vertical cliff, 10 feet above Witch Creek. The back wall of the crater is fluted like a clam shell. The geyser's play splashes against this wall, so most of the water falls back into the pool. Some more violently spraying bursts escape the crater and reach as high as 10 feet. Rare, nearly horizontal jets that arch outward as far as 15 feet may actually arise from a second, independent geyser vent near the back of the crater. The intervals between eruptions are usually 2 to 3 minutes long, and the duration is seldom more than 10 seconds. Shell sometimes regresses to perpetual spouting, but its only known complete dormancy was in 1997.

32. HOODED SPRING is a perpetual spouter. It lies at the north end of the rift that gave the Fissure Group its name. Hooded Spring was named because of a sinter projection that extends partially out over the crater, deflecting the larger bursts of the eruption. This "hood" appears to be the remains of an old geyserite cone. The eruption is mostly 1 foot or less high, but relatively frequent surges spray up to 3 or 4 feet. Just behind Hooded Spring, hidden back within the fissure, is another perpetual spouter; it plays 1 foot high.

33. SHELF SPRING is a long, oval pool on a sinter platform near the upper end of the Fissure Group. It used to be a beautiful, pale-blue color that made it one of the prettiest features at Heart Lake. In early 1992 a mud flow dumped debris into the crater. On the basis of heavily washed areas surrounding the crater, Shelf Spring apparently had some eruptions (or perhaps only heavy overflow) later in 1992, possibly because of plumbing system changes caused by the slide debris.

34. UNNG-HFG-P51 plays from a small cone between Shelf Spring (33) and the hillside. It behaves as a perpetual spouter, weakly splashing less than 1 foot high. Its general appearance is very similar to HFG-P52 (35) on the opposite side of Shelf, but unlike that spring, HFG-P51 has never been known to undergo truly periodic eruptions.

35. UNNG-HFG-P52 ("SIPHON GEYSER") has a cone almost identical in appearance to that of HFG-P51 (34). No eruptions were witnessed until 1986, but erosion then revealed a splash zone of beaded sinter and a runoff channel that could only have been formed by earlier eruptive activity. "Siphon" was named because its eruptions cause the water level in Shelf Spring (33) to drop slightly. The first intervals were a few hours long and consisted of vigorous splashing 1 to 2 feet high with heavy overflow for durations of 2 to 3 minutes. However, the activity gradually grew weaker, and by 2007 it had regressed to weak, near-constant boiling.

36. UNNG-HFG-P91 is a small double pool with jagged craters created by one or more steam explosions. Sometimes active as a perpetual spouter, as a geyser it shows both intervals and durations of 4 to 8 minutes. The 1- to 2-foot splashes would be higher were it not for a roof of old sinter that extends over the craters.

Fissure Springs (numbers 37 through 41)

The final five members of the Fissure Group are situated along the south-central portion of the fissure itself, within about 30 feet of each other. These geysers plus some non-erupting springs comprise the Fissure Springs. The geysers are closely related to one another, and just a brief period of observation reveals their competitive activity. Any one of them may be much reduced in frequency and vigor because of a corresponding increase in the action elsewhere along the fissure.

37. UNNG-HFG-P70 is usually the most active geyser among the Fissure Springs, nearly but not quite a perpetual spouter, with pauses that last only a few seconds. The vent is about 8 feet long and is nothing more than a wide zone along the fracture. The fan-shaped eruption may be continuous for as long as 2 minutes, although a more typical duration is only 5 to 10 seconds. The jetting reaches several feet high, and exceptional surges to 12 feet have been seen. When HFG-P69 (38) has its stronger form of eruptions, this geyser will briefly be quiet.

38. UNNG-HFG-P69 spouts from the next opening along the fissure, south of HFG-P70 (37). Its eruptions generally recur every minute and last 5 to 10 seconds. The normal play is only a foot or two high, but at highly irregular times HFG-P69 has stronger eruptions. Then, the water can spray as high as 4 feet above the ground. It is only during these "major" eruptions that a cone adjacent to the fissure vent can also splash up to 1 foot.

39. UNNG-HFG-P68 lies in a very narrow opening in the rift. The vent is nowhere more than 2 inches wide, but it is more than 6 feet long. The activity of HFG-P68 is closely tied to that of HFG-P67 (40). When HFG-P67 is dormant, which is the usual case, HFG-P68 erupts frequently. The most powerful bursts are in concert with eruptions by HFG-P69 (38), adjacent to it along the rift. Some of the surges reach 3 feet above the ground. Both the intervals and durations are usually only seconds long, but this geyser sometimes declines, so the brief play is only 1 foot high at intervals as long as 5 minutes. Similar weak play is the rule when HFG-P67 is active.

40. UNNG-HFG-P67, a small round vent slightly offset from the fissure, is rarely active. The only observed major eruptions took place during 1973 and 1999. Erratic splashing of minor scale was seen in 1978 and 1986. During the active phases, this geyser will overflow steadily between eruptions. The flow suddenly becomes very heavy as the play begins, and it quickly builds into steady jetting that reaches 4 to 8 feet high throughout the 15-second duration.

Intervals were 11 to 17 minutes in 1999. Nearby geyser HFG-P68 (39) is all but dormant during HFG-P67's active phases. When HFG-P68 is active, as is usual, HFG-P67 surges upward and sometimes reaches overflow during the other geyser's quiet interval.

41. "FISSURE SPRINGS GEYSER" is the southernmost of the geysers among the Fissure Springs. Its only known major eruptions took place in 1986 and 2000, when the action showed regular intervals of 3½ to 5 minutes. The play was steady jetting at a sharp angle in the downslope direction. The durations were 27 to 36 seconds in 1986, when the strongest jets reached 12 to 15 feet outward. The eruptions in 2000 were of much shorter, 6- to 11-second durations, but the play was substantially stronger, with some water reaching more than 20 feet away from the vent. "Fissure Springs Geyser" has acted as a small perpetual spouter during all years other than 1986 and 2000.

UPPER GROUP

The Upper Group (Map 12.5, Table 12.4, numbers 45 through 50) is a short distance up Witch Creek from the Fissure Group. It covers a large area, but most of its activity is in the form of relatively

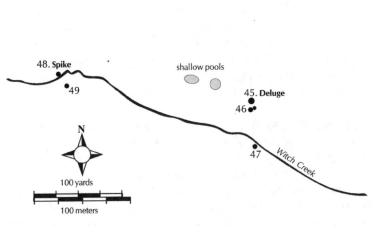

Map 12.5. *Upper Group, Heart Lake Geyser Basin*

Table 12.4. Geysers of the Upper Group, Heart Lake Geyser Basin

Name	Map No.	Interval	Duration	Height (ft)
Deluge Geyser	45	3–6 min*	1–2 min	boil
Spike Geyser	48	steady	steady	1–2
UNNG-HUG-P11	47	steady	steady	1
UNNG-HUG-P16	46	4–7 min	40 sec	1–2
UNNG-HUG-P51	50	see text	long	1–3
Yellow Funnel Spring	49	[1984?]	unrecorded	1

* When active.
[ ] Brackets enclose the year of most recent activity for rare or dormant geysers. See text.

cool and muddy acid springs. Alkaline springs with clear water and high temperatures are confined mostly to two small clusters immediately next to Witch Creek and several hundred feet apart from one another. At least seven geysers are historically known, but, as with the Fissure Group, the water levels have dropped and the activity has significantly decreased during the 2000s. A portion of the Upper Group known as White Gulch extends far up the slope to the north. This is the area immediately below the Heart Lake trail at the Paycheck Pass viewpoint.

45. DELUGE GEYSER is the largest clear spring in the Upper Group. Named in 1878, it and Columbia Spring in the Rustic Group are the only deep blue pools in the Heart Lake Geyser Basin. The crater has massive geyserite shoulders and internal formations and is surrounded by a wide sinter platform studded with smaller hot springs. It shows all signs of large eruptions, but the only party of observers known to have seen bursting play was the one in 1878. Without citing any details, they reported it to be 10 to 15 feet high. More recently, there possibly were large eruptions in 1976, 1997, 1998, and 2000, when the surroundings were found cleanly washed and grasses around the edges of the platform had been scalded. However, the typical activity consisted of frequent boiling periods, during which the level of the quiet pool rose to heavy overflow as superheated water reached the surface, resulting in a sizzling, sputtering eruption. The boiling commotion was seldom higher than 1 foot. The usual interval was 3 to 6 minutes and the duration 1 to 2 minutes. Unfortunately, by 2003 some "eruptions" included nothing but quiet overflow, and in 2007 Deluge was found to be a cool, completely quiet pool 2 feet below overflow.

46. UNNG-HUG-P16 erupts from a ragged vent at the edge of the runoff from Deluge Geyser (45). It appears to have formed quite recently, and in fact it was neither mapped nor reported prior to 1982. The play recurs every 4 to 7 minutes, when splashing 1 to 2 feet high lasts around 40 seconds. Similar UNNG-HUG-P14 is just a few feet away, but it has not erupted for several years. There is no apparent relationship between the timing of these geysers and that of Deluge, despite their proximity.

47. UNNG-HUG-P11 plays from a crater just above and directly across Witch Creek from Deluge Geyser (45). The eruptions have excavated a small alcove in the slope, implying considerable activity, but the only eruptions actually seen were in 1985. At that time the eruptions were frequent, lasted about 1 minute, and reached 2 to 4 feet high. Since then the crater has been filled with debris, and HUG-P11 acts as a sizzling spouter 1 foot high.

48. SPIKE GEYSER lies in the second cluster of alkaline springs, about 600 feet upstream from Deluge Geyser (45). In several ways, Spike is one of the most intriguing geysers in Yellowstone. Starting with the earliest description in 1878, it has probably received more attention than most Heart Lake springs because of its strange cone. Although only 2 feet tall, the cone resembles the dead and dying cones of the Monument Geyser Basin (see Chapter 9). It stands near one end of a geyserite bridge that spans Witch Creek and is surrounded by an assortment of tiny vents that spout when Spike is active. Spike was reported in eruption by some of the early explorers of Yellowstone but never again until after the 1959 earthquake. Either Spike itself or at least some of the minor vents have been active during every known observation since then. The play is as curious as the cone. Very little water issues from the tiny vent at the top of the cone; what does is squirted about 6 inches high. The tallest spout comes from a pencil-sized vent near the base of the cone. It squirts 2 feet high every second or so. A small pool at the back side of the cone surges steadily and boils, throwing some water to about 1 foot. It has taken Spike a very long time to form its cone, yet the surrounding area is badly weathered. It looks as though Spike is a remnant of more vigorous prehistoric activity and as though long active periods are separated by equally long dormancies. Spike has been minimally active since 1998.

The odd formations of Spike Geyser include small geyserite cones and a sinter bridge that spans Witch Creek.

49. YELLOW FUNNEL SPRING is a small, quiet pool that is appropriately named. Its temperature is often far below boiling, but on occasion it is much hotter and bubbles quite vigorously. A single report of splashing eruptions dates to 1984, but that might refer to a smaller, funnel-like crater a few feet away. That spring, UNNG-HUG-P28, boiled violently in 1987 and 1991.

Upstream from Yellow Funnel Spring and Spike Geyser, Witch Creek flows through a relatively narrow canyon where there are some acid springs surrounded by treacherous thermal mud. Explore this area with caution.

50. UNNG-HUG-P51. Scattered along White Gulch, a branch of the Upper Group that extends up the hillside toward the hiking trail, are many hot springs. Most are small and muddy, and there are numerous mud pots and steam vents among them. Nevertheless, this unlikely setting produced at least two geysers during the 1980s. Both may have resulted from chance seasonal changes in normally non-eruptive acid springs, as neither lasted for more than a few weeks.

CHAPTER THIRTEEN

Other Yellowstone Geysers

Geysers have been known to occur in at least seventeen other areas of Yellowstone National Park (Map 13.1). Most of these places are relatively inaccessible, but this book would not be complete without mention of them. In each case, geysers are small in number and secondary to other types of hot springs.

1. SEVEN MILE HOLE

Seven Mile Hole, roughly 7 miles downstream from the Lower Falls of the Yellowstone River, is the only place where the bottom of the Grand Canyon of the Yellowstone can be reached by trail. The trail follows the canyon rim from near Inspiration Point and then abruptly drops into the canyon. The elevation difference is about 1,200 feet in the last 1½ miles.

Hot springs occur in several small, scattered groups within Seven Mile Hole. One of the first reached is Halfway Spring, a small

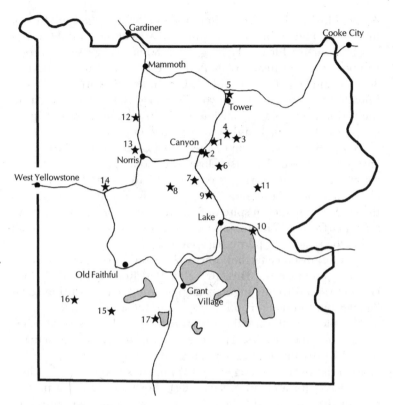

Map 13.1. *Other Yellowstone Geysers*

perpetual spouter that plays atop an impressively large geyserite cone. On occasion it may act as a geyser. Within another group of springs about halfway to the actual bottom of the canyon are two geysers. One, surprisingly, has an official name. Safety-valve Geyser erupts every 10 to 20 minutes for durations of 2 to 3 minutes and splashes about 3 feet high. Nearby is a second geyser whose height is not more than 2 feet. At and near the canyon floor are several perpetual spouters.

2. THE GRAND CANYON OF THE YELLOWSTONE

The main part of the Grand Canyon is brilliantly colored from the Lower Falls about 2 miles downstream. The coloration is the result of hydrothermal rock alteration that has formed a mixture of numerous oxide minerals. The thermal area at the Grand Canyon has been active for a very long time. Some old geyserite deposits above the rim near Artists Point are estimated to be between 300,000 and 600,000 years old. Today's hot springs are largely confined to the bottom of the canyon. Several geysers can be seen from the various viewpoints along the rim. Most are small and require binoculars to see well, and none is reached by trail.

On the southeast side of the river just downstream from the base of the Lower Falls (so close as to sometimes be within the spray) is a tiny hot spring group. Uncle Tom's Trail (now partly closed because of rockslides) used to extend to these springs, so they received names in spite of their very small sizes. Among them are Fairy Geyser, Watermelon Geyser, and Tom Thumb Geyser, all of which are actually perpetual spouters.

A short distance farther downstream and across the river is the site of another geyser. Directly below Red Rock Point, Red Rock Geyser erupted 15 feet high at an angle into the river during 1947. Smaller but frequent eruptions were also recorded in 1991.

Visible in both directions from Artists Point are several other hot springs and geysers. The most important of these is actually named Upstream Artists Point Geyser. It plays from three or more vents at the river's left edge about ¼ mile upstream from the viewpoint—look for a cloud of steam just above the river at a narrow spot in the canyon. One vent jets water at a low angle into and sometimes completely across the river, while the other openings splash weakly. Frequent and regular during the 1930s, with intervals of 15 minutes and durations around 2 or 3 minutes, this geyser is now essentially a perpetual spouter. There are additional geysers and perpetual spouters downstream from Artists Point, but they are difficult to spot except on cold, steamy days.

3. JOSEPH'S COAT HOT SPRINGS

Joseph's Coat lies within the trail-less Mirror Plateau east of the Grand Canyon, and access requires both excellent wilderness skills and a special backcountry permit. Topographic maps show a Whistler Geyser at Joseph's Coat, but it is a perpetual spouter that has probably never had periodic activity. Nearby, however, is Broadside Geyser. Its best-known activity was during 1978, when play up to 15 feet high recurred every 20 to 40 minutes for durations of 5 to 6 minutes. Not geysers but interesting because of their unusual mineral deposits are the yellow-orange Ochre Spring (hydrous iron oxide) and olive-green Scorodite Spring (hydrous iron arsenate).

There are other thermal groups in the same general area as Joseph's Coat Hot Springs. The Rainbow Springs are located at the point where Bellow Spring jets almost horizontally into the adjacent creek; it is probably a perpetual spouter. The Coffee Pot Hot Springs include no named features, but several features spout from geyserite-lined vents and might be geysers.

4. "FAIRYLAND BASIN"

"Fairyland Basin" is tremendously difficult to get to, requiring not only a full day's hike from Joseph's Coat Hot Springs but also a nearly sheer drop into a deep canyon. It is so remote that it has probably been visited fewer than ten times. Whether any true geyser exists here is uncertain. One part of "Fairyland Basin," the "Gnome Group," is a sinter shield studded with numerous odd, tall, thin cones believed to be composed of a mixture of geyserite and travertine. Many of the cones are completely sealed in. In 1992 only one was flowing water, but fresh deposits implied intermittent activity and possibly true eruptions from at least two of the others. Only one feature here has an unofficial name, "Magic Mushroom."

5. CALCITE SPRINGS

The Calcite Springs emerge from the steep cliffs above the Yellowstone River, about midway between Tower Junction and Tower Fall in the northeastern part of Yellowstone. These springs are unusual in that they sometimes discharge organic liquids (essentially, crude oil) along with water, and they are depositing minerals such as calcite (calcium carbonate), gypsum (calcium sulphate), and pure sulphur in addition to silica. The water contains unusually large amounts of carbon dioxide, and this gas is no doubt responsible for the small eruptions occasionally reported. Although one spring was seen to intermittently erupt as high as 10 feet during 1985, it is unlikely that there are any true geysers at Calcite Springs.

6. BOG CREEK HOT SPRINGS

Southeast of Canyon Village and east of Hayden Valley are some extensive thermal areas along Bog Creek. There is no trail, and access requires advanced cross-country orienteering abilities. Dominated by steam vents, frying pans, and mud pots, this is the site of three geysers. Enigma Geyser, the first of the three to be discovered, was named because of its unlikely location among acid springs; some of its eruptions reach 6 to 10 feet high. Discovered a few years later, "Vitriol Geyser" and a smaller geyser a few feet away might be the most acid geysers anywhere, with a measured pH value of less than 1.0.

7. CRATER HILLS

The Crater Hills area is within Hayden Valley. Although it is only about 1 mile west of the highway, reaching the hot springs involves some danger. This is prime bear country, where several people have been mauled by grizzly bears even when walking within a few yards of the road. Visitors are advised to hike in large, noisy groups. There are numerous hot springs among the Crater Hills, only one of which is depositing geyserite. Sulphur Spring (better known as Crater Hills Geyser) is in eruption for a greater percentage of time than it is quiet. The play reaches as much as 8 feet high. Other named springs in this intriguing area are Foam Spring and Turbid Blue Mud Spring.

8. WESTERN HAYDEN VALLEY

Along Alum Creek in the far northwestern portion of Hayden Valley are several hot spring groups. One of these is Highland Springs. Located in the hills west of the valley proper, it contains some perpetual spouters and a single geyser. All of the features here are acid, and the well-named Miniature Geyser might be as acidic as the geysers at Bog Creek. Although tiny, it apparently has been continuously active since first described in 1887.

Along the south fork of Alum Creek is Glen Africa Basin. It contains many perpetual spouters, some of which may act as geysers at times. Glen Africa is not indicated on modern maps, but it gained considerable attention in the early days of Yellowstone because the only east-west road across the Park passed nearby. Among Glen

Mud Geyser, as photographed by William H. Jackson in 1871. Credit, USGS Online Photographic Library.

Africa's features with official names are Pseudo Geyser, a cyclic spouter up to 15 feet high; Red Jacket Spring, which plays steadily to 8 feet; and Flutter-Wheel Spring, a drowned but vigorous fumarole that plays as high as 3 feet within an echoing cavern.

A single small geyser exists at the Alabaster Springs, about 2 miles north of Glen Africa Basin, and small spouters have been observed at Violet Springs, still farther to the northeast.

9. MUD VOLCANO AREA

The Mud Volcano thermal area consists primarily of muddy, sulfurous springs that are strongly acid in character; Sulfur Cauldron's pH has been measured as low as 0.7, which is battery acid strength. Springs such as Dragons Mouth Spring, Black Dragon's Caldron, Sizzling Basin, and Mud Volcano itself are the primary attractions. Nevertheless, the area was the site of a powerful geyser during the 1870s, and it now boasts at least two cold water "soda pop" geysers.

Mud Geyser was named in 1870 when it underwent eruptions as high as 50 feet. The fact that it was a true geyser was confirmed by the Hayden Survey in 1871. Intervals as short as 3½ hours separated durations of 15 to 30 minutes. Then the activity gradually decreased, and Mud Geyser was dormant by 1879. Brief reactivations occurred in 1889 and the 1890s, possibly in 1905, and in 1922. Mud Geyser is the southernmost feature west of the road and is now nearly inactive.

Cold Water Geyser is near the Yellowstone River some distance southeast of the Mud Volcano area proper. Erupting cold water as high as 1½ to 3 feet, the action is powered by carbon dioxide gas rather than steam. Its activity has been erratic in recent years, with long dormant periods separating brief active phases. Nearer the east side of the road about ¼ mile south of Mud Volcano is a smaller cold water feature that acts as a 6-inch-high perpetual spouter.

10. SEDGE BAY OF YELLOWSTONE LAKE (BUTTE SPRINGS)

In early 1990, at a time when the water level of Yellowstone Lake was exceptionally low, a spring exposed on the shoreline of Sedge Bay played as a geyser. The eruptions recurred every 1½ minutes, lasted 40 seconds, and splashed up to 3 feet high. The activity only persisted for a few weeks until the spring thaw raised the water level of the lake. Small eruptions have also been seen among the Butte Springs, which lie across the highway northeast of Sedge Bay.

11. PELICAN CREEK VALLEY

Along the upper reaches of Pelican Creek are numerous thermal areas. Almost without exception the water is acid, and the typical springs are frying pans and mud pots, such as The Mushpots and The Mudkettles. However, West Pelican Geyser was described in 1878 at a location across the creek east of The Mudkettles. The eruption vigorously jetted "slightly turbid" water to heights as great as 15 feet, with activity said to be "almost constant." Although the spring's modern identity is uncertain, it is probably the feature labeled on modern topographic maps as "hot spring" just south of The Mushpots.

Semi-Centennial Geyser at Clearwater Springs underwent major eruptions only during 1922, Yellowstone's fiftieth anniversary year. Credit, NPS photo by J. E. Haynes.

East of the trail about 3 miles north of The Mushpots is the Pelican Creek Mud Volcano. In 1879 it behaved as a geyser when it underwent intermittent eruptions that threw "cartloads" of mud as high as 75 feet. Now it is a gigantic and violently active black mud pot.

12. CLEARWATER SPRINGS

The Clearwater Springs are just north of Roaring Mountain, along the road between Norris and Mammoth. Geyser activity was first seen here in 1918, when there were a few eruptions about 50 feet high. However, it was on August 14, 1922, the year of Yellowstone's fiftieth anniversary, when Semi-Centennial Geyser gained widespread notice. Some of the initial eruptions reached at least 300 feet high, and play 150 to 200 feet high continued until late 1922. There have been no eruptions since then. Minor intermittent bubbling has been seen during some recent seasons, but since

Obsidian Creek flows directly through the crater, no eruptions can be expected.

13. UNNAMED AREA NORTHWEST OF NORRIS GEYSER BASIN

Roughly 2 miles west-northwest of the Norris Museum is a thermal area that was almost never visited until the 1980s, when some very large mud explosions took place. Access is difficult, involving trekking across wet meadows, fording high streams, and bushwhacking dense forest made worse since the fires of 1988. The area contains a variety of springs, from mud pots to clear pools to geysers. The geysers have been known for years, but primarily because of distant, intermittent steam clouds visible on cold days. The largest of the geysers erupted from a long crack within a steep stream bank, sending a fan-shaped spray of water as high as 30 feet; some early reports that might refer to this spring imply eruptions much higher than that, and during the early 1970s the steam clouds were routinely reported by the ranger at the fire lookout atop Mt. Holmes. Unfortunately, this geyser now appears to be dead, or at least dormant and severely altered, because of a landslide that thoroughly buried most of the fracture. Two other geysers play from murky pools, their highest splashes reaching perhaps 3 feet, and there are several perpetual spouters. The mud explosions of the 1980s created two impressive craters 15 to 20 feet deep within the forest where there previously had never been hot springs of any kind. In 2004 one of these was weakly active as a mud pot, and the other was nearly full of bubbling water.

14. TERRACE SPRINGS

The Terrace Springs lie next to the road just north of Madison Junction. Whether true geyser activity ever existed there is doubtful. All of the features are rather cool, and the largest pool, Terrace Spring itself, is only about 140°F (60°C). Its apparent boiling is caused by carbon dioxide and other gases rather than steam. No activity other than this has been reported since 1871. However, when travelers C. C. Clawson, R. W. Raymond, and others visited these springs in 1871, Clawson described Terrace Spring (his "the Cauldron") as a geyser that threw "a column of hot water from 15 to 20 feet high." Nearby were other smaller springs "puffing and blowing and squirting hot water," to which Raymond gave names such as Kettle, Safety-Valve, and Reservoir. Of course, we will never know for sure, but those 1871 visitors likely let their imaginations exaggerate reality, as they had not yet visited the geyser basins.

15. FERRIS FORK OF THE BECHLER RIVER

Within a deep canyon along the Ferris Fork of the Bechler River, extending from Three River Junction upstream about 1 mile, is a thermal area where at least two geysers and one large perpetual spouter are known. The geysers are small but frequent in their action. The spouter is larger and may act as a geyser at times, reaching up to 10 feet high.

16. BOUNDARY CREEK

In the southwestern portion of Yellowstone are several clusters of hot springs along Boundary Creek. Because they are the closest thermal groups within Yellowstone to the Island Park area outside the park where geothermal drilling was proposed, activity in the Boundary Creek area was closely monitored during the 1980s. The fact that the area included two small geysers was a surprise. Eruptions were frequent then, but neither geyser has been reported since 1993.

17. LEWIS LAKE HOT SPRINGS

Barely mentioned in published reports, a 2001 survey of the hot springs near the west shore of Lewis Lake revealed a deep blue pool, several perpetual spouters, and one small geyser. The geyser, which is hidden up a small canyon away from the lake, erupts frequently, reaching 1 or 2 feet high for durations of several seconds.

Geyser Fields of the World

This appendix presents information about all the geyser fields of the world, known as of 2007. These locations are shown on the Appendix Map and Table. The larger of these geyser fields are given full discussions, while many of the smaller areas are only summarized.

The sources of this information are diverse. Geysers have been noted in the course of numerous geologic studies—geysers are rare enough to make their existence an important part of any geologic study—but few of these studies have been published for wide distribution. Such reports have been obtained whenever possible, but the most valuable sources of information about geysers in the modern world have been personal communications, mostly letters from geologists around the world, plus personal journeys to several of the geyser fields around the Pacific Rim.

It is important to note that the word "geyser" is often used loosely, especially in recent travel descriptions posted on the World

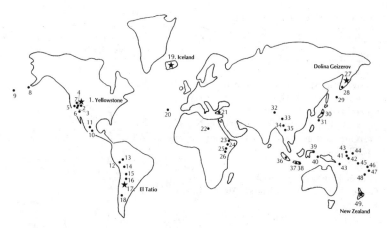

Appendix Map. *The Geyser Fields of the World*

Wide Web. There, "geyser" may refer to almost any variety of hot spring, including things like small pools that are cool enough to bathe in. Even in professional literature, however, the term is often misused in reference to steam vents and mud pots and almost always in describing what are actually perpetual spouters. Thus, some of the areas listed here might, in fact, not be geyser fields after all.

These studies have made one thing especially clear: true geysers are extraordinarily rare features, restricted with few exceptions to small numbers in special geologic environments. Yellowstone National Park is far and away the greatest geyser field anywhere. This book enumerates over 700 Yellowstone hot springs that have erupted during recorded times. Almost all of these springs are true geysers, and most remain active today. The small area of the Upper Geyser Basin alone contains at least 20 percent of all the geysers in the world, more than in any other geyser field on Earth, and in total Yellowstone encompasses well over half the world's total number of geysers.

According to the best current information, the combination of *all* other geyser fields of the world, listed here as forty-eight in number, probably includes fewer than 500 geysers total.

Just why geysers are so common in Yellowstone is not known with certainty. It could be chance as much as anything. The existence of geysers is a complex matter, involving details such as water temperature, water supply, degree of self-sealing within the geothermal system, and, perhaps of utmost importance, the extent to

Appendix Table. Summary of the Geyser Fields of the World

Name	Text and Map Number	Number of Geysers[†]
North America		
Yellowstone National Park, Wyoming, U.S.A.	1	500
Beowawe, Nevada, U.S.A.	2	0*
Steamboat Hot Springs, Nevada, U.S.A.	3	0*
Great Boiling Springs, near Gerlach, Nevada, U.S.A.	4	3
Morgan Springs, near Lassen Volcanic National Park, California, U.S.A.	5	3
Long Valley, California, U.S.A.	6	3*
Mickey Hot Springs, Oregon, U.S.A.	7	1
Geyser Valley, Umnak Island, Alaska, U.S.A.	8	9
Kanaga Island, Alaska, U.S.A.	9	several
Comanjilla, Guanajuato, Mexico	10	11
Ixtlan de los Hervores, Michoacan, Mexico	11	2
South America		
Calacoa (Carumas), Moquegua, Peru	12	8*
Puentebello, Puno, Peru	13	1
Calientes de Candarave, Tacna, Peru	14	3
Suriri (Polloquere), Tarapaca, Chile	15	1
Puchuldiza and Tuja, Tarapaca, Chile	16	8
El Tatio, Antofagasta, Chile	17	85
Volcan Domuyo, Neuquén, Argentina	18	5
Mid–Atlantic Ocean		
Iceland	19	30*
Volcan Furnas, Azores Islands, Portugal	20	0*
Europe		
There are no true geysers in Europe.		
Asia Minor and Africa		
Ayvacik, Turkey	21	1
Soborom Hot Springs, Tarso Voon (volcano), Chad	22	reported
Allallobeda, Danakil, Ethiopia	23	2
Geyser Island, Lake Langano, Ethiopia	24	1
Unnamed spring group, Lake Abaya, Ethiopia	24	1
Logipi and other areas, northern Kenya	25	reported
Lake Bogoria, Kenya	26	16
Asia		
Dolina Geizerov, Kamchatka Peninsula, Russia	27	100(?)
Mutnovka, Kamchatka Peninsula, Russia	28	unknown
Pauzhetsk, Kamchatka Peninsula, Russia	28	2*
Shiaskotan Island, Kuril Islands, Russia	29	many?
Onikobe, Miyagi Prefecture, Honshu Island, Japan	30	2
Beppu, Kyushu Island, Japan	31	artificial?
Peting Chuja and Naisum Chuja, Tibet, People's Republic of China	32	14

continued on next page

Name	Text and Map Number	Number of Geysers[†]
Tagajia, Tibet, People's Republic of China	32	4
Chapu, Tibet, People's Republic of China	32	2
Gulu, Tibet, People's Republic of China	32	reported
Yangpachen, Tibet, People's Republic of China	32	several
Chaluo, Sichuan, People's Republic of China	33	4
Rehai, Yunnan, People's Republic of China	33	19 (?)
Balazhang, Yunnan, People's Republic of China	33	1
Southeast Asia		
Lashio, Shan District, Myanmar (Burma)	34	1
Pai, Tavoy District, Myanmar (Burma), possibly Thailand	34	1
Chiang Mai and Mae Hong Son Provinces, Thailand	35	19*
Lampung-Semangko, Sumatra, Indonesia	36	3
Tapanuli, Sumatra, Indonesia	36	3
Kerinci, Sumatra, Indonesia	36	1
Pasaman, Sumatra, Indonesia	36	1
Cisolok, Java, Indonesia	37	2
Gunung Papandayan, Java, Indonesia	38	reported
Minahasa District, Sulawesi (Celebes), Indonesia	39	1
Bacan Island, Maluku Group, Indonesia	40	1
Pacific Ocean Rim		
Garua Harbour, New Britain Island, Papua–New Guinea	41	2
Koimumu (Kasiloli), New Britain Island, Papua–New Guinea	42	14
Narage Island, Papua–New Guinea	43	1
Waramung and Kapkai, Ambitle Island, Papua–New Guinea	44	8
Lihir Island, Papua–New Guinea	44	0
Tongonan, Leyte Island, Philippines	none	0*
Deidei and Iamelele, Fergusson Island, Papua–New Guinea	45	12
Vutusuala Hot Springs, Savo Island, Solomon Islands	46	several
Sladen Boiling Springs, Gaua Island, Vanuatu	47	reported
Mt. Sereama, Vanua Lava Island, Vanuatu	47	reported
Nakama Springs, Savusavu, Vanua Levu, Fiji	48	5
New Zealand	49	70*

[†] The number of active geysers according to the best information currently available, which is sometimes many years old. See the text for details. As noted in Chapter 1 of this book, different authorities may derive different counts depending on their personal definition of "geyser."

* Areas in which significant human developments have affected the activity, reducing the number of active geysers. The five areas listed as 0 (zero) formerly contained significant numbers of geysers but now have none because of geothermal exploitation or natural changes.

which water has been both over-pressured beyond normal hydrostatic pressure and over-heated (or superheated) above the normal boiling point for a given depth and pressure. Yellowstone is a near-perfect combination of factors scarcely matched elsewhere. For example, Iceland contains more hot springs than Yellowstone does, yet the geysers there number only about thirty.

This does not mean these other geysers aren't special. Icelanders and New Zealanders are justifiably proud of their geysers and have recently taken steps to reverse exploitation and assure their preservation. A growing concern for the environment in Chile may limit development of those geyser fields. The Russians created a national park on Kamchatka, in part because of the geysers, just as happened at Yellowstone. That, however, ends the list of protected areas. Many of the world's geyser fields are not what they once were. Humankind's ever-widening search for energy has destroyed several of these geothermal systems, and several of those that remain untapped are presently being explored for potential development. Just as "you can't have your cake and eat it, too," it is impossible to exploit and simultaneously preserve hot springs. Geysers are disappearing from our world. Were they biological organisms, they would be on the endangered species list.

NORTH AMERICA

1. YELLOWSTONE NATIONAL PARK, WYOMING, U.S.A. In terms of the number of geysers, the power of the activity, and any other category one would like to consider, Yellowstone is far and away the world's largest and best geyser field. Its more than 700 historically known geysers—as many as 60 percent of all the geysers in the world—are scattered among nine major geyser basins plus several smaller areas, as detailed in this book.

2. BEOWAWE, EUREKA COUNTY, NEVADA, U.S.A. Beowawe is a small community, little more than a whistle-stop on the railroad, halfway between Battle Mountain and Elko in northern Nevada. The geysers were found about 4 miles southwest of the town, near the head of Whirlwind Valley. The area was first reported in literature to contain geysers in 1869—interestingly, geysers were confirmed at Beowawe before they were in Yellowstone—but the Beowawe Geysers did not receive scientific attention until 1934. Further studies during the late 1940s and early 1950s revealed that as many as thirty of its springs were capable of erupting, although generally only a few were active at any given time. The area generated enough attention that it was proposed as a national monument and later as a Nevada state park. Those proposals failed in part because of land ownership issues.

All descriptions of the Beowawe Geysers must be written in the past tense, however. New technology has allowed small geothermal systems to be exploited. In 1986 the Whirlwind Valley power plant began producing electricity. The production wells are located away from the Main Terrace, where most of the geysers were located, but the spent fluid is re-injected into the ground nearby. For reasons geologists cannot thoroughly explain, many of the geysers rejuvenated briefly in late 1986, shortly after the first production, but by 1987 all of the hot spring activity had quit. One of the most concentrated geyser fields anywhere is dead.

3. STEAMBOAT HOT SPRINGS, WASHOE COUNTY, NEVADA, U.S.A.
Steamboat Springs is almost too accessible. The area is astride U.S. Highway 395 just 9 miles south of downtown Reno, Nevada. It suffered severe damage from a lack of management while most of the local population was completely unaware of its rare importance. The geyser activity at Steamboat Springs was continuously changing, but there was always some degree of eruptive activity, and more than twenty-five of the springs had histories as geysers.

Most of the action was on the Main Terrace, west of the highway. Many different geysers had been active there. In 1860 one of them was very regular and played as high as 75 feet, and two years later different vents reached 50 to 60 feet. More recently, Geyser #42w was known to reach over 20 feet high, and #40 played up to 10 feet from a water level well belowground.

The same technology that led to the destruction of Beowawe did the same for Steamboat Springs. The operators of the power plant, which began full commercial production in 1988, were supposed to monitor the natural springs and assure their preservation. That was not done, and Steamboat Hot Springs has joined the list of thoroughly destroyed areas.

4. GREAT BOILING SPRINGS, GERLACH, WASHOE COUNTY, NEVADA, U.S.A.
The Great Boiling Springs were first reported by John C. Fremont in 1845. He wrote about "a circular space about fifteen

"Geyser 40" at Steamboat Hot Springs played frequently and vigorously as much as 20 feet above its pool level and 10 feet aboveground during the early 1980s. It and all the other geysers at Steamboat stopped playing shortly after a geothermal power plant started operating in 1987.

feet in diameter, entirely occupied by boiling water. It boils up at irregular intervals, and with much noise." It is certain that some of the springs act as true geysers on occasion. Eruptions up to 5 feet high were reported in a 1984 travel magazine, and a pool underwent intermittent boiling in 1988. Now, however, the area is closed as fenced private property.

A few miles north of Gerlach are the Ward's Hot Springs, at Fly Ranch. Near the graded dirt road is the "Fly Ranch Geyser," which has appeared in numerous magazine articles. However, it is not a geyser but a well drilled in 1916 that acts as a perpetual spouter.

5. MORGAN SPRINGS, TEHAMA COUNTY, AND TERMINAL GEYSER, PLUMAS COUNTY, CALIFORNIA, U.S.A. Located just outside the south entrance to Lassen Volcanic National Park, Morgan Springs used to be the site of a Forest Service campground. This small collection of very hot, flowing springs and pools included one geyser in 1916 and at least three (possibly five) geysers during the 1950s. The Morgan Springs now lie within private property closed to public access, and their current state is unknown.

The largest of the geysers at Hot Creek, in California's Long Valley Caldera, occasionally reached as high as 40 feet shortly after the earthquakes of May 1980. Credit, Photo by Paul Strasser.

Terminal Geyser is a few miles away within the national park. It currently acts as a perpetual spouter, constantly jetting a few feet high from several openings. But in the 1870s Terminal was reported to spout intermittently as high as 20 feet, and in 1925 a publication about the geology of Lassen Park referred to Terminal as "The Geyser," with eruptions 8 feet high.

6. LONG VALLEY CALDERA, MONO COUNTY, CALIFORNIA, U.S.A.
The Long Valley Caldera is a volcanic explosion crater very much like that of Yellowstone. Although it is 140,000 years older, it is logical that it should contain similar hot springs. It does, but all the hot springs groups are small.

The Casa Diablo Hot Springs, near the junction of U.S. Highway 395 with State Highway 203, was the site of a geyser that reached as high as 30 feet until the development of geothermal power plants. At the swimming area at Hot Creek, several springs have histories as geysers, most notably following earthquakes in 1980. One of these pools erupted up to 40 feet high, but the activity gradually died down. No significant eruptions were reported after 1990 until

June 2006, when several geysers up to 7 feet high made temporary appearances. Little Hot Creek is a tiny area on an unimproved back road where there are at least two geysers that play frequently as high as 2 feet.

7. MICKEY HOT SPRINGS, HARNEY COUNTY, OREGON, U.S.A. The fact that Mickey Hot Springs possesses a single geyser was only discovered in 1986. The best year on record for "Mickey Geyser" was 1991, when it had intervals of 4 to 5 minutes, durations of about 1 minute, and heights up to 6 feet.

8. UMNAK ISLAND, ALEUTIAN CHAIN, ALASKA, U.S.A. Umnak Island is one of the first and largest of the Aleutian Islands, quite close to the mainland of the Alaska Peninsula. At the Geyser Bight Geothermal Area in the narrow midsection of the island are several groups of hot springs, three of which contain geysers. Although all are small and mostly of the boiling bubble-shower type, their activity is vigorous. Most recorded intervals have ranged between 3 and 10 minutes, with durations of about 3 minutes and heights of 1 to 10 feet. One of the geysers is capable of larger eruptions, occasionally reaching up to 20 feet high.

9. KANAGA ISLAND, ALEUTIAN CHAIN, ALASKA, U.S.A. Kanaga Island is in the outer Aleutian Chain, near Adak Island. During a general survey of Alaska's geothermal resources, it was discovered that Kanaga's hot springs include small geysers. No statistics were reported except that no geyser played more than about 2 feet high.

10. COMANJILLA, GUANAJUATO, MEXICO. The Hotel Refugio at Comanjilla is a hot spring resort between the cities of Guanajuato and Leon, northeast of Guadalajara. The first description of these geysers was written in 1910. The resort uses the springs to supply its pools and spas, but the developments are low-grade, and most of the eleven geysers described in 1910 could still be identified in 1981. The largest, Geyser Humboldt, plays up to 10 feet high when active. Another had frequent eruptions to 6 feet, and one hidden deep within a small cavern next to a swimming pool played very regularly every 2 hours.

As impressive as it might be, an erupting drilled well is one of the few signs that a major geyser field used to exist at Ixtlan de los Hervores, Mexico.

11. IXTLAN DE LOS HERVORES, MICHOACAN, MEXICO. The name roughly translates as "boiling pools," and Ixtlan was once promoted as "Mexico's Little Yellowstone." In 1906 as many as fourteen geysers occurred among several hot spring groups scattered along a 2-mile stretch of the valley. Unfortunately, Ixtlan is another site of geothermal drilling. One of the geothermal wells at Ixtlan spouts continuously, occasionally reaching over 100 feet high. It has been promoted as "the geyser," but all of the natural springs have disappeared except in a small area behind the village of Salitre. At least two geysers were active there in 1981, and it is probably the Salitre area where a Mexican tourism Web page cited "active geysers" during 2005.

SOUTH AMERICA

12. CALACOA, MOQUEGUA DEPARTMENT, PERU. The vicinity of Calacoa (also known as Carumas), along the Rio Putina between 9,500 and 12,000 feet above sea level in the Andes Mountains of southern Peru, is filled with hot springs. All are of significantly high temperatures, and at least eight geysers have been observed among three of the six most important thermal groups. The waters of all these areas are used for bathing and cooking, and evidence of Inca usage is found in numerous places.

13. PUENTE BELLO, MOQUEGUA DEPARTMENT, PERU. The geysers of Puente Bello are adjacent to the dirt road between the cities of Puno and Moquegua. They occur both inside and in front of a natural bridge named Cueva Ccollo (a native Indian name) that extends more than 400 feet through the hillside. Within the cave, dangerously accessible because of the hot water and inhospitable atmosphere, are "many" geysers. Playing 3 to 6 feet high, these may be perpetual spouters. However, outside the cave is a truly periodic geyser with heavy discharge. The usual play consists of quick jets typically just 4 to 5 feet high, but the geyser is said to occasionally reach as high as 80 feet. In 2002, Puente Bello was proposed as a new natural monument within the Peruvian national park system.

14. CALIENTES DE CANDARAVE, TACNA DEPARTMENT, PERU. The springs near Candarave, the "Place of the Condor," are on the east flank of Volcan Yucamani and higher than 14,400 feet above sea level. There are at least forty-four boiling springs, three of which erupt intermittently as geysers to heights of 1 to 4 feet. In 1979, however, one erupted with extremely regular 55-minute intervals; the duration was 2 minutes, and the height was over 20 feet. This area lies within a proposed new Peruvian natural monument.

14a. ANCOCOLLO, TACNA DEPARTMENT, PERU. A small cluster of boiling springs deep within the gorge of Rio Ancocollo was only revealed as a geyser locality in February 2005. Four of the sixteen springs splash intermittently a few feet high. Nothing more is known.

15. POLLOQUERE AT SALAR DE SURIRI, PARINACOTA PROVINCE, TARAPACA REGION, CHILE. Nearly 15,000 feet high in the Andes of far northern Chile, Polloquere is within the Salar de Suriri Natural Monument. Access is by way of poor dirt roads that are often closed during the November-to-March wet season. One small geyser was observed in 1944. In a geothermal survey report published in 1972, no geyser was specifically noted, but several of the 230 high-temperature springs were noted as "boiling up quite high."

The geyser field of El Tatio, Chile, a remote area high in the Andes Mountains, boasts of at least eighty-five geysers. Credit, NPS photo.

16. PUCHULDIZA, IQUEQUE PROVINCE, TARAPACA REGION, CHILE. Puchuldiza is located in northern Chile, high in the Andes Mountains at an elevation just under 14,000 feet. Four geysers were described in 1972 and five in 1980, when some of the eruptions reached 12 feet high. Unfortunately, the area has been damaged by both geothermal and gold exploration drilling. The drill holes now act as perpetual spouters, and they probably account for the "geysers" promoted in 2002 by a tour company.

About 3 miles downstream along the Rio Puchuldiza is a smaller thermal area known as Tuja. In 2002 it apparently included one small, natural geyser.

17. EL TATIO, EL LOA PROVINCE, ANTOFAGASTA REGION, CHILE. El Tatio is one of the world's premier geyser fields. Although exploratory geothermal drilling was done in 1967, the geysers remain largely unaltered, and a local organization based in the city of Calama, the "Movement in Defense of Atacaman Water and Life," is working toward permanent protection. Meanwhile, El Tatio is visited almost daily by four-wheel-drive tours that operate from the town of San Pedro de Atacama, about 50 miles away. Rental cars are also available, and people properly equipped with high-quality gear can camp among the geysers. Even though the bone-dry Atacama Desert is nearby, winter conditions are the rule at the altitude of 13,800 feet. There are no commercial facilities.

The geysers occur among scattered groups of hot springs in a field that spans several square miles. Of the more than 300 hot springs at El Tatio, probably over 100 have been active as geysers. The record for a single observation was the 85 true geysers observed in March 2002, when 30 perpetual spouters were also seen. None of the eruptions is of significantly large size, the tallest reaching perhaps 20 feet, but the action is vigorous, and several geysers rise out of massive geyserite cones.

Given the difficulties in making long-term observations, little more is known about El Tatio. However, with no fewer than 85 geysers, it ranks as the third largest existing geyser field on Earth, trailing only Yellowstone and Dolina Geizerov on Kamchatka (#27).

18. DOMUYO PROTECTED NATURAL AREA, NEUQUÉN TERRITORY, ARGENTINA. Hot springs are common along a fracture zone that extends down the southwest flank of Volcan Domuyo, near the border with Chile in northern Patagonia. Geysers apparently exist in at least two of the thermal areas. Five geysers have been described at Villa Aguas Calientes. They erupt frequently to heights as great as 6 feet. At nearby Los Olletas there are "intermittent steam emanations," probable geysers that play as high as 3 feet. At El Humazo there are two springs that spout as high as 10 feet, but they are probably perpetual spouters.

MID-ATLANTIC OCEAN

19. ICELAND. Most discussions of the geysers of Iceland seem to generalize, giving the impression that there is a single geyser basin. Reality is that while the large geysers of fame are confined to one small area, hot spring groups with geysers are scattered widely about the island. They total a large enough number to make Iceland the fifth largest geyser field in the world. Most famous of all is Geysir, the namesake of all the world's geysers. Translated, the name means "gusher" or "spouter." The Icelandic people are proud of Geysir, and the name is copyrighted so it can never be applied to any other geyser, anywhere.

Geysir ("Spouter"), in Iceland, is the namesake of all, but it is rarely active. This eruption was in 1968. Credit, NPS photo.

The names of the geysers and the basins are fascinating. Icelandic is a Viking tongue, largely unrelated to other modern languages, and many of these names are difficult for English-speaking people to pronounce. The geyser basins bear such names as Torfastathir, Hruni, Reykir in Ölfus (now called Hveragerdi), Hveravellir, and Borgarfjartharsysla. The geysers themselves have names such as Sturlüreykir, Svathi, Eyvindarhver, and Opherrishola. Each name has a specific meaning, and, like those of Yellowstone, they are often very descriptive.

Geysir is at Haukadalur. Although active, it rarely undergoes the 200-foot eruptions historically known. Why this is so is uncertain; perhaps it is simply because of an exchange of function with other springs in the area, or perhaps the geyser was damaged after suffering through too many artificially induced eruptions (a practice illegal in Yellowstone and now banned at Geysir). The modern play usually fails to exceed 30 feet high. This takes place several times per day. Another important geyser at Haukadalur is Strokkur, whose name means "churn." It gushes forth every 8 to 20 minutes, the brief bursts of most eruptions reaching 75 feet high. Three or four other, smaller geysers occur at Haukadalur,

Geysir Strokkur ("The Churn") is the largest remaining geyser in Iceland. It erupts about every 20 minutes and reaches as high as 75 feet. Credit, Photo by Alan Glennon.

which is maintained with a park-like atmosphere and served by a hotel complex.

Another concentration of geysers was at Hveragerdi. Most of these geysers can no longer be identified because of geothermal drilling for the space heating of greenhouses. Gone are Gryla, which played every 2 to 3 hours as high as 40 feet; Littli, which could reach 80 feet; and more than a dozen other geysers. At least four small geysers and some perpetual spouters persist, however. A trail system threads up the thermal valley to Svarth, Bathstafuhver, and an unnamed geyser, each of which erupts a few feet high.

In far northern Iceland is Reykjadalur. Two geysers there have been active with little change for at least 200 years. Ystihver plays every 2 to 3 minutes, reaching 9 (rarely 20) feet high. Uxahver is less frequent and 8 feet high. Odiphtong, a large pool, has rare eruptions, and Sythstihver has been dormant since 1918.

Probably the most active basin in Iceland now is at Hveravellir, almost dead center on the island. In a barren region far from any town, it is seldom visited and minimally developed. It contains at least five geysers, one of which is known to reach 15 feet high. In the 1800s, Gamli Strokk played as high as 150 feet, but apparently not even a trace of its crater remains visible.

Iceland has virtually no energy sources other than geothermal, so the exploitation of the hot springs is vital to the country's economy. Although geysers have been observed in at least fifteen other areas in Iceland, many of them have been altered or destroyed by geothermal and other developments. Until recently, there were few attempts to preserve the geysers in places other than Haukadalur. But a new attitude is beginning to appear, and there are now interpretive centers at Haukadalur and Hveragerdi and new boardwalk trails at Hveravellir. Perhaps most of Iceland's remaining geyser basins, where there are about thirty active geysers, will survive.

20. VOLCAN FURNAS, ISLA SAN MIGUEL, AZORES ISLANDS, PORTUGAL. The name "Furnas" translates just as it sounds—"furnace." It is a volcano, probably only dormant rather than dead, within which are numerous hot springs. At the turn of the twentieth century there were several geysers, one of which erupted frequently to "several feet." The activity is now restricted to a few boiling springs, one or two of which may qualify as intermittent bubble-shower geysers.

CONTINENTAL EUROPE

There are no true geysers anywhere in Europe, despite many reports to the contrary about Lardarello, Italy. Cold water soda pop geysers, whose eruptions are powered by carbon dioxide gas, are known in France, Germany, Slovakia, and Serbia.

ASIA MINOR AND AFRICA

21. AYVACIK, TURKEY. Hot springs occur throughout Turkey, and some near the country's northwest coast are of very high temperatures. One occurs close to the town of Ayvacik. Gayzer Suyu is the single geyser in the area, reported in a 1968 geothermal report to erupt frequently as high as 6 feet.

22. TIBESTI MOUNTAINS, CHAD. The Tibesti Mountains are little known to most westerners, but they comprise the highest mountain massif in the Sahara Desert. The major peaks are volcanoes with summit elevations higher than 10,000 feet. The area is largely considered off limits because of unstable governments and rebel activity along the border between Chad and Libya. Within the small Yirrigue Caldera near Pic Tousside volcano are hot springs long reported to include small geysers. No details are known, and they might only be fumaroles. More definite would seem to be the "geysers, hot springs, and mud pots" at Soborom Hot Springs, near the volcano Tarso Voon, 50 miles east of Pic Tousside. Again, details are not known.

23. ALLALLOBEDA, DANAKIL DEPRESSION, ETHIOPIA. The Allallobeda Thermal Area is in eastern Ethiopia, within the northern part of the East African Rift System. Several hot springs are aligned along a fault zone. The largest spring is the Allallobeda Spouter, a perpetual spouter that plays to about 20 feet; on very rare occasions it apparently bursts as high as 100 feet. At least one other spring is a true geyser, although its eruption reaches only about 1 foot, while four other features undergo intermittent boiling.

24. THE LAKES DISTRICT, ETHIOPIA. Numerous geothermal areas are found in the highlands of southern Ethiopia, where several large lakes occupy low areas in the East African Rift Valley. Geysers are known in at least two of the hot spring groups, and they very possibly exist in others as well. In Lake Langano's North Bay is Edo Laki Island, also called Geyser Island. Following an earthquake in 1906, one spring erupted powerfully, some of the play reaching over 60 feet high. The play was extremely brief, perhaps a single burst, but it recurred every 30 seconds. Through the years the activity declined, and in 1966 the spring had regressed to intermittent overflow. A geyser active in 1993 "at Lake Langano" may be this one. The Chokore Hot Springs lie along the northwest shore of Lake Abaya. The geyserite platform there is dotted with small cones and beaded deposits, and in 1979 there was one active geyser. Its action was erratic but frequent and up to 6 feet high.

25. NORTHERN KENYA. The East African Rift Valley extends southward from Ethiopia through Kenya. Geothermal areas occur everywhere along the valley, and geysers are known to exist at three of these places. Numerous hot springs and high-temperature steam

"Geyser KL-19d" is one of the geysers scattered around the shoreline of Lake Bogoria, Kenya. Credit, Photo by Robin Renaut.

vents exist on North Island, a long-dormant volcano in northern Lake Turkana, where a geyser was described in 2004; details are not known. Logipi Geyser (also spelled Logkippi, Lokippi, and so on) is at Elboitong, in the Suguta Valley south of Lake Turkana. (It is not at Lake Logipi, as previously reported here.) Active in the 1930s as a "regular" geyser reaching 4 feet high, it was later described as a perpetual spouter. A geothermal survey of that area found boiling springs but no active geyser in 1990. In 1902, springs that "throw out only an intermittent jet" existed on Ol Kokwe, a volcanic island in Lake Baringo. No natural geyser has been seen there in recent years, although "Baringo Geyser" played over 200 feet high from a drilled well during 2004.

26. LAKE BOGORIA, KENYA. Lake Bogoria (formerly Lake Hannington) in central Kenya hosts the largest geyser field in Africa. There are at least 200 hot springs distributed among five named hot spring groups. The 16 known geysers include 3 unknown before 2005, when the lake's water level dropped and exposed additional geothermal ground. The name "Loburu" has been used for two areas and at least 2 individual springs.

The northernmost thermal group is the modern Loburu, where 3 of the largest Bogoria geysers have been seen. Each can erupt over 10 feet high, but usually they are smaller and perpetual in their action.

Chemurkeu (the "Loburu" of older literature) includes 40 hot springs and 4 geysers. One reaches about 4 feet high, while the others play to less than 1 foot.

At Ng'wasis are numerous flowing springs and steam vents. A 1975 report by a Kenyan geologist identified this as the site of Loburu Geyser, but no eruptive activity has been seen in this group since at least 1988.

Koibobei is the location of at least 3 geysers. Koibobei Geyser is the largest at Lake Bogoria, reaching over 15 feet high but only when the lake level is low enough to expose its vent. The 2 other geysers play 1 to 2 feet high.

Finally, Losaramat boasts 3 geysers among its 17 springs. Each is about 2 feet high.

Previous editions of this book cited geysers or perpetual spouters at nearby Maji ya Moto. In fact, it includes only some warm springs where eruptions have never been observed.

NORTHERN AND CENTRAL ASIA

27. DOLINA GEIZEROV (GEYSER VALLEY), KRONOTSKY NATURE PRESERVE, KAMCHATKA PENINSULA, RUSSIA. The Kamchatka Peninsula is that extension of Siberia that points south into the North Pacific Ocean. One of the closest parts of the former Soviet Union to the United States, it was closed to outsiders until 1991. This author is proud to have served as the leader of the first American expedition allowed to spend time in Dolina Geizerov, eight days in June–July 1991 that revealed the valley to be the second largest extant geyser field in the world.

Kamchatka is an intensely volcanic place. Dolina Geizerov lies just outside the Uzon Caldera, a large Yellowstone-style volcanic explosion and collapse crater, and is low on the flank of Kikhpinych Volcano, which had some small, ashy eruptions in the 1890s. Several other active volcanoes are in the vicinity. The Kronotsky Nature Preserve is the descendant of an area that was closed to hunting in the 1880s to preserve sables, but it is such a remote and rugged region that Dolina Geizerov wasn't discovered until 1941. World War II postponed explorations until the late 1940s, but within a

Geizer Gorizontalnyi ("Horizontal Geyser") is but one of the vigorously active gey-sers in Russia's "Valley of Geysers" on the Kamchatka Peninsula. Gorizontalnyi was not affected by the 2007 landslide that devastated part of the valley.

few years the valley had been thoroughly mapped and its existence revealed to the outside world. With the development of limited tourism since 1991, trails and boardwalks have been established, but there are no commercial facilities. Access is via a 120-mile helicop-ter flight from the Yelizovo airport near the city of Petropavlovsk-Kamchatskii, and it is apparently available only for one-day tours. Overnight camping is not allowed except to researchers.

Prior to 1991, published reports about Dolina Geizerov lim-ited the discussion to about 23 geysers, and since the same features were always described, the impression was that they were the only geysers present. In fact, in 1991 there were at least 200 geysers. Most are relatively small, playing only a few feet high, but the activ-ity is intense. Almost all the geysers of significant size have inter-vals of less than 1 hour. Fontan (Fountain), Malyi (Small), Bolshoi (Great), Troynoy (Three), and Pervenets (First Born) all reach 40 to 80 feet high, while Skalistyi (Rocky), Conus Khrustalnyi (Crystal Cone), Malenkii Prinz (Little Prince), and Tschell (Crack) play to lesser heights but with great frequency and regularity. Unknown prior to 1991 is Grot Yubileinyi (Jubilee Grotto), which can send its angled water jet as high as 100 feet and as far outward as 250

Geizer Sosed ("The Neighbor") had vigorous eruptions from a crater near several other small geysers in Kamchatka's "Valley of Geysers." It was deeply buried by a huge landslide in June 2007.

feet. Velikan Geyser (Giant) is the largest geyser that had received publicity. Although impressive in size at 90 to 120 feet high, the play lasts less than 1 minute, and the 5-hour interval is one of the longest in the valley. In addition to the geysers, Dolina Geizerov is studded with many small perpetual spouters, pools, mud pots, and steam vents.

Unfortunately, some of the geysers mentioned here must be considered to be in the past. In June 2007 a massive landslide buried nearly half the valley. In addition, the lake that backed up behind the muddy dam inundated many more geysers. The status of the area is entirely uncertain, but perhaps as few as 100 geysers remain. Nevertheless, Dolina Geizerov still ranks as the second largest geyser field in the word, and the fact that the Kronotsky Nature Preserve is listed as a World Heritage Site will probably assure the preservation of what remains.

28. MUTNOVKA AND PAUZHETSK, KAMCHATKA PENINSULA, RUSSIA.
Mutnovka is about 90 winding road miles south of Petropavlovsk-Kamchatsky. Since the partial destruction of Dolina Geizerov in the Kronotsky Nature Preserve, this much smaller area has been

promoted by tourism agencies as "Kamchatka's Little Valley of Geysers." Little is known about Mutnovka, except that its geysers are noted as "not as numerous" and "smaller" than those of Dolina Geizerov. The road exists because of a growing geothermal power plant complex, and its operators openly state that the production of electricity is more important than the preservation of hot springs. Mutnovka is likely to be destroyed very soon.

The geothermal field of Pauzhetsk is much farther south, near the southern tip of the Kamchatka Peninsula. Despite geothermal developments and a small (11 megawatt) power plant, two of its geysers remain active. The eruptions generally reach only a few feet high, but in 1991 a geothermal geologist stated that eruptions 30 feet high have been seen.

29. SHIASKOTAN ISLAND, KURIL ISLANDS, RUSSIA. Similar to the Aleutians, the Kurils are the volcanic island chain between Kamchatka and Japan. Shiaskotan is a small island in the middle of the chain. Although it was nicknamed "The Island of a Thousand Geysers" in a 1971 book, only one geyser was actually described. It apparently erupted frequently and was "several meters" high. In 1992, two Russian volcanologists stated that there are also geysers on another (unnamed) island in the southern Kuriles (probably Kunashir).

30. ONIKOBE SPRINGS, HONSHU ISLAND, JAPAN. Considering the number of volcanic zones in Japan, many of which contain rhyolite rock, geysers are few in number and presently exist in only one or two places in the country. The Onikobe Springs are in Miyagi Prefecture, about 200 miles north of Tokyo. Part of an extensive zone in which there are several resorts and sulfur mines, Onikobe includes as many as five geysers among its hundred hot springs. The largest and the only one that is regularly active is Onikobe Geyser itself, which in 2006 erupted as high as 50 feet on intervals of 6 to 20 minutes for durations of 1 to 1½ minutes. Megama and Ogama each can play 6 to 10 feet high but are active mostly as non-eruptive intermittent springs. There are also several pools whose geyserite deposits show signs of large-scale geyser eruptions in the past. Drilled wells named Benten and Miyazawa also erupt, reaching 45 to 60 feet high as frequently as every 30 minutes. The Onikobe thermal area is operated as a park, and visitors who have paid the entry fee can also use a geothermal swimming pool.

Elsewhere in Japan, there once was a large but infrequent geyser at the city of Atami, but it has been inactive since the 1950s. The geysers reported at Tsuchiya, Shikabe, Yunotani, and Tamatsukuri are all artificial, erupting from drilled wells, and it is unlikely that natural geysers ever existed in those places. The same is true of Suwako Geyser, which erupts hourly as high 160 feet at a recreation area 100 miles west of Tokyo.

31. BEPPU, KYUSHU ISLAND, JAPAN. The city of Beppu is a major hot spring resort. Perpetual spouters are common, and some boiling pools might qualify as bubble-shower geysers. Tatsumaki jigoku, whose name translates as something like "water spout from Hell," has frequent eruptions several feet high but is probably a well drilled in 1929. Another spring, commonly called "Dragon's Breath," sprays water out of a cavern to a distance of 10 or 12 feet into a pool, but it is probably also a drilled well.

32. TIBETAN PLATEAU, TIBET (XIZANG) AND QINGHAI, PEOPLE'S REPUBLIC OF CHINA. "We marched for three successive days without coming to tents. Then we saw in the distance a great column of smoke rising into the sky. We wondered if it came from a chimney or a burning house, but when we got near we saw it was the steam rising from hot springs. We were soon gazing at a scene of great natural beauty. A number of springs bubbled out of the ground, and in the middle of the cloud of steam shot up a splendid little geyser fifteen feet high. After poetry, prose! We all naturally thought of a bath."

So wrote Heinrich Harrer in his book, *Seven Years in Tibet*, as he neared the end of his journey to Lhasa. Unfortunately, there is no way to positively identify his geyser, but the best candidate area is Yangpachen. At least eight separate geyser fields have been described in Tibet and one in Qinghai, and it is entirely possible that geysers occur in a number of other geothermal areas as well.

One of the difficulties in understanding these places is that the early descriptions used Tibetan language for geographic names (in which "chuja" refers to hot springs), whereas most modern references use Chinese names. However, the Tibet Map Institute has published a series of relief maps (some as recently as 2002) that use the classical Tibetan names. When these are compared with Chinese geographic coordinates, some confusion has been eliminated.

The first geysers described in Tibet were identified by Montgomerie in 1871–1872 in remote localities along the Lahu Chu River. At an elevation of nearly 17,000 feet, this remains a very remote place served only by a primitive road. The larger of the two geyser basins is Peting Chuja, where more than a dozen geysers were described among 100 other hot springs atop a geyserite terrace. In 1872 and when seen again in 1912, several of these geysers reached 40 to 50 feet high, and additional spouts rose from the bed of the river. About 15 miles farther up the river is Naisum Chuja. It reportedly included just 2 geysers, but both were said to spout 50 to 60 feet high. It is probable that one of these places is the same as the modern Chinese "Bibiling," the site of 3 perpetual spouters, the largest of which erupts at an angle as high as 60 feet.

Dagyel Chuja (Chinese "Dagajia") is probably the most significant geyser field in Tibet. It is near the northern route of the east-west "highway" through Tibet, near the headwaters of the Lagetzangbo River at an altitude of almost 16,000 feet. The Tibetan name is translated as "King Tiger Hot Springs." A 2000 geothermal summary claimed the existence of about 100 geysers, 4 of significant size. The 2 largest erupted to heights of 130 to 165 feet. The interval of one was about 3 hours. Another was cyclic, undergoing series of several 10-minute eruptions separated by quiet of at least 36 hours; its eruption was jetted at a 45-degree angle. (It should be noted that no other report about Dagyel Chuja cites more than a few hot springs in total or any eruption higher than 20 or 30 feet.)

Another geyser area is Chabu, also known as "Capu." Geysers such as Quzun and Semi are reported to have increased the frequency of their eruptions by up to five times following an earthquake in 1959 (not the same as Yellowstone's earthquake the same year, of course). One of them is highly regular in its action, erupting every 7 minutes for about 30 seconds and reaching as high as 25 feet.

The hot springs at Buxiunglanggu ("Gudui") consist mostly of small pools, but eruptions as high as 3 feet have been described. The main attraction at Kurme ("Gurma") is a series of explosion craters, several of which contain small geysers or perpetual spouters. Nearby is Kau ("Kew"), where a single geyser plays a few feet high. With no details beyond a tabulated listing of geothermal areas, two geysers are reported at a place called Gulu (or Guhu).

In the far west of Tibet, in a region sometimes closed to western explorers, Qupu has recently experienced large hydrothermal explosions. In 1975 the creation of one of these craters threw rocks as far as a mile away. Craters with diameters as great as 250 feet are filled with boiling pools and are surrounded by "spouting vents," but whether there are actual geysers at Qupu is uncertain.

The closest large geothermal field to the city of Lhasa is Yangpachen ("Yangbajing"). To the native people of Tibet, hot springs were considered a gift from Buddha, sacred and inviolate. The Chinese feel differently. Yangpachen is the site of a geothermal plant that provides electrical power to Lhasa. Since 2000, the power plant has suffered a decline in production as a result of severe mineral deposition within the drilled wells. It is likely that the natural springs have been altered by the developments, but the China Ministry of Culture claims "fountains, geysers, [and] boiling springs" are still active. A tourist area charges an admission fee of about US$2.50 (20 Chinese yuan), which allows the use of concrete-lined bathing pools at the Geothermal Spring Holiday Resort that offers food and lodging.

Geysers have been reported at Fenghuoshan ("Wind-fire Mountain"), in Qinghai north of Tibet. Near the new railroad line between Golmud and Lhasa, it is at an altitude of over 16,000 feet. Satellite photos indicate an extensive thermal area, but nothing is known about the activity.

In 2002, a Chinese geological survey of the remote mountain region of far northern Tibet reportedly located a geyser field to rival Yellowstone. That doubtful claim was supported by no details.

Worth mention is that a Chinese book published in 1982 expressly states that "geyser eruptions must reach 3 meters [almost 10 feet] in height to be counted." This is a strange restriction not followed elsewhere in the world. Since boiling springs are described in many places other than those cited as containing geysers—indeed, over 300 individual geothermal districts have been mapped in Tibet—one wonders how many geysers less than 10 feet high might have been ignored by Chinese researchers.

33. SICHUAN AND YUNNAN, PEOPLE'S REPUBLIC OF CHINA. Not far outside the eastern boundary of Tibet in Sichuan (Szechwan) and extending south through Yunnan toward Burma and Thailand is a zone of high geothermal heat flow and numerous hot spring basins. Near Chaluo, in Sichuan, are several hot spring groups that include at least four geysers in addition to numerous fumaroles, boiling springs, and pools. The largest geyser is cyclic in its activity—a series of minor eruptions is punctuated about every 3 hours by major play that reaches as high as 15 feet for 15 to 20 minutes.

Rehai ("Hot Sea"), in Yunnan, is 6 miles from the city of Tengchong and has been declared a "National Scenic Geological Park." Admission is 20 Chinese yuan (about US$2.50), or free for those who stay in the adjacent Rehai Hotel. As many as "19 large geysers and numerous smaller ones" at Rehai have been described. Most of them are probably perpetual spouters; however, a recent geological report described Dagunguo ("Big Boiler") as erupting frequently but intermittently, "with bounding streams and a thunderous roar." Another is called Snatching Birds Pool because its jets of water are said to kill birds in mid-flight. Other springs in the park bear names that translate as Small Boiler, Pearl Spring, Toad's Throat, Drumbeat Spring, and even Pregnancy-inducing Pool. Perched on a mountainside 5 miles southwest of Rehai is Reshuitang, where at least two pools undergo intermittent episodes of boiling.

Farther south in Yunnan, next to the major highway to Burma and touted as a tourist attraction, is a pool called Balazhang Hot Spring. It began frequent eruptions following a 1976 earthquake, with play that repeats every few minutes and reaches 4 to 8 feet high.

A Chinese source states that there are 86 significant hot spring localities in Sichuan and 179 in Yunnan. Many of these geothermal areas have springs with near-boiling temperatures, so geysers might well occur in some of the geothermal areas other than Chaluo, Rehai, and Balazhang.

34. BURMA (MYANMAR). Northeastern Burma lies along the same geological province that includes the geysers of Sichuan and Yunnan, China (area 33, above), and the northern parts of Thailand (area 35, below), so it stands to reason that it includes geysers among the hot springs. However, the only locality described seems questionable. Near the small city of Lashio in Fang State is a small park where "among the sites to see is the geyser." A publicity photo disseminated on-line in 2004 showed a boiling spring surrounded by a concrete wall.

A geyser at Pai, in the "Tavoy District of Burma," was described in 1864. Its eruption reached as high as 6 feet. It is believed that this actually refers to the town of Pai in northern Thailand, where there are numerous boiling springs, or to Thailand's Pa Pai Hot Springs, which definitely boasts small geysers.

35. CHIANG MAI AND MAE HONG SON PROVINCES, NORTHERN THAILAND. Thailand is a well-explored place, yet the fact that there are geysers in northern Chiang Mai and Mae Hong Son was only revealed by geothermal studies that date to the late 1970s. At least nineteen geysers in five localities have been described.

The Pong Hom Hot Springs at San Kamphaeng are only a few miles from the city of Chiang Mai. Several geysers with eruptions up to 20 feet high resemble the steady jets that often spout from drilled wells, but local authorities state that no drilling has taken place there.

A few miles from Fang, near the border with Burma and not far from Laos, are the hot springs of Ban Muang Chom. A small geothermal power plant has been operating there for several years, but somehow the nearby geysers had survived as of 1991. One, next to the access road, had intervals of about 10 minutes, playing up to 6 feet high for a few minutes. A number of smaller geysers existed nearby. The more recent status of these geysers is unknown.

Pa Pai is a tiny cluster of only seven hot springs, but four are geysers. All have intervals of 15 to 45 seconds, durations of 5 to 10 seconds, and reach 2 feet high. There is one small geyser at the Thepanom Hot Springs, and near the campground at the Pong Duad Hot Springs Park are bubble-shower geysers that boil up as high as 3 feet.

In Mae Hong Son Province, a geyser splashes several feet high at Muang Paeng.

36. SUMATRA, INDONESIA. Sumatra, the largest and northwestern-most island of the Indonesian Archipelago, is intensely volcanic, and geysers have been reported in four localities. In only two, however—Waipanas and Tapanuli—are true geysers positively known to exist. Taken from southeast to northwest, these fields include the following.

A caldera within the Lampung-Semangko District at the far southeastern end of Sumatra used to contain geysers in a number of places, including one basin that reportedly boasted a dozen or more geysers of substantial size. A few decades ago the entire region was disrupted by a series of powerful phreatic (steam-powered) explosions, and most of the geysers were destroyed. Such activity is now restricted to two small hot spring groups. At Waipanas is a single geyser, which in late 1971 had intervals of 10 seconds, durations of 5 seconds, and heights up to 12 feet. Nearby Waimuli might have two geysers, both of which play about 1 foot high.

Kerinci is Sumatra's most active volcano, centerpiece of a large national park established for the preservation of the Sumatran tiger and rhinoceros. The Geyser Gao Gadang is probably the largest in Indonesia—if it is a true geyser. The activity reportedly is dependent on atmospheric conditions, the eruptions occurring only when the air temperature is exceptionally high and/or the barometric pressure very low. Then it may erupt frequently (or perhaps steadily) more than 70 feet high. Such eruptions are rare. Under other conditions this spring acts as a perpetual spouter only 3 feet high.

Among the small cluster of Pasaman Hot Springs near the town of Panti is one possible geyser, playing 5 feet high.

Within an area of a few square miles, geysers apparently occur at three hot spring groups and a large perpetual spouter in another, places collectively known as Tapanuli. Near the village of Sibanggor Jae is a single geyser, erupting 8 feet high. Tarutung has a single geyser that reaches perhaps 10 feet. The most extensive of these groups is at Silangkitang, where a geyser 8 feet high plays from a large pool and several smaller geysers may exist. Finally, a muddy perpetual spouter at Sipirok jets a steady stream of acid water to about 60 feet.

37. CISOLOK, JAVA, INDONESIA. Cisolok is on the southwestern coast of Java. Most of the hot springs rise from the bed of the Cipanas River, and only those whose deposits have grown above the water level erupt. Two geysers have been described. One reaches 12 to 17 feet high, while the other plays to just 1 foot.

38. GUNUNG PAPANDAYAN, JAVA, INDONESIA. Papandayan is one of Indonesia's most active volcanoes. Hot springs are common on and near the slopes of the mountain. They certainly include boiling pools and perpetual spouters, and a travel guidebook noted "small geysers" as one attraction in the area.

39. MINAHASA DISTRICT, CELEBES (SULAWESI), INDONESIA. The northeastern arm of Celebes Island contains the volcanic area of Minahasa, near the main city of Manado. Although hot springs are extremely common throughout the district, geysers are known to exist only near the village of Toraget, where a deep, clear pool erupts every 3 to 5 minutes. The play lasts only a few seconds and reaches 1 to 2 feet high. Other geysers have been recorded in this area, but most of their activity has been restricted to short periods of time following earthquakes.

40. BACAN ISLAND, MALUKU GROUP, INDONESIA. The hot springs on Bacan Island (also spelled Bactian, Batjan, and Bacjan) are of boiling temperatures, and several were apparently active as geysers during the 1800s. The largest was named Atoe Ri. No modern geothermal survey has been conducted on the island, and the present status of these springs in a rebel-held area is unknown.

PACIFIC OCEAN RIM

41. GARUA HARBOUR, NEW BRITAIN ISLAND, PAPUA–NEW GUINEA. New Britain is the large island immediately east of New Guinea. Garua Harbour, about midway along the north coast, is probably a caldera, and hot springs occur all around its shore. Near the village of Pangalu north of the bay is an extensive thermal area in Wabua ("Hot Water") Valley that includes spouting springs in four of its eleven hot spring groups. The best known of these is Rabili, where there are two geysers. Both come and go with time, sometimes being completely inactive and at others acting as perpetual spouters but most commonly erupting as periodic geysers. The larger of these can play as high as 25 feet; the other typically reaches 6 feet. Hudi has boiling springs within sinter-lined craters; Vavua has been known as the site of "periodic eruptions in muddy pools"; and Narera includes "clear pools that throw jets of water into the air," one of which has erupted as high as 10 meters (32 feet).

Across the harbor is the town of Talasea. Scattered throughout the settlement are boiling springs and extensive geyserite deposits as well as vigorous mud pots. Actual geysers have never been described, but their existence is possible.

42. KOIMUMU (KASILOLI), NEW BRITAIN ISLAND, PAPUA–NEW GUINEA. The Koimumu thermal area (historically also called Kasiloli, Kasoli, Livigi, Namagura, and Lotatolo) is about 50 miles southeast of Garua Harbour and a few miles inland from Cape

Hoskins. The active area is small, measuring less than 1,000 feet in any dimension. Specifics about the geyser activity were published in 1956, when Koimumu contained at least fourteen geysers, enough to rank it among the largest geyser fields in the world. Tabé Geyser was cyclic in its action, with intervals that ranged from 15 minutes to several hours in length; the duration was 3 minutes and the height in excess of 30 feet. The other geysers, none of which was named, all played between 1 and 4 feet high. A restricted area because of the presence of megapode birds (whose eggs are incubated when buried in the warm ground), guided visits are available only to organized tours. The current status of Koimumu is uncertain, since a violent eruption of nearby Mount Pago volcano in 2002 blanketed the region, including the hot springs, in a deep layer of volcanic ash.

43. NARAGE (GAROVE) ISLAND, WITU GROUP, PAPUA–NEW GUINEA. Narage Island is a speck of dormant volcanism in the tiny Witu Island Group, about 50 miles north of New Britain. Hot springs occur at a number of places around the island, and among them is one geyser. A report from the 1880s talks of eruptions as high as 30 feet. The geyser was much weaker in 1970 when it exhibited intervals of 2 to 3 minutes, durations of 20 to 30 seconds, and a height of around 3 feet.

44. LIHIR AND AMBITLE ISLANDS, NEAR NEW IRELAND ISLAND, PAPUA–NEW GUINEA. These islands lie off the east coast of New Ireland, forming the easternmost lands of the sprawling nation of Papua–New Guinea.

Lihir (or Lir) Island is small, measuring about 5 by 3½ miles. There are—or were—several hot spring groups, and two of them may have contained geysers. However, it is likely that none of the natural springs exists now. The near-surface geothermal system contains one of the world's largest gold deposits (containing an estimated 1,000 tons, or 24 million troy ounces, of gold). It is being mined, and much of the island has been devastated.

Ambitle (or Anir, or Feni) Island is larger and contains more hot springs than Lihir did, and geysers positively exist here. At a low elevation, on the grounds of the Waramung Plantation, are numerous hot springs. Several show intermittent activity, and in 1996 one was clearly a geyser, erupting irregularly but frequently to about 1½ feet high. Inland 2 miles is another thermal area, called Kapkai. It contains several very hot, geyserite-lined pools. One of these pools is a large but infrequent geyser that has been seen to erupt 30 feet high. Elsewhere in this group are as many as five other geysers plus a 10-foot perpetual spouter. Unfortunately, a gold deposit similar to that on Lihir has been discovered on Ambitle, and the geyserite at Kapkai contains as much as 1 ounce per ton—high-grade ore, to be sure, but fortunately limited in volume. Mining had not taken place as of 2005.

45. FERGUSSON ISLAND, D'ENTRECASTEAUX ARCHIPELAGO, PAPUA–NEW GUINEA. Fergusson is the largest of the d'Entrecasteaux Islands, located near the southeastern tip of New Guinea. The thermal areas of interest, Deidei and Iamelele, are just a few miles apart. At Deidei there are several geysers. Most of the eruptions are small, but in 1954 one geyser frequently reached as high as 15 feet, while another played less often 5 to 10 feet high. Since 2000, Deidei Village has been a regular stop for regional tourist cruises to allow visitors to "watch boiling water spout up to 20 meters [65 feet]" after hiking through a forest "filled with birds of paradise."

Iamelele covers a larger area than Deidei, but geyser activity there (if any) is much less intense. Small eruptions from just two or three of the springs were reported in 1956, and they were probably perpetual spouters. The most notable feature in 1956 was a steadily boiling pool called Italautaliagu, whose action reached up to 3 feet high.

46. SOLOMON ISLANDS. Hot springs are known to exist on most of the volcanic Solomon Islands, but true geysers probably exist in only one thermal area. The extensive Paraso Thermal Area lies along the Ulo River on western Vella Lavella Island. Although most of those springs are mud pots, the existence of small geysers in a limited part of the field was confirmed by a Solomon Islands volcanologist in early 2006. Geysers have long been reported on Savo Island, but the evidence is that most, if not all, of those springs are fumaroles that occasionally are drowned by surface water. The same is true for the geysers reported on Ranongga Island.

47. VANUATU. The islands of Vanuatu (formerly known as the New Hebrides Islands) lie along the same volcanic arc just south of the Solomon Islands. The existence of true geysers on two islands in the Banks Group was confirmed in January 2006 by the Vanuatu

Department of Geology, Mines and Water, although no details were provided. These are the "erupting thermal geysers" at Sladen Boiling Springs on Gaua Island and unnamed hot springs at Mt. Sereama on Vanua Lava Island reported as "steaming pools and geysers." The geysers reported on several other islands of Vanuatu are probably drowned fumaroles and perpetual spouters.

48. NAKAMA SPRINGS, SAVUSAVU, VANUA LEVU, FIJI. The Nakama Springs are actually within the town limits of Savusavu, adjacent to the school grounds and near the Hot Springs Hotel. The thermal tract is tiny, measuring only about 100 by 60 feet. The first time distinct eruptions were recorded was during 1878, when two of the springs began to play and attracted considerable attention by jetting columns of water up to 60 feet high; the intervals were as short as 10 minutes, and the durations were about as long. No further eruptions were recorded until the 1950s. Since then, only small geysers have been seen, and it is not known if any of the modern features play from the same craters as the geysers of 1878 did. In 1993 the Nakama Springs included five small geysers, none of which played higher than 2 feet.

49. NEW ZEALAND. The thermal areas in the central part of New Zealand's North Island used to comprise the second largest geyser field in the world, trailing only Yellowstone. As many as 300 or more geysers have been known within the Rotorua-Taupo Volcanic Zone. Because of recent geothermal and hydroelectric developments, three-quarters of those geysers have disappeared.

The best-known of the geyser basins is at Whakarewarewa, just outside the city of Rotorua. The area is operated by native Polynesians as part of the Maori Arts and Crafts Institute. A number of these people live in the area, using the hot water for bathing, cooking, space heating, and so on. Although there is no geothermal development actually within "Whaka," extensive drilling in the city has had a serious impact on the geysers. About a dozen geysers are active. By far the largest are Pohutu (Big Splash) and the adjacent Prince of Wales Feathers geysers. For a brief time after Pohutu's eruption, this complex is quiet, but in short order Prince of Wales Feathers begins jetting an angled, steady column. This gradually grows in strength, and about the time it reaches 30 feet high, Pohutu joins in. At full force, Pohutu can reach up to 100 feet high, although 60 to 70 feet is more typical. Nearby Waikorohihi

Pohutu Geyser, joined by the arching play of Prince of Wales Feathers Geyser, is now the largest of New Zealand's active geysers, with some play reaching over 100 feet high at intervals as short as a few minutes.

and Mahanga geysers both show cyclic eruption patterns; when active, they can be frequent and up to 20 feet high, but both have been dormant since 2003. Kereru Geyser is very erratic, often splashing but rather uncommonly bursting up to 10 feet. Near the Maori village, several large pools have rare bursting eruptions, and one small geyser plus several perpetual spouters erupt in an area away from the tourist trails. In addition to the geysers, Whaka boasts several large mud pots and a few pools. Directly across the highway, the Arikikapakapa area has mud pots and pools as unique golf course hazards.

The largest concentration of geysers presently active in New Zealand is at comparatively little-known Orakei Korako. Until 1961, well over 100 geysers and intermittent springs were known, 20 of which were of major proportions. Orakeikorako Geyser would sometimes erupt to 180 feet. When it was dormant, nearby Minguini would play, infrequently reaching a measured 295 feet high! Rameka played to 30 feet, Terata to 50, Porangi to 80, and Te Mimi-a-homai-te-rangi to 75. Most of these geysers are gone now. In 1961 they were covered by the water of Lake Ohakuri. Government geologists have publically called the inundation "New Zealand's worst case of environmental vandalism." However, above the lake level about 35 small geysers remain. The largest of the existing geysers are Diamond, which reaches 8 to 25 feet; Cascade, which has strong bursts within a cavern; and Sapphire, which is frequent and as much as 10 feet high. Orakei Korako is operated as a tourist attraction, and shuttle boats cross the lake from a gift shop and café.

The country around Lake Rotomahana was once the site of two of the largest geyserite formations in the world. The Pink and White Terraces were over 50 feet high and hundreds of feet broad, but they were completely blown away by the explosive Mt. Tarawera–Rotomahana-Waimangu volcanic eruptions of 1886. In the aftermath of those destructive blasts, new and equally fascinating thermal features were formed. Star of them all—star of all the world's known geysers—was the great Waimangu. Beginning in 1900 and lasting four years, Waimangu Geyser (Black Water) played like no other geyser has ever done. Eruptions were frequent and predictable. During some stretches of time, it was actually in eruption nearly as often as it was quiet. All eruptions were several hundred feet high. Jets reaching 600 to 1,000 feet were common, and some approached 1,500 feet high! Waimangu's death in 1904 came not because of changes to its plumbing system but because of groundwater changes as a result of nearby landslides. The dominant features at Waimangu now are Frying Pan Lake, around whose shores are several small geysers, and Inferno Crater Lake, which varies its water level by more than 30 feet over regular cycles of about 38 days.

A few geysers also exist at Waiotapu (5 geysers), Waikite (2), Te Kopia (1), and Tokaanu (5), but they are all quite small and largely secondary to other geothermal attractions, such as mud pots and some of the largest geothermal pools in the world.

Once upon a time, Wairakei would have headed New Zealand's list. Its Geyser Valley was the site of at least 100 geysers, some of them large and spectacular. New Zealand's first geothermal power plant spelled their end and also destroyed the geysers at The Spa, a few miles away near Taupo. A similar demise befell the single geyser at Ohaaki.

The distance between Rotorua and Taupo is about 50 miles. Within this zone there are still about 70 geysers, enough to rank New Zealand high on the list of world geyser fields. But so much has been destroyed. New Zealand is a nation of limited energy resources such as coal and oil, so there is pressure for additional geothermal

New Zealand's Waimangu Geyser ("Black Water") was the tallest geyser ever observed anywhere. Active between 1900 and 1904, some of its most powerful bursts reached 1,500 feet high.

development. However, some geologists and private citizens think there must be a better way, and the situation is improving. Legal action has limited geothermal production in Rotorua, and as a result the water levels and geyser activity are slowly recovering at Whakarewarewa. The Waimangu Valley–Lake Rotomahana area has been designated as a research area and preserve. There has even been a proposal to remove the dam and drain Lake Ohakuri at Orakei Korako (although this is deemed unlikely to happen for several years). However, no new geothermal developments are planned near any of the other geyser basins, so New Zealand stands to remain one of the major geyser fields on Earth.

OTHER POSSIBLE GEYSER FIELDS

The forty-nine areas just described are those positively known to contain geysers at this time or in the recent past. In addition, a number of other localities around the world have been known or reported to contain geysers. Several of the following used to contain geysers, but for one reason or another they disappeared many years ago. For others, the answer as to whether geysers are actually present will have to wait for future exploration. These places are described briefly.

THE GEYSERS, SONOMA COUNTY, CALIFORNIA, U.S.A. In spite of the name, most professional opinion has it that The Geysers never included geysers—it was named during California's early history because of the billowing steam clouds that rose from the springs in a deep valley. However, a relatively specific series of short articles and reply letters that appeared in the *San Francisco Chronicle* newspaper in 1851 seems to describe true geysers, complete with pools within craters lined with geyserite. These geysers had names like Agassiz's Maelstrom, Silliman's Fountain, Pluto's Cauldron, and Panther Geyser, and one of the reports cited eruptions 20 to 30 feet high. In any case, the development of what is presently the largest complex of geothermal power plants in the world has long since destroyed any such springs.

AMEDEE HOT SPRINGS, LASSEN COUNTY, CALIFORNIA, U.S.A. The balance of evidence shows that the Amedee area would never have contained geysers, as the appropriate water chemistry and geyserite deposits are lacking. However, the railroad town of Amedee (now long gone) once had a newspaper, *The Amedee Geyser*, and a clock at the hotel supposedly moved forward 38 seconds with each eruption of "the geyser." There is now a geothermal power plant at Amedee.

BRADY'S HOT SPRINGS, CHURCHILL COUNTY, NEVADA, U.S.A. A geyser at Brady's was described in 1848 by a surviving member of the ill-fated Donner Party, who complained about the water being hot and unpalatable. In the 1920s, eruptions 4 to 6 feet high were seen. No geyser has been reported at Brady's since then, but mud pots and steam vents continued to exist as late as 1988.

HELL'S KITCHEN, SALTON SEA, IMPERIAL COUNTY, CALIFORNIA, U.S.A. The hot springs at Hell's Kitchen, near the volcanic rhyolite dome of Mullett Island, were described as early as 1853. They included a few small geysers that survived until about 1950, when they were inundated by the Salton Sea's rising water level.

CERRO PRIETO, BAJA CALIFORNIA NORTE, MEXICO. Following major earthquakes in 1852, 1915, and 1934, large but temporary geysers erupted among the hot springs at Cerro Prieto, a few miles south of Mexicali. Some of the eruptions approached 200 feet high, but none of the geysers persisted for longer than a few weeks. Cerro Prieto is now the site of one of the world's largest geothermal power plants, but one 35-foot geyser, questionably said to be natural, was active in 2001.

ARARO, MICHOACAN, MEXICO. In 1952 this area near the eastern shore of Lake Cuitzeo north of Morelia was reported to contain clear intermittent springs and one geyser among its sinter-lined pools and mud pots. A 1981 investigation was unable to find such springs.

QUETZALTENANGO DISTRICT, GUATEMALA. Near the bathing resorts of Aguas Georginas and Aguas Amargas are said to be "fumaroles and geysers for the traveler to see," and another thermal area southeast of Zunil was reported to contain "geysers and fumaroles." These eruptions probably rose from drowned fumaroles. Zunil is now the site of a geothermal power plant.

VOLCAN PURACE NATIONAL PARK, COLOMBIA. The hot springs at Termales de San Juan have been called geysers, and there are eruptions a few feet high. However, the water is sulfurous and strongly acid, several degrees cooler than boiling, and "never ending." These features are probably gassy perpetual spouters.

PINCHOLLO, VALLE DEL COLCA, DEPARTMENTO AREQUIPA, PERU. At least one geyser used to exist on the flank of Volcan Sabancaya, at an altitude of 13,200 feet and "a four-hour hike" above the village of Pinchollo. It had eruptions as high as 100 feet until a 1999 earthquake "contained it." What that phrase means is unclear, but the fact that the geyser is still active in some fashion is implied by its listing in 2002 as an attraction within the proposed Valle del Colca Natural Monument.

ULUCAN (OR UBINAS), DEPARTMENTO MOQUEGUA, PERU. Near the village of Omate about 35 miles northwest of the geysers at Calacoa (see Area #12), about twenty springs described as geysers are scattered along a sinter terrace adjacent to Rio Vagabundo. The largest spring of the group, known as "Los Meaderos," is said to be intermittent in its overflow, but it apparently does not actually erupt. Five other vents spout between 1 and 4 feet high, but they are probably gas-driven perpetual spouters since their water temperature is only 160°F (70°C).

QUIGUATA, IQUIQUE PROVINCE, TARAPACA REGION, CHILE. A reference published in 1967 included Quiguata as a geyser locality in a table that also listed Suriri, Puchuldiza, and El Tatio. It is not mentioned in any other known geological report, and a Chilean geologist denied that there is a thermal area by that name. However, an Internet gazetter of Chilean place names locates Quiguata about 35 miles southeast of Puchuldiza and notes that its "geysers are curious and worthy of study."

BOLIVIA. Geysers have been reported at numerous localities in the Andes Mountains of Bolivia, but the majority of reports refer to places where the water temperatures in the springs are well below boiling. Near the village of Sajama are hot springs that erupt as high as 3 feet within sinter-lined craters, but in 2005 a German geophysicist stated that these springs are "constant" in their boiling.

Sol de Mañana includes numerous mud pots, perpetual spouters, and perhaps some small geysers. Termales Rio Quetana lies deep in a canyon where a geologist's report noted "steam vents and small geysers on a sintered alluvial terrace above the river." At Towa, on the slopes of a volcano with the same name, are said to be "hot pools and geysers among small deposits of siliceous sinter."

BANOS VILLAVINCENCIO, MENDOZA PROVINCE, ARGENTINA. A scientific paper that emphasized the biological resources of endangered environments in Argentina was published under the auspices of the World Wildlife Fund in 2001. It briefly noted the geysers at Volcan Domuyo and also stated that geysers exist at Banos Villavincencio. However, other sources state that the water at Villavincencio is only 104°F (40°C), a temperature far too low to support geyser activity.

COPAHUE AND CAVIAHUE, NEUQUÉN TERRITORY, ARGENTINA. Volcan Copahue is in Chile, but the resort towns of Copahue and Caviahue on the northeastern and eastern flanks of the mountain, respectively, are in Argentina. Copahue is a hot springs resort where the main attraction is bathing in thermal swimming pools at hotels or in natural springs such as Lago Verde. About 2½ miles south of town are the hot springs at Las Maquinas ("The Machines") and Las Maquinitas. Both include fumaroles as hot as 203°F (95°C) and boiling pools. An on-line photograph showed a jet of water, so although never described as such, geysers or perpetual spouters may exist at Copahue. About 11 miles south of Copahue is the smaller resort town of Caviahue, where there definitely are a few small spouters and possible geysers. Between the two towns is El Anfiteatro ("The Amphitheater"), where spouting springs are probably drowned fumaroles rather than geysers.

HELLS GATE, KENYA. Hells Gate, also known as Njorowa Gorge, is a long canyon traversed only by a trail south of Lake Naivasha. It certainly contains hot springs that have been referred to as geysers, with water jets several feet high. In reality, however, they are apparently fumaroles that occasionally spray a little liquid water. The Olkaria Geothermal Power Station is nearby.

KERGUELEN ISLAND, SOUTH INDIAN OCEAN. Kerguelen, a large but virtually unknown island, is a French territory in the sub-Antarctic.

Several groups of hot springs occur in the southern part of the island and are said to include "bubbling, splashing pools."

SHIKABE, HOKKAIDO ISLAND, JAPAN. A geyser has long been reported to erupt about 6 feet high at Shikabe, but it probably does so from a drilled well.

BAISHUI TERRACE, ZHONGDIAN, YUNNAN, CHINA. The primary attraction near the town of Zhongdian, Yunnan, is the Baishui ("White Water") Terrace. Tourist literature promotes it as having been formed by active "geysers and hot springs," but the fact that the terrace is composed of calcium carbonate (travertine) implies that true geysers probably do not occur there.

TONGONAN, LEYTE ISLAND, PHILIPPINES. It was only in 1997 that a geological publication reported several geysers at Tongonan, but in that same year the Philippines' largest geothermal power plant began production there. A follow-up report in 2001 stated that the geysers had "ceased to operate."

MAGEKABO, FLORES ISLAND, INDONESIA. Geysers have been rumored to exist in a large thermal area at Magekabo, in the Ende District of western Flores. Numbers cited were "several" and heights "considerable." However, according to the director of a geothermal survey, the area is entirely acid, containing steam vents and mud pots but no geysers or spouters.

BUKAPETING, ALOR ISLAND, INDONESIA. Small geysers were reported during geothermal surveys in the late 1980s, and a follow-up publication in 1998 referred in passing to "spouting springs."

AIRMADIDI, MINAHASA DISTRICT, CELEBES (SULAWESI), INDONESIA. Most reports about the Ranopaso ("Boiling Water") Hot Springs at Airmadidi implied that the springs were all acid and muddy. Two small geysers were reported, but they likely were destroyed by the 1993 construction of a geothermal power plant.

ALA RIVER, NEW BRITAIN ISLAND, PAPUA–NEW GUINEA. Along the Ala River deep in the interior of western New Britain is a thermal area that definitely does include spouting springs, but their temperatures are cooler than boiling, so they are probably gassy perpetual spouters rather than true geysers.

Finally, the next localities have been reported as containing geysers, but they definitely do not now and probably never have.

COSO HOT SPRINGS, CALIFORNIA, U.S.A. High-temperature springs with siliceous sinter deposits plus some large mud pots exist, but no geysers were ever observed at Coso.

AHUACHAPAN, EL SALVADOR. An acid area of mud pots and fumaroles, sometimes called "Hell's Half Acre," has been related to Yellowstone by several travel guides, but no geysers have ever existed.

VOLCAN ALCEDO, ISLA ISABELA, GALAPAGOS ISLANDS, ECUADOR. Several published references state that geysers occur within this caldera, but they are actually fumaroles that occasionally become drowned by surface water.

LARDARELLO, ITALY. Lardarello is the site of the world's oldest geothermal power plant, in operation since 1905. There used to be boiling springs at Lardarello, but true geysers were never observed. A geyser described in recent travel literature is only a weak steam vent.

ZAMBIA AND ZIMBABWE, AFRICA. Several spots in each of these countries have been reported to contain geysers. Two in Zimbabwe, officially named Chimanimani Geyser and Zongola Geyser, are steadily spouting artesian springs with hot but non-boiling water. At the site of the Chilambwa Geyser, in Zambia, there is no thermal activity whatsoever; it was probably a cold artesian spring that stopped spouting decades ago.

MASY RIVER, MADAGASCAR. Lukewarm springs along the Masy River in central Madagascar undergo intermittent eruptions as high as 10 feet. These are soda pop geysers, however, with eruptions powered by carbon dioxide gas.

ARTIFICIAL "GEYSERS" IN THE WESTERN UNITED STATES. Highway maps often indicate geysers here and there around the American

West. They are all artificial, drilled wells that now undergo natural, sometimes spectacular, eruptions. Among them are the "Old Faithful of California" at Calistoga; "Old Perpetual" near Lakeview, Oregon; and "Crump Geyser" near Adel, Oregon. The "Captive Geyser" at Soda Springs, Idaho, is controlled so as to erupt warm, gas-charged water as high as 100 feet every hour. The Crystal, Woodside, Tumbleweed, Ten Mile, and Champagne geysers near Green River, Utah, are abandoned wildcat oil wells that act as soda pop, or "cold water," geysers. They play cold water, Crystal to well over 100 feet, because of the evolution of dissolved carbon dioxide gas—the effect is similar to vigorously shaking a can of soda pop before popping it open. A "geyser" near Afton, Wyoming, erupts several feet high because of a natural siphoning action along a stream that briefly flows underground.

"Old Perpetual" is a drilled well that now undergoes natural eruptions in front of a motel near Lakeview, Oregon. Credit, Photo by Alan Glennon

Algae. Colonial, single-celled plants; in geothermal areas, true algae are common in cold stream and lake water but not in geothermal water unless it is acid and/or thoroughly cooled. See also *cyanobacteria.*

Bubble-shower spring. A geyser whose eruption consists of intermittent episodes of violent boiling as *superheated* water surges to the surface of a *pool.* Considered by some to be a variety of intermittent spring and not a geyser.

Burst. Applicable to fountain-type geysers, a burst may be a single throw of water; or, in the case of geysers having a series of short, closely spaced eruptions, a burst may be one of the periods of spouting.

Calorie. The amount of heat energy necessary to raise the temperature of 1 gram of water by 1° Celsius; by conversion, it takes about

250 calories to raise the temperature of 1 pound of water by 1° Fahrenheit.

Complex. A cluster of springs or geysers so intimately associated that the activity of any one member will affect that of the others. Compare with *group.*

Concerted eruptions. Simultaneous eruptions by two or more geysers within a *complex* where there are direct connections between their *plumbing systems,* so the eruption of one geyser triggers the eruption of the other. Compare with *dual eruptions.*

Cone. With respect to geysers, a built-up formation of geyserite on the Earth's surface, within which is the geyser's *vent.*

Cone-type geyser. A geyser whose eruption is jetted as a steady column of mixed water and steam from a small vent with little or no surface pool. The vent is often but not necessarily at the top of a built-up *cone* of geyserite.

Crater. May be synonymous with *vent;* more often, the crater is either a broad, shallow depression within which is centered a comparatively small vent or a wide, deep hole that contains a pool.

Cyanobacteria. A variety of primitive, *thermophilic* chlorophyll-bearing bacteria that survives in alkaline geothermal water at temperatures below 167°F (75°C). Previously known as "blue-green algae" but distinct from true algae in that a cyanobacterium has no nucleus in its cell. See also *algae.*

Cycle. The time span from the start of one active phase of a geyser, through the following dormancy, to the start of the next active phase. Sometimes incorrectly used as a synonym for *period.*

Dead. A geyser or hot spring that will never again undergo active cycles because of a permanent loss of its supply of water. This term must be used with extreme caution, as most seemingly dead features are only in a *dormancy.*

Diurnal variations. Changes in geyser or other hot spring activity because of the physical differences between daytime and nighttime.

Dormancy (or Dormant). A span of time, ranging from days to years, during which a geyser temporarily ceases to erupt, usually because of *exchange of function.* Many geysers will be active for long periods of time and then go dormant. Note that dormant does not mean *dead.*

Dual eruptions. Eruptions by two or more geysers that occur simultaneously only by chance and not because of any physical relationship between them. Compare with *concerted eruptions*.

Duration. How long a geyser eruption lasts; that is, the time from the beginning of an eruption to the end of the same eruption.

Eruption. The spouting action of a geyser.

Exchange of function. The subterranean shift of energy and/or water from one geyser or hot spring group to another, resulting in a decline of activity in the first and an increase in the other.

Extinct. Inactive and without any possible activity in the future. Synonymous with *dead* and often incorrectly used as a synonym for *dormancy*.

Fountain-type geyser. A geyser whose eruption is a series of separate explosive bursts of water, usually issuing from a pool whose *vent* is within a *crater*.

Frequent. A general term for eruptions that are irregularly spaced in time yet occur often, usually separated by only a few minutes.

Fumarole. A steam vent; that is, a hot spring in which all available water is converted to steam at depth before reaching the surface.

Geothermal. The term applied to any geological system or process that relies on the Earth's internal heat as the source of energy.

Geyser. A hot spring that erupts because of the boiling of water at depth within the confining space of a plumbing system, which forcibly ejects water out of the vent in an intermittent fashion.

Geyser basin. A portion of a *geyser field* within which groups of hot springs including geysers are found.

Geyser field. A geographical region that contains one or more *geyser basins*. Yellowstone National Park, for example, is one geyser field within which are nine major and several small geyser basins; by contrast, many of the world's other geyser fields consist of a single geyser basin.

Geyserite. The variety of opal, technically amorphous hydrated silica, deposited by geysers and perpetual spouters, usually with a beaded surface; also known as *siliceous sinter*. Distinct from *travertine*, which is composed of calcium carbonate.

Group. An assortment of hot springs and geysers considered as a unit on the basis of some geographical separation from other nearby groups.

Height. The distance from the ground to the top of a geyser's erupted water. The height listed in the tables and descriptions invariably expresses the maximum height per eruption and is not necessarily characteristic of the entire eruption.

Hydrogen sulfide. A volcanic gas emitted from the hot springs that causes the "rotten egg odor" of the thermal areas.

Infrequent. A general term for eruptions that are very irregular, usually with intervals days to weeks long.

Intermittent spring. A hot spring that undergoes occasional episodes of quiet overflow without a bursting or jetting eruption.

Interval. The amount of time from the start of one geyser eruption to the start of the next. In some parts of the world, this start-to-start time is called the "*period*" of the eruption, whereas the "interval" is the quiet time between eruptions; that is, from end to start.

Irregular. The term for eruptions that show no evident pattern of distribution, with intervals ranging from minutes to days in length.

Mud pot. An acid hot spring with a limited water supply; not enough water is present to carry away clay mud that forms from the chemical and bacterial alteration of the rocks within the crater.

Period. The span of time from the start of one eruption to the start of the next; that is, the interval plus the duration. Nearly synonymous with *interval*.

Perpetual spouter. An erupting spring that resembles a geyser except that the eruption does not stop. Although included as geysers throughout this book, important mechanical differences mean that a perpetual spouter is not a true geyser.

Play. The eruptive activity of a geyser; a synonym for *eruption*.

Plumbing system. The subsurface network of tubes, cavities, and channels that makes up the water supply system of any hot spring; it is especially important for geysers in that it must contain a near-surface constriction, be pressure tight, and be accessible to large volumes of superheated water.

Pool. A non-eruptive hot spring that has an open body of water within a *crater*; also, the body of water that occupies the crater of a *fountain-type* geyser during the interval between eruptions.

Preplay. Any activity such as heavy overflow or minor splashing that precedes a geyser's eruption; useful in that preplay is usually an indication that the time of eruption is near.

Rare. The general term for eruptions or active phases that almost never occur, months to years sometimes passing between them.

Seldom. The general term for eruptions that are very widely spaced, several weeks to a few months often passing between them.

Siliceous sinter. Another term for the opaline silica or *geyserite* of the geyser basins.

Sinter. A general term for any mineral deposit formed by hot water. In Yellowstone and most other geyser fields, it is the *siliceous sinter* or *geyserite*, composed of hydrated silicon dioxide (a form of opal), that is of greatest importance, but thermal areas may also include calcareous sinter or *travertine*, composed of calcium carbonate, and *ferric sinter*, made largely of iron oxide.

Sput. The term for a small geyser or perpetual spouter whose activity is relatively insignificant compared to surrounding features but whose action can have a deleterious effect on nearby geyser activity; also, a general term for the members of a large group of small erupting features.

Superheated. Referring to the water of a geyser, hotter than the surface boiling temperature (which is about 198°F [90°C] in Yellowstone).

Thermophilic. Literally "heat loving," the term used in reference to the communities of primitive bacteria, algae, and animals that live in or on hot spring water.

Travertine. A hot spring deposit formed of calcium carbonate rather than silica; also called "calcareous sinter."

Uncommon. A synonym for *infrequent*.

Vent. The surface opening of the plumbing system of any hot spring; in geysers, it is the point from which the eruption issues, usually located within a *cone* or at the bottom of a *crater*.

Allen, E. T., and A. L. Day. 1935. *Hot Springs of the Yellowstone National Park*. Washington, D.C.: Carnegie Institute of Washington, Publication Number 466.

Brock, T. D., and M. L. Brock. 1994. *Life at High Temperatures*. Yellowstone National Park: Yellowstone Association.

Bryan, T. S. 2005. *Geysers: What They Are and How They Work*. Missoula, Mont.: Mountain Press.

Christiansen, R. L. 2001. *The Quaternary and Pleistocene Yellowstone Plateau Volcanic Field of Wyoming, Idaho, and Montana*. Washington, D.C.: U.S. Geological Survey Professional Paper 729-G.

Duckworth, C., ed. 2004. "Thermophiles," in *Yellowstone Resources and Issues 2004* (an annual compendium of information about Yellowstone National Park). Yellowstone National Park: Division of Interpretation.

Geyser Observation and Study Association, The (GOSA). 1989, 1990, 1992, 1993, 1998, 2002, 2005. GOSA Transactions. *The Journal of the Geyser Observation and Study Association* (a nonprofit corporation), Palmdale, Calif.

Haines, A. L. 1977. *The Yellowstone Story: A History of Our First National Park.* 2 vols. Yellowstone National Park: Yellowstone Library and Museum Association, with Colorado Associated University Press.

Keefer, W. R. 1972. *The Geologic Story of Yellowstone National Park.* Washington, D.C.: U.S. Geological Survey Bulletin 1347; republished by Yellowstone National Park: Yellowstone Association.

Marler, G. D. 1951. Exchange of Function as a Cause of Geyser Irregularity. *American Journal of Science* 249: 329–342.

———. 1964. *Effects of the Hebgen Lake Earthquake of August 17, 1959 on the Hot Springs of the Firehole Geyser Basins, Yellowstone National Park.* Washington, D.C.: U.S. Geological Survey Professional Paper 435.

———. 1973. *Inventory of Thermal Features of the Firehole River Geyser Basins, and Other Selected Areas of Yellowstone National Park.* Washington, D.C.: National Technical Information Service Publication Number PB-221289; republished by The Geyser Observation and Study Association.

Marler, G. D., and D. E. White. 1975. Seismic Geyser and Its Bearing on the Origin and Evolution of Geysers and Hot Springs of Yellowstone National Park. *Geological Society of America Bulletin* 86: 749–759.

Sheehan, K. B., B. L. Dicks, and J. M. Henson. 2005. *Seen and Unseen: Discovering the Microbes of Yellowstone.* Helena, Mont.: Globe Pequot Press (a Falcon Guide).

White, D. E., R. O. Fournier, L.J.P. Muffler, and A. H. Truesdell. 1975. *Physical Results of Research Drilling in Thermal Areas of Yellowstone National Park, Wyoming.* Washington, D.C.: U.S. Geological Survey Professional Paper 892.

White, D. E., R. A. Hutchinson, and T.E.C. Keith. 1988. *The Geology and Remarkable Thermal Activity of Norris Geyser Basin, Yellowstone National Park, Wyoming.* Washington, D.C.: U.S. Geological Survey Professional Paper 1456.

Whittlesey, L. H. 1989. *Wonderland Nomenclature: A History of the Place Names of Yellowstone National Park.* Helena: Montana Historical Society Press; typescript manuscript, original of 2,242 pages, cooperatively republished at no profit by The Geyser Observation and Study Association.

INDEX OF GEYSER AND HOT SPRING NAMES

*In cases of duplicate and similar names, the geyser basin of
the location is given within parentheses.*

Verdant Spring, 194
Vermillion Springs, 277
Veteran Geyser, 297
Victory Geyser, 117
Vitriol Geyser, 398
Vixen Geyser, 295

Wall Pool, 144
Washing Machine, 274
Washtub Spring, 63
Watermelon Geyser, 396
Wave Spring, 79
Wedge Spring, 359
West Flood Geyser, 180
West Geyser, 150
West Pelican Geyser, 400
West Round Spring, 93

West Sentinel Geyser, 126
West Sprinkler Geyser, 235
West Trail Geyser, 163
West Triplet Geyser, 72
Whirligig Geyser, 272
Whistle Geyser (Upper), 155
Whistler Geyser (Josephs Coat), 396
White Bubbler, 277
White Dome Geyser, 199
White Geyser, 167
White Hot Spring, 356
White Pyramid Geyser Cone, 105
White Sand Spring, 154

Wisp Geyser, 386
Witches Cauldron, 80

Yellow Bubbler Geyser, 152
Yellow Funnel Spring (Heart Lake), 393
Yellow Funnel Spring (Norris), 293
Yellow Sponge Spring, 362
YF-305, 211
YM-210, 135
Young Hopeful Geyser, 213

Zig Zag Spring, 96

GROUP ABBREVIATIONS FOR UNNAMED SPRINGS

BBG, Biscuit Basin (Upper)
BSB, Black Sand Basin (Upper)
CDG, Cascade Group (Upper)
CGG, Castle, Grand, and Sawmill Groups (Upper)
CLC, Chain Lakes Complex (Upper)
CLX, Culex Basin (Lower)
DSG, Daisy Group (Upper)
FCG, Fairy Creek Groups (Lower)
FLG, Black Warrior (Firehole Lake) Group (Lower)
FTN, Fountain Group (Lower)
GFG, Great Fountain Group (Lower)
GHG, Geyser Hill Group (Upper)
GIB, Gibbon Geyser Basin
GNT, Giant Group (Upper)
GRG, Grotto Group (Upper)
HFG, Fissure Group (Heart Lake)
HLG, Lower Group (Heart Lake)
HMG, Middle Group (Heart Lake)
HRG, Rustic Group (Heart Lake)
HUG, Upper Group (Heart Lake)
IMP, Imperial Group (Lower)
KLD, Kaleidoscope Group (Lower)
LSG, Lake Shore Group (West Thumb)
LST, Lone Star Geyser Basin
MGB, Midway Geyser Basin
MGG, Morning Glory Group (Upper)
MHG, Marshall's Hotel Group (Lower)

MMG, Morning Mist Group (Lower)
MMS, Morning Mist Springs (Lower)
MYR, Myriad Group (Upper)
NBK, Back Basin (Norris)
NPR, Porcelain Basin (Norris)
OFG, Old Faithful Group (Upper)
ORG, Old Road Group (Upper)
OSG, Orange Spring Group (Upper)
PBG, Punch Bowl Group (Upper)
PIN, Pine Springs (Upper)
PMG, Pipeline Meadows Group (Upper)
PNK, Pink Cone Group (Lower)
POT, Potts Hot Spring Basin (West Thumb)
QAG, Quagmire Group (Lower)
RSG, Round Spring Group (Upper)
RVG, River Group (Lower)
SDP, Serendipity Springs (Lower)
SHO, Shoshone Geyser Basin
SMG, Sentinel Meadow Group (Lower)
SPR, Sprinkler Group (Lower)
TGG, Tangled Geysers (Lower)
THD, Thud Group (Lower)
UPG, Upper Springs (Upper)
WCG, White Creek Group (Lower)
WDG, White Dome Group (Lower)
WSG, Westside Group (Upper)
WTL, Lower Group (West Thumb)

T. Scott Bryan was a seasonal employee at Yellowstone National Park from 1970 through 1986, working in the maintenance division at Canyon for four summers and as a ranger-naturalist at Norris and Old Faithful after that. He held other National Park Service positions in Glacier National Park, Death Valley National Monument, Glen Canyon National Recreation Area, and the Los Angeles Field Office.

After serving in the U.S. Navy, during which time he was able to visit geysers in Japan and New Zealand, he received his B.S. in geology at San Diego State University. His education continued at the University of Montana, where he received an M.S. in 1974. In addition to the summers in Yellowstone, he was a professor of geology, astronomy, and general physical sciences and the director of the planetarium at Victor Valley College in Victorville, California. Retired in 2001, he and his wife, Betty, now reside in West Yellowstone, Montana, during the summer and in Oro Valley, Arizona, in the winter.

Photo by Bill Warnock.

The Geysers of Yellowstone was Bryan's first book. He is also the author of *Geysers: What They Are and How They Work* (Mountain Press), several articles on the natural history and geology of the American West, and scientific journal reports. He and his wife are coauthors of *The Explorer's Guide to Death Valley National Park.*

Bryan is associated with The Geyser Observation and Study Association (GOSA), a nonprofit 501(c)(3) corporation devoted to furthering the study and understanding of geysers worldwide. In addition to his studies in Yellowstone, he has been to geyser fields throughout the contiguous United States, Mexico, Japan, Fiji, and New Zealand and to the "Valley of Geysers" on the Kamchatka Peninsula of Russia, when he led the first-ever U.S. study group there in 1991.